Particle Accelerator Physics

Helmut Wiedemann

Particle
Accelerator Physics

Basic Principles and Linear Beam Dynamics

With 160 Figures

Springer-Verlag
Berlin Heidelberg New York
London Paris Tokyo
Hong Kong Barcelona
Budapest

PHYSICS

 d6859835

Professor Dr. Helmut Wiedemann

Applied Physics Department and Stanford Synchroton Radiation Laboratory,
Stanford University, Stanford, CA 94305, USA

ISBN 3-540-56550-7 Springer-Verlag Berlin Heidelberg New York
ISBN 0-387-56550-7 Springer-Verlag New York Berlin Heidelberg

Library of Congress Cataloging-in-Publication Data. Wiedemann, Helmut, 1938- Particle accelerator physics:
basic principles and linear beam dynamics/Helmut Wiedemann. p. cm. Includes bibliographical references and
index. ISBN 0-387-56550-7 (New York: alk. paper). – ISBN 3-540-56550-7 (Berlin: alk. paper) 1. Beam dynamics.
2. Linear accelerators. I. Title. QC793.3.B4W54   1993   539.7'3–dc20   93-10040

The use of general descriptive names, registered names, trademarks, etc. in this publication does not imply, even
in the absence of a specific statement, that such names are exempt from the relevant protective laws and regulations
and therefore free for general use.

Typesetting: Camera-ready copy from the author using a Springer TEX macro package
54/3140 - 5 4 3 2 1 0 – Printed on acid-free paper

To my sons
Frank and Martin

Preface

The purpose of this textbook is to provide a comprehensive introduction into the physics of particle accelerators and particle beam dynamics. Particle accelerators have become important research tools in high energy physics as well as sources of incoherent and coherent radiation from the far infra red to hard x-rays for basic and applied research. During years of teaching accelerator physics it became clear that the single most annoying obstacle to get introduced into the field is the absence of a suitable textbook. Indeed most information about modern accelerator physics is contained in numerous internal notes from scientists working mostly in high energy physics laboratories all over the world.

This text intends to provide a broad introduction and reference book into the field of accelerators for graduate students, engineers and scientists summarizing many ideas and findings expressed in such internal notes and elsewhere. In doing so theories are formulated in a general way to become applicable for any kind of charged particles. Writing such a text, however, poses the problem of correct referencing of original ideas. I have tried to find the earliest references among more or less accessible notes and publications and have listed those although the reader may have difficulty to obtain the original paper. In spite of great effort to be historically correct I apologize for possible omissions and misquotes. This situation made it necessary to rederive again some of such ideas rather than quote the results and refer the interested reader to the original publication. I hope this approach will not offend the original researchers, but rather provide a broader distribution of their original ideas, which have become important to the field of accelerator physics.

This text is split into two volumes. The first volume is designed to be self contained and is aimed at newcomers into the field of accelerator physics, but also to those who work in related fields and desire some background on basic principles of accelerator physics. The first volume therefore gives an introductory survey of fundamental principles of particle acceleration followed by the theory of linear beam dynamics in the transverse as well as longitudinal phase space including a detailed discussion of basic magnetic focusing units. Concepts of single and multi particle beam dynamics are introduced. Synchrotron radiation, its properties and effect on beam dynamics and electron beam parameters is described in considerable detail followed by a discussion of beam instabilities on an introductory level, beam lifetime and basic lattice design concepts.

The second volume is aimed specifically to those students, engineers and scientists who desire to immerse themselves deeper into the physics of particle accelerators. It introduces the reader to higher order beam dynamics, Hamiltonian particle dynamics, general perturbation theory, nonlinear beam optics, chromatic and geometric aberrations and resonance theory. The interaction of particle beams with rf fields of the accelerating system and beam loading effects are described in some detail relevant to accelerator physics. Following a detailed derivation of the theory of synchrotron radiation, particle beam phenomena are discussed while utilizing the Vlasov and Fokker Planck equations leading to the discussion of beam parameters and their manipulation and collective beam instabilities. Finally design concepts and new developments of particle accelerators as synchrotron radiation sources or research tools in high energy physics are discussed in some detail.

This text grew out of a number of lecture notes for accelerator physics courses at Stanford University, the Synchrotron Radiation Research Laboratory in Taiwan, the University of Sao Paulo in Brazil, the International Center for Theoretical Physics in Trieste and the US Particle Accelerator School as well as from interaction with students attending those classes and my own graduate students.

During almost thirty years in this field I had the opportunity to work with numerous individuals and accelerators in laboratories around the world. Having learned greatly from these interactions I like to take this opportunity to thank all those who interacted with me and have had the patience to explain their ideas, share their results or collaborate with me. The design and construction of new particle accelerators provides a specifically interesting period to develop and test theoretically new ideas, to work with engineers and designers, to see theoretical concepts become hardware and to participate in the excitement of commissioning and optimization. I have had a number of opportunities for such participation at the Deutsches Elektronen Synchrotron DESY in Hamburg, Germany and at the Stanford University at Stanford, California and am grateful to all colleagues who hosted and collaborated with me. I wished I could mention them individually and apologize for not doing so.

A special thanks goes to the operators of the electron storage rings SPEAR and PEP at the Stanford Linear Accelerator Center, specifically to T. Taylor, W. Graham, E. Guerra and M. Maddox, for their dedicated and able efforts to provide me during numerous shifts over many years with a working storage ring ready for machine physics experimentation.

I thank Mrs. Joanne Kwong, who typed the initial draft of this texts and introduced me into the intricacies of TEX typesetting. The partial support by the Department of Energy through the Stanford Synchrotron Radiation Laboratory in preparing this text is gratefully acknowledged. Special thanks to Dr. C. Maldonado for painstakingly reading the manuscript. Last but not least I would like to thank my family for their patience in dealing with an "absent" husband and father.

Palo Alto, California   *Helmut Wiedemann*
April 1993

Contents

1 Introduction

The development of charged particle accelerators and it's underlying principles builds on the theoretical and experimental progress in fundamental physical phenomena. While active particle accelerator experimentation started only in this century, it depended on the basic physical understanding of electromagnetic phenomena as investigated both theoretically and experimentally mainly during the nineteenth and beginning twentieth century. In this introduction we will recall briefly the history leading to particle accelerator development, applications and introduce basic definitions and formulas governing particle beam dynamics.

1.1 Short Historical Overview

The history and development of particle accelerators is intimately connected to the discoveries and understanding of electrical phenomena and the realization that the electrical charge comes in lumps carried as a specific property by individual particles. It is reported that the greek philosopher and mathematician *Thales of Milet*, who was born in 625 BC first observed electrostatic forces on amber. The greek word for amber is electron or $\eta\lambda\epsilon\kappa\tau\rho\sigma\nu$ and has become the origin for all designations of electrical phenomena and related sciences. For more than two thousand years this observation was hardly more than a curiosity. In the eighteenth century, however, electrostatic phenomena became quite popular in scientific circles and since have been developed into a technology which by now completely embraces and dominates modern civilization.

It took another two hundred years before the carriers of electric charges could been isolated. Many systematic experiments were conducted and theories developed to formulate the observed electrical phenomena mathematically. It was Coulomb, who in 1785 first succeeded to quantify the forces between electrical charges which we now call *Coulomb forces*. As more powerful sources for electrical charges became available, glow discharge phenomena were observed and initiated an intensive effort on experimental observations during most of the second half of the nineteenth century. It was the observations of these electrical glow discharge phenomena that led the scientific community, for example, to the discovery of elementary particles and electromagnetic radiation which are both basic ingredients for particle acceleration.

Research leading to the discovery of elementary particles and to ideas for the acceleration of such particles is dotted with particularly important milestones which from time to time set the directions for further experimental and theoretical research. It is obviously somewhat subjective to choose which discoveries might have been the most directive. Major historical discoveries leading to present day particle accelerator physics started to happen more than a hundred and fifty years ago:

1815 W. Proust, a british physician and chemist, postulated initially anonymous, that all atoms are composed of hydrogen atoms and that therefore all atomic weights come in multiples of the weight of a hydrogen atom.

1839 Faraday [1.1] published his experimental investigations of electricity and described various phenomena of glow discharge.

1858 Plücker [1.2] reported on the observation of *cathode rays* and their deflection by magnetic fields. He found the light to become deflected in the same spiraling direction as Ampere's current flows in the electromagnet and therefore postulated that the electric light, as he calls it, under the circumstances of the experiment must be magnetic.

1867 L. Lorenz working in parallel with J.C. Maxwell on the theory of electromagnetic fields formulated the concept of retarded potentials although not yet for point charges.

1869 Hittorf [1.3], a student of Plücker started his paper with the statement (translated from german): *"The undisputed darkest part of recent theory of electricity is the process by which in gaseous volumes the propagation of electrical current is effected"*. Obviously observations with glow discharge tubes displaying an abundance of beautiful colors and complicated reactions to magnetic fields kept a number of researchers fascinated. Hittorf conducted systematic experiments on the deflection of the light in glow discharges by magnetic fields and corrected some erroneous interpretations by Plücker.

1871 C.F. Varley postulated that cathode rays are particle rays.

1874 H. von Helmholtz postulated atomistic structure of electricity.

1883 J.C. Maxwell publishes his *Treatise on Electricity and Magnetism.*

1883 T.A. Edison discovers thermionic emission.

1886 E. Goldstein [1.4] observed positively charged rays which he was able to isolate from a glow discharge tube through channels in the cathode. He therefore calls these rays *Kanalstrahlen.*

1887 H. Hertz discovered electromagnetic waves and photoelectric effect.

1891 G.J. Stoney introduced the name *electron.*

1895 H.A. Lorentz formulated electron theory, *Lorentz force* equation and *Lorentz contraction.*

1894 P. Lenard built a discharge tube that allows cathode rays to exit to atmospheric air.

1895 W. Röntgen discovered *x-rays*.

1895 Wiedemann [1.5] reported on a new kind of radiation studying electrical sparks. Thomson [1.6] explained the emission of this radiation as being caused by acceleration occurring during the collision of charged particles with other atoms and calculated the energy emitted per unit time to be $(2e^2 f^2)/(3V)$, where e is the charge of the emitting particle, f the acceleration and V the velocity of light.

1897 J.J. Thomson measured the ratio e/m for kanal and cathode rays with electromagnetic spectrometer and found the e/m ratio for cathode rays to be larger by a factor of 1700 compared to the e/m ratio for kanal rays. He concluded that cathode rays consist of free electricity giving evidence to free electrons.

1897 J. Larmor [1.7] formulated concept of *Larmor precession*

1898 Liènard [1.8] calculated the electric and magnetic field in the vicinity of a moving point charge and evaluated the energy loss due to *electromagnetic radiation* from a charged particle travelling on a circular orbit.

1900 Wiechert [1.9] derived expression for *retarded potentials* of point charges.

1907 Schott [1.10, 11] formulated the first theory of *synchrotron radiation* in an attempt to explain atomic spectra.

1905 A. Einstein published *theory of special relativity*.

1901 W. Kaufmann first alone and in 1907 together with A.H. Bucherer measured increase of electron mass with energy. First experiment in support of theory of special relativity.

1909 R.A. Millikan started measuring electric charge of electron.

1913 First experiment by J. Franck and G. Hertz to excite atoms by accelerated electrons.

1914 Marsden produced first proton beam irradiating paraffin with alpha particles.

1920 H. Greinacher [1.12] built first *cascade generator*.

1922 R. Wideroe as a graduate student sketched *ray transformer*.

1924 Ising [1.13] invented *electron linac* with drift tubes and spark gap excitation.

1928 Wideroe [1.14] reported first operation of linear accelerator with potassium and sodium ions. Discusses operation of *betatron* and failure to get beam for lack of focusing.

1928 P.A.M. Dirac predicted existence of *positrons*.

1931 Van de Graaff [1.15] built first high voltage generator.

1932 Lawrence and Livingston [1.16] accelerated first proton beam from 1.2 MeV *cyclotron*.

1932 Cockcroft and Walton [1.17] used technically improved cascade generator to accelerate protons and initiated first artificial atomic reaction: Li + p $\rightarrow$ 2 He.

1932 C.D. Anderson discovered *positron, neutrons* were discovered by J. Chadwick and H.C. Urey discovered *deuterons.*

1939 W.W. Hansen, R. Varian and his brother S. Varian invented *klystron* microwave tube at Stanford.

1941 Kerst and Serber [1.18] completed first functioning betatron.

1941 Touschek and Wideroe formulated *storage ring principle.*

1944 Ivanenko and Pomeranchuk [1.19] and Schwinger [1.20] pointed out independently an energy limit in circular electron accelerators due to *synchrotron radiation* losses.

1945 Veksler [1.21] and McMillan [1.22] independently discovered the principle of *phase focusing.*

1945 Blewett [1.23] experimentally discovered synchrotron radiation by measuring the energy loss of electrons.

1947 Alvarez [1.24] designed first *proton linear accelerator* at Berkeley.

1947 Ginzton et al. [1.25] accelerated electrons to 6 MeV with Mark I *disk loaded linac* at Stanford.

1949 E.M. McMillan et al. commissioned 320 MeV electron synchrotron.

1950 Christofilos [1.26] formulated concept of *strong focusing.*

1951 Motz [1.27] built first *wiggler magnet* to produce quasi monochromatic synchrotron radiation

1952 Livingston et al. [1.28] described design for 2.2 GeV *Cosmotron* in Brookhaven.

1952 Courant et al. [1.29] published first paper on strong focusing.

1952 Chodorow et al. [1.30] completed 600 MeV MARK III *electron linac.*

1954 R.R. Wilson et al. operated first AG *electron synchrotron* in Cornell at 1.1 GeV. Lofgren et al. accelerated protons to 5.7 GeV in *Bevatron.*

1955 Sands [1.31] defined limits of phase focusing due to *quantum excitation.*

1958 Courant and Snyder [1.32] published their paper on the *Theory of the Alternating-Gradient Synchrotron.*

Research and development in accelerator physics blossomed significantly during the fifties supported by the development of high power radio frequency sources and the increased availability of government funding for accelerator projects. Parallel with the progress in accelerator technology, we also observe advances in theoretical understanding, documented in an increasing number of publications. It is beyond the scope of this text to only try to give proper credit to all major advances in the past thirty years and refer the interested reader to more detailed references.

1.2 Particle Accelerator Systems

Particle accelerators come in many forms applying a variety of technical principles. All are based on the interaction of the electric charge with static and dynamic electromagnetic fields and it is the technical realization of this interaction that leads to the different types of particle accelerators. Electromagnetic fields are used over most of the available frequency range from *static electric fields* to ac magnetic fields in betatrons oscillating at 50 or 60 Hz, to *radio frequency fields* in the MHz to GHz range and ideas are being developed to use *laser beams* to generate high field particle acceleration.

In this text we will not discuss the different technical realization of particle acceleration but rather concentrate on basic principles which are designed to help the reader to develop technical solutions for specific applications meeting basic beam stability requirements. For particular technical solutions we refer to the literature cited in the bibliography and throughout this text.

Nevertheless to discuss basic principles of particle acceleration and beam dynamics it is desirable to stay in contact with technical reality and reference practical and working solutions. We will therefore repeatedly refer to certain types of accelerators and apply theoretical beam dynamics solutions to exhibit the salient features and importance of the theoretical ideas under discussion. In these references we use mostly such types of accelerators which are commonly used and are extensively publicized.

1.2.1 Basic Components of Accelerator Facilities

In the following paragraphs we will describe components of particle accelerators in a rather cursory way to introduce the terminology and overall features. Particle accelerators consist of two basic units, the *particle source* or *injector* and the main accelerator. The particle source comprises all components to generate the desired type of particles.

Generally *glow discharge columns* are used to produce *proton* or *ion beams* which then are first accelerated in electrostatic accelerators like a *Van de Graaff* or *Cockcroft-Walton* accelerator and then in an *Alvarez-type linear accelerator*. To increase the energy of heavy ion beams the initially singly charged ions are, after some acceleration, guided through a thin metal foil to strip more electrons off the ions. More than one stripping stage may be used at different energies to reach the maximum *ionization* for most efficient acceleration.

Much more elaborate measures must be used to produce *antiprotons*. Generally a high energy proton beam is aimed at a heavy metal *target* where through hadronic interactions with the target material among other particles antiprotons are generated. Emerging from the target these antiprotons are collected by strong focusing devices and further accelerated.

Electrons are commonly generated from a heated *cathode*, also called a *thermionic gun*, which is covered on the surface by specific alkali oxides or any other substance with a low work function to emit electrons at technically practical temperatures. Another method to create a large number of electrons within a short pulse uses a strong laser pulse like that from a YaG laser directed at the surface of a *photo cathode*. Systems where the cathode is inserted directly into an accelerating rf field are called *rf guns*. Positrons are created the same way as antiprotons by aiming high energy electrons on a heavy metal target where, through an electromagnetic shower and pair production, positrons are generated. These positrons are again collected by strong magnetic fields and further accelerated.

Whatever the method of generating particles may be, in general they do not have the time structure desired for further acceleration or special application. Efficient acceleration by rf fields occurs only during a very short period per oscillation cycle and most particles would be lost without proper preparation. For high beam densities it is desirable to compress the continuous stream of particles from a thermionic gun or a glow discharge column into a shorter pulse with the help of a *chopper* device and a *prebuncher*. The chopper may be a mechanical device or a deflecting magnetic or rf field moving the continuous beam across the opening of a slit. At the exit of the chopper we observe a series of beam pulses, called *bunches*, to be further processed by the prebuncher. Here early particles within a bunch are decelerated and late particles accelerated. After a well defined drift space, the bunch length becomes reduced due to the energy dependence of the particle velocity. Obviously this compression works only as long as the particles are not relativistic while the particle velocity can be modulated by acceleration or deceleration.

No such compression is required for antiparticles, since they are produced by high energetic particles having the appropriate time structure. Antiparticle beams emerging from a target have, however, a large beam size and beam divergence. To make them suitable for further acceleration they are generally stored for some time in a *cooling* or *damping ring*. Such cooling rings are circular "accelerators" where particles are not accelerated but spend just some time circulating. Positrons circulating in such *storage rings* quickly lose their transverse momenta and large beam divergence through the emission of synchrotron radiation. In the case of antiprotons external fields are applied to damp the transverse beam size.

Antiparticles are not always generated in large quantities. On the other hand, the accelerator ahead of the conversion target can often be pulsed at a much higher rate than the main accelerator can accept injection. In such cases, the antiparticles are collected from the rapid cycling injector into an *accumulator ring* and then transferred to the main accelerator when required.

Particle beams prepared in such manner may now be further accelerated in linear or circular accelerators. A linear accelerator consists of a linear sequence of many accelerating units where accelerating fields are generated and timed such that particles absorb and accumulate energy from each acceleration unit. Most commonly used *linear accelerators* consist of a series of cavities excited by radio frequency sources to high accelerating fields. In the *induction accelerator*, each accelerating unit consists of a transformer which generates from an external electrical pulse a field on the transformer secondary which is formed such as to allow the particle beam to be accelerated. Such induction accelerators can be optimized to accelerate very high beam currents at medium beam energies.

For very high beam energies linear accelerators become very long and costly. Such practical problems can be avoided in circular accelerators where the beam is held on a circular path by magnetic fields in bending magnets passing repeatedly every turn through accelerating sections, similar to those in linear accelerators. This way the particles gain energy from the accelerating cavities at each turn and reach the maximum energy while the fields in the bending magnets are raised in synchronism. Such a circular accelerator is called a *synchrotron*.

The basic principles to accelerate particles of different kind are similar and we do not need to distinguish between protons, ions, and electrons. Technically the accelerators differ more or less to adjust to the particular beam parameters which have mostly to do with the particle velocities. For highly relativistic particles the differences in beam dynamics vanish. Protons and ions are more likely to be nonrelativistic and therefore vary the velocity as the kinetic energy is increased, thus generating problems of synchronism with the oscillating accelerating fields which must be solved by technical means.

After acceleration in a linear accelerator or synchrotron the beam can be directed onto a target, mostly a target of liquid hydrogen, to study high energy interactions with the target protons. Such *fixed target* experimentation dominated nuclear and high energy particle experimentation from the first applications of artificially accelerated particle beams far into the seventies and is still a valuable means of basic research. Obviously, it is also the method in conjunction with a heavy metal target to produce secondary particles like antiparticles for use in colliding beam facilities and mesons for basic research.

To increase the center-of-mass energy for basic research, particle beams are aimed not at fixed targets but to collide head on with another beam. This is one main goal for the construction of *colliding beam facilities* or storage rings. In such a ring, particle and antiparticle beams are injected in opposing directions and made to collide in specifically designed *interaction regions*. Because the interactions between counter orbiting particles is very rare, storage rings are designed to allow the beams to circulate for many turns with beam life times of several hours to give the particles ample opportunity to collide with another counter rotating particle. Of course, beams can counter rotate in the same magnetic fields only if one beam is made of the antiparticles of the other beam while two *intersecting storage rings* must be employed to allow the collision of particles with equal charge.

The circulating beam in an electron storage ring emits synchrotron radiation due to the transverse acceleration during deflection in the bending magnets. This radiation is highly collimated in the forward direction, of high brightness and therefore of great interest for basic research, technology, and medicine.

Basically the design of a storage ring is the same as that for a synchrotron allowing some adjustment in the technical realization to optimize the desired features of acceleration and long beam lifetime, respectively. Beam intensities are generally very different in a synchrotron from that in a storage ring. In a synchrotron the particle intensity is determined by the injector and this intensity is much smaller than desired in a storage ring. The injection system into a storage ring is therefore designed such that many beam pulses from a linear accelerator, an accumulator ring or a synchrotron can be accumulated. A synchrotron serving to accelerate beam from a low energy preinjector to the injection energy of the main facility, which may be a larger synchrotron or a storage ring, is also called a *booster synchrotron* or a *booster* for short.

Although a storage ring is not used for particle acceleration it often occurs that a storage ring is constructed long after and for a higher beam energy than the injector system. In this case, the beam is accumulated at the maximum available injection energy. After accumulation the beam energy is slowly raised in the storage ring to the design energy.

Electron positron storage rings have played a great role in basic high-energy research. For still higher collision energies, however, the energy loss due to synchrotron radiation has become a practical and economic limitation. To avoid this limit, beams from two opposing linear accelerators are brought into head on collision at energies much higher than is possible to produce in circular accelerators. To match the research capabilities in colliding beam storage rings, such *linear colliders* must employ sophisticated beam dynamics, focusing arrangements and technologies which do not exist as yet and are under development.

1.2.2 Applications of Particle Accelerators

Particle accelerators are mainly known for their application as research tools in nuclear and high energy particle physics requiring the biggest and most energetic facilities. Smaller accelerators, however, have found broad applications in a wide variety of basic research and technology, as well as medicine. In this text we will not discuss the details of all these applications but try to concentrate only on the basic principles of particle accelerators and the theoretical treatment of particle beam dynamics and instabilities. An arbitrary and incomplete listing of applications for charged particle beams and their accelerators is given for reference to the interested reader:

Nuclear physics	*Synchrotron radiation*
Electron/proton accelerators	Basic atomic and
Ion accelerators/colliders	molecular physics
Continuous beam facility	Condensed matter physics
	Earth sciences
High-energy physics	Material sciences
Fixed target accelerator	Chemistry
Colliding beam storage rings	Molecular and cell biology
Linear colliders	Surface/interface physics
Power generation	*Coherent radiation*
Inertial fusion	Free electron lasers
Reactor fuel breeding	Microprobe
	Holography
Industry	*Medicine*
Radiography by x-rays	Radiotherapy
Ion implantation	Health physics
Isotope production/separation	Digital subtraction angiography
Materials testing	Microsurgery with tunable FEL
Food sterilization	
X-ray lithography	

This list is by no means exhaustive and additions must be made at an impressive pace as the quality and characteristics of particle beams become more and more sophisticated, predictable and controllable. Improvements in any parameter of particle beams create opportunities for new experiments and applications which were not possible before. More detailed information on specific uses of particle accelerators as well as an extensive catalogue of references has been compiled by Scharf [1.33].

1.3 Basic Definitions and Formulas

Particle beam dynamics can be formulated in a variety units and it is therefore prudent to define the units used in this text to avoid confusion. In addition we recall fundamental relations of electromagnetic fields and forces as well as some laws of special relativity to the extend that will be required in the course of discussions.

1.3.1 Units and Dimensions

A set of special physical units, selected primarily for convenience, are most commonly used to quantify physical constants in accelerator physics. The use of many such units is often determined more by historical developments than based on the choice of a consistent set of quantities useful for accelerator physics. In order to ease cross references to existing literature and adjust to common use of units in accelerator physics the author decided against using strictly the international set of units.

Generally, accelerator physics theory is formulated in either the metric MKS system of units or in gaussian cgs units, but neither system is strictly adhered to. While one observes a majority of theoretical papers using the cgs system, *stat* units are not used to measure physical particle beam quantities. In this text we use the cgs system of units for theoretical discussions while quantitative formulas will be expressed in *practical* units as they are widely used in the accelerator physics community. By this choice of units, we follow the majority of the accelerator physics community and thus facilitate the reading of subject papers in this field.

To measure the energy of charged particles neither the unit *joule* nor *erg* is used. The basic unit of energy in particle accelerator physics is the *electron Volt* eV which is the energy a particle with one basic unit of electrical charge *e* would gain while being accelerated between two conducting plates at a potential difference of one Volt. Specifically we will often use derivatives of the basic units to express actual particle energies in a convenient form:

$$1 \text{ keV } = 10^3 \text{ eV}, \qquad\qquad 1 \text{ MeV } = 10^6 \text{ eV},$$

$$1 \text{ GeV } = 10^9 \text{ eV}, \qquad\qquad 1 \text{ TeV } = 10^{12} \text{ eV}.$$

To describe particle dynamics we find it necessary to sometimes use the *particle's momentum* and sometimes the *particle's energy*. The effect of the Lorentz force from electric or magnetic fields is inversely proportional to the momentum of the particle. Acceleration in rf fields, on the other hand, is most conveniently measured by the increase in kinetic or total energy.

In an effort to simplify the technical jargon used in accelerator physics the term *energy* is used for all three quantities although mathematically the momentum is then multiplied by the velocity of light for dimensional consistency. There are still numerical differences which must be considered for all but very highly relativistic particles. Where we need to mention the pure particle momentum and quote a numerical value we generally use the unit eV/c. With this definition a particle of *energy* $cp = 1$ eV would have a momentum of $p = 1$ eV/c.

An additional complication arises in the case of composite particles like *heavy ions*, consisting of protons and neutrons. In this case, the particle energy is not quoted for the whole ion but in terms of the kinetic energy per nucleon.

The particle beam current is measured generally in Amperes, no matter what general system of units is used but also occasionally in terms of the total charge or number of particles. In circular accelerators this beam current I relates directly to the *beam intensity* or the number of circulating particles N. If βc is the velocity of the particle and Z the charge multiplicity, we get for the relation of *beam current* and *beam intensity*

$$I = e\,Z\,f_{\text{rev}}\,N\,, \tag{1.1}$$

where the *revolution frequency* $f_{\text{rev}} = \beta c/C$ and C is the circumference of the circular accelerator.

For a linear accelerator or beam transport line where particles come by only once, the definition of the beam current is more subtle. We still have a simple case if the particles come by in a continuous stream and the beam current is proportional to the particle flux $\dot{N}$ or $I = eZ\,\dot{N}$. This case, however, occurs very rarely since particle beams are generally accelerated by rf fields. As a consequence there is no continuous flux of particles reflecting the time varying acceleration of the rf field. The particle flux therefore is better described by a series of equidistant *particle bunches* separated by an integral number of wavelengths of the accelerating rf field. Furthermore, the acceleration often occurs only in bursts or pulses producing either a single bunch of particles or a string of many bunches. In these cases we distinguish between different current definitions. The peak current is the peak instantaneous beam current for a single bunch, while the average current is defined as the particle flux averaged over the duration of the beam pulse.

Magnetic fields are quoted either in Gauss or Tesla. Similarly, field gradients and higher derivatives are expressed in Gauss per centimeter or Tesla per meter. Frequently we find the need to perform numerical calculations with parameters given in units of different systems. Some helpful numerical conversions are compiled in Table 1.1.

Table 1.1. Numerical conversion factors

quantity:	replace cgs parameter:	by practical units:
potential	1 esu	300 V
electrical field	1 esu	3×10^4 V/m
current	1 esu	$0.1 \cdot c$ A
charge	1 esu	0.3333×10^{-9} C
force	1 dyn	10^{-5} N
energy	1 eV	1.602×10^{-19} J
	1 eV	1.602×10^{-12} erg

Similar conversion factors can be derived for electromagnetic quantities in formulas by comparisons of similar equations in the cgs and MKS system. Table 1.2 includes some of the most frequently used conversions. The absolute dielectric constant is

$$\epsilon_\text{o} = \frac{10^7}{4\pi c^2}\frac{\text{C}}{\text{V m}} = 8.854 \times 10^{-12}\,\frac{\text{C}}{\text{V m}} \tag{1.2}$$

and the absolute magnetic permeability is

Table 1.2. Conversion factors for equations

quantity:	replace cgs parameter:	by MKS parameter:
potential	V_{cgs}	$\sqrt{4\pi\epsilon_\mathrm{o}}\, V_{\mathrm{MKS}}$
electric field	$\mathbf{E}_{\mathrm{cgs}}$	$\sqrt{4\pi\epsilon_\mathrm{o}}\, \mathbf{E}_{\mathrm{MKS}}$
current	I_{cgs}	$\frac{1}{\sqrt{4\pi\epsilon_\mathrm{o}}}\, I_{\mathrm{MKS}}$
current density	$\mathbf{j}_{\mathrm{cgs}}$	$\frac{1}{\sqrt{4\pi\epsilon_\mathrm{o}}}\, \mathbf{j}_{\mathrm{MKS}}$
charge	q_{cgs}	$\frac{1}{\sqrt{4\pi\epsilon_\mathrm{o}}}\, q_{\mathrm{MKS}}$
charge density	ρ_{cgs}	$\frac{1}{\sqrt{4\pi\epsilon_\mathrm{o}}}\, \rho_{\mathrm{MKS}}$
conductivity	σ_{cgs}	$\frac{1}{4\pi\epsilon_\mathrm{o}}\, \sigma_{\mathrm{MKS}}$
inductance	L_{cgs}	$4\pi\epsilon_\mathrm{o}\, L_{\mathrm{MKS}}$
capacitance	C_{cgs}	$\frac{1}{4\pi\epsilon_\mathrm{o}}\, C_{\mathrm{MKS}}$
magnetic field	H_{cgs}	$\sqrt{4\pi\mu_\mathrm{o}}\, H_{\mathrm{MKS}}$
magnetic induction	B_{cgs}	$\frac{\sqrt{4\pi}}{\sqrt{\mu_\mathrm{o}}}\, B_{\mathrm{MKS}}$

$$\mu_\mathrm{o} \;=\; 4\pi \times 10^{-7}\,\frac{\mathrm{V\,s}}{\mathrm{A\,m}} \;=\; 1.2566 \times 10^{-6}\,\frac{\mathrm{V\,s}}{\mathrm{A\,m}}\,. \tag{1.3}$$

Both constants are related by

$$\epsilon_\mathrm{o}\,\mu_\mathrm{o}\,c^2 \;=\; 1\,. \tag{1.4}$$

Using these conversion factors it is possible to convert all formulas in cgs units in this text into the equivalent form for MKS units.

1.3.2 Basic Relativistic Formalism

In accelerator physics the dynamics of particle motion is formulated for a large variety of energies from nonrelativistic to highly relativistic values and the equations of motion obviously must reflect this. Relativistic mechanics is therefore a fundamental ingredient of accelerator physics and we will recall a few basic relations of relativistic particle mechanics from a variety of more detailed derivations in generally available textbooks.

Beam dynamics is expressed in a fixed laboratory system of coordinates but some specific problems are better discussed in the moving coordinate system of a single particle or of the center of charge for a collection of particles. Transformation between the two systems is effected through a *Lorentz transformation*

$$x = x^*, \quad y = y^*, \quad s = \frac{s^* + \beta_s ct^*}{\sqrt{1-\beta_s^2}}, \quad ct = \frac{ct^* + \beta_s s^*}{\sqrt{1-\beta_s^2}}, \tag{1.5}$$

where $\beta_s = v_s/c$ and where we have assumed that the particle moves with the velocity v_s along the s-axis with respect to the fixed laboratory system S. The primed quantities are measured in the moving system S^*.

Characteristic for relativistic mechanics is the *Lorentz contraction* and *time dilatation*, both of which become significant in the description of particle dynamics. The Lorentz contraction is the result of a *Lorentz transformation* from one coordinate system to another moving relative to the first system. We consider a rod at rest along the s-coordinate with the length $\Delta s = s_2 - s_1$ in the coordinate system S. In the system S^* moving with the velocity v_s along the s-axis with respect to the system S, this rod appears to have the length $\Delta s^* = s_2^* - s_1^*$ and with (1.5) we get the relationship

$$\Delta s = \gamma(s_2^* + v_s t^*) - \gamma(s_1^* + v_s t^*) = \gamma \Delta s^* \tag{1.6}$$

or

$$\Delta s = \gamma \Delta s^*, \tag{1.7}$$

where γ is the total particle energy E in units of the particle rest energy mc^2

$$\gamma = \frac{E}{m c^2} = \frac{1}{(1 - \beta_s^2)^{1/2}}. \tag{1.8}$$

The rod appears shorter in the moving particle system by the factor γ and longest in the system where it is at rest.

Because of the Lorentz contraction the volume of a body at rest in the system S appears also reduced in the moving system S^* and we have for the volume of a body in three dimensional space

$$V = \gamma V^*. \tag{1.9}$$

Only one dimension of this body is Lorentz contracted and therefore the volume scales only linearly with γ. As a consequence, the charge density ρ of a particle bunch with the volume V is lower in the laboratory system S compared to the density in the system moving with this bunch and becomes

$$\rho = \frac{\rho*}{\gamma}. \tag{1.10}$$

Similarly, we may derive the *time dilatation* or the elapsed time between two events occurring at the same point in both coordinate systems. Applying the Lorentz transformations we get from (1.5) with $s_2^* = s_1^*$

$$\Delta t = t_2 - t_1 = \gamma \left(t_2^* + \frac{\beta_s s_2^*}{c} \right) - \gamma \left(t_1^* + \frac{\beta_s s_1^*}{c} \right) \tag{1.11}$$

or

$$\Delta t = \gamma \Delta t^*. \tag{1.12}$$

13

For a particle at rest in the moving system S^* the time t^* varies slower than the time in the laboratory system. This is the mathematical expression for the famous *twin brother paradox* where one of the brothers moving in a space capsule at relativistic speed would age slower than his twin brother staying back. This phenomenon becomes reality for unstable particles. For example, high-energy pion mesons, observed in the laboratory system, have a longer lifetime by the factor γ compared to low-energy pions with $\gamma = 1$. As a consequence we are able to transport high-energy pion beams a longer distance than low energy pions.

The total energy of a particles is given by

$$E = \gamma E_o = \gamma mc^2,\tag{1.13}$$

where $E_o = mc^2$ is the rest energy of the particle. The kinetic energy is defined as the total energy minus the rest energy

$$E_{\mathrm{kin}} = E - E_o = (\gamma - 1)mc^2.\tag{1.14}$$

The change in kinetic energy during acceleration is equal to the product of the accelerating force and the path length over which the force acts on the particle. Since the force may vary along the path we use the integral

$$\Delta E_{\mathrm{kin}} = \int_{L_{\mathrm{acc}}} \mathbf{F}\, ds\tag{1.14}$$

to define the energy increase. The length L_{acc} is the path length through the accelerating field. In discussions of energy gain through acceleration we consider only energy differences and need therefore not to distinguish between total and kinetic energy. The particle *momentum* finally is defined by

$$c^2 p^2 = E^2 - E_o^2\tag{1.16}$$

or

$$cp = \sqrt{E^2 - E_o^2} = m c^2 \sqrt{\gamma^2 - 1} = \gamma \beta m c^2 = \beta E,\tag{1.17}$$

where $\beta = v/c$. The simultaneous use of the terms energy and momentum might seem sometimes to be misleading as we discussed earlier. In this text, however, we will always use physically correct quantities in mathematical formulations even though we sometimes use the term energy for the quantity cp. In electron accelerators the numerical distinction between energy and momentum is insignificant since we consider in most cases highly relativistic particles. For proton accelerators and even more so for heavy ion accelerators the difference in both quantities becomes, however, significant.

Often we need differential expressions or expressions for relative variations of a quantity in terms of variations of another quantity. Such relations can be derived from the definitions in this section. By variation of $cp = mc^2 \sqrt{\gamma^2 - 1}$ in (1.17) for example we get

$$dcp = \frac{mc^2}{\beta}\,d\gamma \;=\; \frac{dE}{\beta} \;=\; \frac{dE_{kin}}{\beta} \tag{1.18}$$

and

$$\frac{dcp}{cp} \;=\; \beta^{-2}\,\frac{d\gamma}{\gamma}\,. \tag{1.19}$$

Varying $cp = \gamma\beta\,mc^2$ in (1.17) and replacing $d\gamma$ from (1.18) we get

$$dcp \;=\; \gamma^3\,mc^2\,d\beta \tag{1.20}$$

and

$$\frac{dcp}{cp} \;=\; \gamma^2\,\frac{d\beta}{\beta}\,. \tag{1.21}$$

In a similar way other relations can be derived.

Electromagnetic fields and the interaction of charged particles with these fields play an important role in accelerator physics. We find it often useful to express the fields in either the laboratory system or the particle system. Transformation of the fields from one to the other system is determined by the Lorentz transformation of electro magnetic fields. We assume again the coordinate system S^* to move with the velocity v_s along the s-axis with respect to a right-handed (x, y, s) reference frame S. The electromagnetic fields in this moving reference frame are identified by primed symbols and can be expressed in terms of the fields in the laboratory frame of reference S:

$$
\begin{aligned}
E_x^* &= \gamma\,(E_x + \beta_s\,B_y), & B_x^* &= \gamma\,(B_x - \beta_s\,E_y),\\
E_y^* &= \gamma\,(E_y - \beta_s\,B_x), & B_y^* &= \gamma\,(B_y + \beta_s\,E_x),\\
E_s^* &= E_s, & B_s^* &= B_s.
\end{aligned}
\tag{1.22}
$$

These transformations exhibit interesting features for accelerator physics, where we often use magnetic or electrical fields, which are pure magnetic or pure electric fields when viewed in the laboratory system. For relativistic particles, however, these pure fields become a combination of electric and magnetic fields.

1.3.3 Particle Collisions at High Energies

The most common use of high-energy particle accelerators has been for basic research in elementary particle physics. Here accelerated particles are aimed at a target, which incidentally may be just another particle beam, and the researchers try to analyze the reaction of high-energy particles colliding with target particles. The available energy from the collision depends on the kinematic parameters of the colliding particles. We define a center of mass coordinate system which is the system that moves with the center of mass of the colliding particles. In this system the vector sum of all momenta is zero and is preserved through the collision. Similarly, the total energy is conserved and we may define a center of mass energy the same way the rest energy of a single particle is defined by

$$E_{\mathrm{cm}}^2 = \left(\sum_i E_i\right)^2 - \left(\sum_i cp_i\right)^2,$$

(1.23)

where the summation is taken over all particles forming the center of mass system. We apply this to two colliding particles with masses m_1 and m_2 and velocities $\mathbf{v_1}$ and $\mathbf{v_2}$, respectively,

$$(m_1, \mathbf{v_1}) \quad \longrightarrow \quad \longleftarrow \quad (m_2, \mathbf{v_2}).$$

The center of mass energy for this system of two colliding particles is then

$$\begin{aligned}
E_{\mathrm{cm}}^2 &= \left[\sum_{i=1}^{2}(E_{\mathrm{kin}} + m\,c^2)_i\right]^2 - \left[\sum_{i=1}^{2} cp_i\right]^2 \\
&= (\gamma_1 m_1 + \gamma_2 m_2)^2\, c^4 - (\gamma_1\, \beta_1\, m_1 + \gamma_2\, \beta_2\, m_2)^2\, c^4 \,.
\end{aligned}$$

(1.24)

We apply these kinematic relations to a proton, where $m = m_{\mathrm{p}}$, of energy γ colliding with a proton at rest in a *stationary target*. For a target proton at rest with $\gamma_2 = 1$, $m_2 = m_{\mathrm{p}}$, $\beta_2 = 0$ and $\beta\gamma = \sqrt{\gamma^2 - 1}$, the center of mass energy is

$$E_{\mathrm{cm}}^2 = (\gamma + 1)^2\, m_{\mathrm{p}}^2 c^4 - (\gamma^2 - 1)\, m_{\mathrm{p}}^2 c^4$$

or after some manipulations

$$E_{\mathrm{cm}} = \sqrt{2(\gamma + 1)}\, m_{\mathrm{p}}\, c^2 \,.$$

(1.25)

The available energy for high-energy reactions after conservation of energy and momentum for the whole particle system is the center of mass energy minus the rest energy of the particles that need to be conserved. If, for example, two protons collide, high-energy physics conservation laws tell us that the hadron number must be conserved and therefore the reaction products must include two units of the hadron number. In the most simple case the reaction will produce just two protons and some other particles with a total energy equal to the available energy

$$E_{\mathrm{avail}} = E_{\mathrm{cm}} - 2\, m_{\mathrm{p}} c^2 = \left[\sqrt{2(\gamma + 1)} - 2\right] m_{\mathrm{p}} c^2 \,.$$

(1.26)

The energy available from such reactions increases only like the square root of the energy of the accelerated particle which makes such *stationary target* physics an increasingly inefficient use of high-energy particles. A significantly more efficient way of using the energy of colliding particles for

elementary particle research can be obtained by *head on collision* of two
equal particles of equal energy. In this case $\gamma_1 = \gamma_2 = \gamma$, the mass of the
colliding particles is $m_1 = m_2 = m_p$, and $\beta_1 = -\beta_2 = \beta$. In this case the
center of mass energy is simply twice the energy of each of the particles

$$E_{cm} = 2\gamma\, mc^2 = 2E. \tag{1.27}$$

In *colliding beam facilities* where particles collide with their antiparticles
no particle type conservation laws must be obeyed and therefore the total
kinetic energy gained though particle acceleration becomes available for
high energy physics experiments and the production of new particles at the
collision point. In a similar way we may calculate the available energy for
a variety of collision scenarios like the collision of an accelerated electron
with a stationary proton, the head on collision of electrons with protons or
collisions involving high-energy heavy ions.

1.4 Basic Principles of Particle-Beam Dynamics

Accelerator physics is to a large extend the description of charged particle
dynamics in the presence of external electromagnetic fields or of fields generated by other charged particles. We use *Maxwell's equations* in a vacuum
environment or in a material with well behaved values of the permittivity
and permeability to describe these fields and the *Lorentz force* to formulate
the particle dynamics under the influence of these electromagnetic fields

$$\nabla\left(\epsilon_r \mathbf{E}\right) = 4\pi\rho, \qquad \nabla \times \mathbf{E} = -\frac{1}{c}\frac{\partial}{\partial t}\mathbf{B},$$
$$\nabla\mathbf{B} = 0, \qquad \nabla \times \frac{\mathbf{B}}{\mu_r} = \frac{4\pi}{c}\rho\mathbf{v} + \frac{1}{c}\frac{\partial}{\partial t}\mathbf{E}. \tag{1.28}$$

Here ρ is the charge density and $\mathbf{v}$ the velocity of the charged particles. In
general, we are interested in particle dynamics in a material free environment and set therefore $\epsilon_r = 1$ and $\mu_r = 1$. For specific discussions we do,
however, need to calculate fields in a material filled environment in which
case we come back to the general form of Maxwell's equations (1.28).

Whatever the interaction of charged particles with electromagnetic fields
and whatever the reference system may be, we depend in accelerator physics
on the invariance of the *Lorentz force equation* under coordinate transformations. All acceleration and beam guidance in accelerator physics will be
derived from the Lorentz force which is defined in the case of a particle with
charge q in the presence of an electric $\mathbf{E}$ and magnetic field $\mathbf{B}$ by

$$\mathbf{F} = q\,\mathbf{E} + \frac{q}{c}\left(\mathbf{v} \times \mathbf{B}\right). \tag{1.29}$$

For simplicity we use throughout this text particles with one unit of electrical charge e like electrons and protons unless otherwise noted. In case of multiply charged ions the single charge e must be replaced by eZ where Z is the *charge multiplicity* of the ion. Both components of the Lorentz force are used in accelerator physics where the force due to the electrical field is mostly used to actually increase the particle energy while magnetic fields are used mostly to guide the particle beams along desired beam transport lines. This separation of functions, however, is not exclusive as the example of the betatron accelerator shows where particles are accelerated by time dependent magnetic fields. Similarly electrical fields are used in specific cases to guide or separate particle beams.

If we integrate the Lorentz force over the time a particle interacts with the field we get a change in the particle momentum,

$$\Delta \mathbf{p} = \int \mathbf{F} \, dt. \tag{1.30}$$

On the other hand, if the Lorentz force is integrated with respect to the path length we get the change in *kinetic energy* E_{kin} of the particle

$$\Delta E_{\mathrm{kin}} = \int \mathbf{F} \, ds. \tag{1.31}$$

Comparing (1.30) and (1.31) we find with $ds = \mathbf{v} \, dt$ the relation between the momentum and kinetic energy differentials

$$c\beta \, d\mathbf{p} = dE_{\mathrm{kin}}. \tag{1.32}$$

Inserting the Lorentz force (1.29) into (1.31) and replacing $ds = \mathbf{v} \, dt$ in the second integral we get

$$\Delta E_{\mathrm{kin}} = q \int \mathbf{E} \, ds + \frac{q}{c} \int (\mathbf{v} \times \mathbf{B}) \mathbf{v} \, dt. \tag{1.33}$$

It becomes obvious that the kinetic energy of the particle increases whenever a finite accelerating electric field $\mathbf{E}$ exists and the acceleration occurs in the direction of the electric field. This acceleration is independent of the particle velocity and acts even on a particle at rest $\mathbf{v} = 0$. The second component of the Lorentz force in contrast depends on the particle velocity and is directed normal to the direction of propagation and normal to the magnetic field direction. We find therefore from (1.33) the well-known result that the kinetic energy is not changed by the presence of magnetic fields since the scalar product $(\mathbf{v} \times \mathbf{B}) \mathbf{v} = 0$ vanishes. The magnetic field causes only a deflection of the particle trajectory. The Lorentz force (1.29) in conjunction with (1.30) is used to derive the *equation of motion* of charged particles in the presence of electromagnetic fields

$$\frac{d}{dt} \mathbf{p} = \frac{d}{dt} (m\gamma \mathbf{v}) = eZ\mathbf{E} + \frac{e}{c} Z (\mathbf{v} \times \mathbf{B}), \tag{1.34}$$

where Z is the charge multiplicity of the charged particle. For simplicity we drop from here on the factor Z since the charge multiplicity is different from unity only for ion beams. For ion accelerators we note therefore that the particle charge e must be replaced by eZ.

The fields can be derived from electrical and magnetic potentials in the well-known way

$$
\text{electric field:} \qquad \mathbf{E} = -\frac{1}{c}\frac{\partial \mathbf{A}}{\partial t} - \nabla \Phi,
$$
$$
\text{magnetic field:} \qquad \mathbf{B} = \nabla \times \mathbf{A},
\tag{1.35}
$$

where Φ is the electric scalar potential and $\mathbf{A}$ the magnetic vector potential. The particle momentum is $\mathbf{p} = \gamma m \mathbf{v}$ and it's time derivative

$$
\frac{\mathrm{d}\mathbf{p}}{\mathrm{d}t} = m\gamma\frac{\mathrm{d}\mathbf{v}}{\mathrm{d}t} + m\,\mathbf{v}\,\frac{\mathrm{d}\gamma}{\mathrm{d}t}.
\tag{1.36}
$$

With

$$
\frac{\mathrm{d}\gamma}{\mathrm{d}t} = \frac{\mathrm{d}}{\mathrm{d}\beta}\frac{1}{\sqrt{1-\beta^2}}\frac{\mathrm{d}\beta}{\mathrm{d}t} = \gamma^3\frac{\beta}{c}\frac{\mathrm{d}v}{\mathrm{d}t}
\tag{1.37}
$$

we get from (1.36) the equation of motion

$$
\mathbf{F} = \frac{\mathrm{d}\mathbf{p}}{\mathrm{d}t} = m\left(\gamma\frac{\mathrm{d}\mathbf{v}}{\mathrm{d}t} + \gamma^3\frac{\beta}{c}\frac{\mathrm{d}v}{\mathrm{d}t}\mathbf{v}\right).
\tag{1.38}
$$

For a force parallel to the particle propagation $\mathbf{v}$ we have $\dot{v}\,\mathbf{v} = \dot{\mathbf{v}}\,v$ and (1.38) becomes

$$
\frac{\mathrm{d}\mathbf{p}_\parallel}{\mathrm{d}t} = m\gamma\left(1 + \gamma^2\beta\frac{v}{c}\right)\frac{\mathrm{d}\mathbf{v}_\parallel}{\mathrm{d}t} = m\gamma^3\frac{\mathrm{d}\mathbf{v}_\parallel}{\mathrm{d}t}.
\tag{1.39}
$$

On the other hand, if the force is directed normal to the particle propagation we have $\mathrm{d}v/\mathrm{d}t = 0$ and (1.38) reduces to

$$
\frac{\mathrm{d}\mathbf{p}_\perp}{\mathrm{d}t} = m\gamma\frac{\mathrm{d}\mathbf{v}_\perp}{\mathrm{d}t}.
\tag{1.40}
$$

It is obvious from (1.39) and (1.40) how differently the dynamics of particle motion is affected by the direction of the Lorentz force. Specifically the dynamics of highly relativistic particles under the influence of electromagnetic fields depends greatly on the direction of the force with respect to the direction of particle propagation. The difference between parallel and perpendicular acceleration will have a great impact on the design of electron accelerators. As we will see later the acceleration of electrons is limited due to the emission of synchrotron radiation. This limitation, however, is much more severe for electrons in circular accelerators where the magnetic forces act perpendicularly to the propagation compared to the acceleration in linear accelerators where the accelerating fields are parallel to the particle propagation. This argument is also true for protons or for that matter, any charged particle, but because of the much larger particle mass the amount of synchrotron radiation is generally negligible small.

1.4.1 Stability of a Charged-Particle Beam

Individual particles in an intense beam are under the influence of strong repelling electrostatic forces creating the possibility of severe stability problems. Particle beam transport over long distances could be greatly restricted unless these *space-charge forces* can be kept under control. First, it is interesting to calculate the magnitude of the problem.

If all particles would be at rest within a small volume, we would clearly expect the particles to quickly diverge from the center of charge under the influence of the repelling forces from the other particles. This situation may be significantly different in a particle beam where all particles propagate in the same direction. We will therefore calculate the fields generated by charged particles in a beam and derive the corresponding Lorentz force due to these fields. Since the Lorentz force equation is invariant with respect to coordinate transformations, we may derive this force either in the laboratory system or in the moving system of the particle bunch.

First we will perform the calculations in the laboratory system and assume a continuous stream of particles moving along the s-axis with the velocity v_s. Assuming a uniform particle density distribution ρ_o within the beam we find for symmetry reasons only a radial electrical field E_r and an azimuthal magnetic field B_φ. The radial electric field E_r at a distance r from the beam axis can be derived from Coulomb's law $\nabla \mathbf{E} = 4\pi \rho_o$ expressed in cylindrical coordinates which becomes after integration

$$E_r = 2\pi \rho_o r. \tag{1.41}$$

Similarly, we get from Ampere's law $\nabla \times \mathbf{B} = \frac{4\pi}{c} \rho_o \mathbf{v}$ the azimuthal magnetic field

$$B_\varphi = 2\pi \rho_o \frac{v}{c} r. \tag{1.42}$$

These field components determine the Lorentz force due to electromagnetic fields generated by the beam itself and acting on a particle within that beam

$$F_r = e \left(E_r - \frac{v}{c} B_\varphi \right) = 2\pi e \frac{\rho_o}{\gamma^2} r. \tag{1.43}$$

Only the radial component of the Lorentz force is finite. The Lorentz force remains repelling but due to a relativistic effect we find that the repelling electrostatic force at higher energies is increasingly compensated by the magnetic field. The total Lorentz force due to space charges therefore vanishes like γ^{-2} for higher energies. Obviously this repelling space charge force is generally much stronger for proton and especially for ion beams because of the smaller value for γ and, in the case of ions, because of the larger charge multiplicity which increases the space-charge force by a factor of Z.

We find the same result if we derive the Lorentz force in the moving system S^* of the particle beam and then transform to the laboratory system. In this moving system we have obviously only the repelling electrostatic force since the particles are at rest and the only field component is the radial electrical field which is from (1.41)

$$F_r^* = e\,E_r^* = e\,2\pi\rho_o^*\,r^* . \tag{1.44}$$

Transforming this equation back into the laboratory system we note that this force is purely radial and therefore acts only on the radial momentum. With $F_r = dp_r/dt$ and $p_r = p_r^*$ we find $F^* = \gamma\,F_r$ since $dt = \gamma\,dt^*$. The charge densities in both systems are related by $\rho^* = \rho/\gamma$, the radii by $r^* = r$, and the Lorentz force in the laboratory system becomes thereby

$$F_r = 2\pi e\,\frac{\rho_o}{\gamma^2}\,r \tag{1.45}$$

in agreement with (1.43).

We obtained the encouraging result that at least relativistic particle beams become stable under the influence of their own fields. For lower particle energies, however, significant diverging forces must be expected and adequate focusing measures must be applied. The physics of such space charge dominated beams is beyond the scope of this book and is treated elsewhere, for example in considerable detail in [1.34, 35].

Problems

Problem 1.1. Use the definition for β, the momentum, the total and kinetic energy and derive expressions $p(\beta, E)$, $p(E_{kin})$, and $E_{kin}(\gamma)$. Simplify the expressions for large energies, $\gamma \gg 1$. Derive from these relativistic expressions the classical nonrelativistic formulas.

Problem 1.2. Protons are accelerated to a kinetic energy of 200 MeV at the end of the Fermilab *Alvarez linear accelerator*. Calculate their total energy, their momentum and their velocity in units of the velocity of light.

Problem 1.3. Consider electrons to be accelerated in the 3 km long SLAC linear accelerator with a uniform gradient of 20 MeV/m. The electrons have a velocity $v = c/2$ at the beginning of the linac. What is the length of the linac in the rest frame of the electron? Assume the particles at the end of the 3 km long linac would enter another 3 km long tube and coast through it. How long would this tube appear to be to the electron ?

Problem 1.4. Plot on log-log scale the velocity β, momentum, and kinetic energy as a function of the total energy for electrons, protons, and gold ions Au^{+14}. Vary the total energy from $0.01\,mc^2$ to $10^4\,mc^2$.

Problem 1.5. A charged pion meson has a rest energy of 139.568 MeV and a mean life time of $\tau_{o\pi} = 26.029$ nsec in its rest frame. What are the life times, if accelerated to a kinetic energy of 20 MeV? and 100 MeV? A pion beam decays exponentially like e^{-t/τ_π}. At what distance from the source will the pion beam intensity have fallen to 50%, if the kinetic energy is 20 MeV? or 100 MeV?

Problem 1.6. The design for the Relativistic Heavy Ion Collider RHIC [1.36] calls for the acceleration of completely ionized gold atoms in a circular accelerator with bending magnets reaching a maximum field of 34.5 kG. What is the maximum achievable kinetic energy per nucleon for gold ions Au^{+77} compared to protons? Calculate the total energy, momentum, and velocity of the gold atoms. $(A_{Au} = 197)$

Problem 1.7. Gold ions Au^{+14} are injected into the Brookhaven Alternating Gradient Synchrotron, AGS, at a kinetic energy per nucleon of 72 MeV/u. What is the velocity of the gold ions? The AGS was designed to accelerate protons to a kinetic energy of 28.1 GeV. What is the corresponding maximum kinetic energy per nucleon for these gold ions that can be achieved in the AGS? The circulating beam is expected to contain 6×10^9 gold ions. Calculate the beam current at injection and at maximum energy assuming there are no losses during acceleration. The circumference of the AGS is $C_{AGS} = 807.1$ m. Why does the beam current increase although the circulating charge stays constant during acceleration?

Problem 1.8. A positron beam accelerated in the linac of problem 1.3 hits a fixed hydrogen target. What is the available energy from a collision with a target electron assumed to be at rest? Compare this available energy with that obtained in a linear collider where electrons and positrons from two similar linacs collide head on at the same energy.

Problem 1.9. The SPEAR colliding beam storage ring has been constructed originally for electron and positron beams to collide head on with an energy of up to 3.5 GeV. At 1.55 GeV per beam a new particle, the Ψ/J-particle, was created. In a concurrent experiment such Ψ/J particle have been produced by protons hitting a hydrogen target. What proton energy was required to produce the new particle? Determine the positron energy needed to create Ψ/J particles by collisions with electrons in a fixed target.

Problem 1.10. Assume you want to produce antiprotons by accelerating protons and letting them collide with other protons in a stationary hydrogen target. What is the minimum kinetic energy the accelerated protons must have to produce antiprotons? Use the reaction $p + p \rightarrow p + p + p + \bar{p}$.

Problem 1.11. A proton with a kinetic energy of 1 eV is emitted parallel to the surface of the earth. What is the bending radius due to gravitational forces? What are the required transverse electrical and magnetic fields to obtain the same bending radius? What is the ratio of electrical to magnetic field? Is this ratio different for a proton energy of say 10 TeV? Why? (gravitational constant $f = 6.6710^{-13} \text{Ncm}^2/\text{g}^2$)

Problem 1.12. Express the equation of motion (1.34) for $Z = 1$ in terms of particle acceleration, velocity and fields only. Verify from this result the validity of (1.39) and (1.40).

Problem 1.13. A circular accelerator with a circumference of 300 m contains a uniform distribution of singly charged particles orbiting with the speed of light. If the circulating current is 1 amp, how many particles are orbiting? We instantly turn on an ejection magnet so that all particles leave the accelerator during the time of one revolution. What is the peak current at the ejection point? How long is the current pulse duration? If the accelerator is a synchrotron accelerating particles at a rate of 100 acceleration cycles per second, what is the average ejected particle current?

Problem 1.14. Prove the validity of the field equations

$$E_r = 2\pi \rho_0 r \quad \text{and} \quad B_\varphi = 2\pi \rho_0 \beta r \tag{1.46}$$

for a uniform cylindrical particle beam with constant charge density ρ_0 within a radius R. Derive the field expressions for $r > R$. Consider a highly relativistic electron bunch of 10^{10} electrons. The bunch has the form of a cylindrical slug, 1 mm long and a diameter of 0.2 μm. What is the electrical and magnetic field strength at the surface of the beam. Calculate the peak electrical current of the bunch. If two such beams in a linear collider with an energy of 500 GeV pass by each other at a distance of 10 μm, what is the deflection angle of each beam due to the field of the other beam?

Problem 1.15. Derive expressions for the electrical and magnetic fields for a radial charge distribution $\rho(r, \varphi, z) = \rho(r)$. Assume a gaussian charge distribution with standard deviation σ

$$\rho(r) = \rho_0 \exp\left(-\frac{r^2}{2\sigma^2}\right). \tag{1.47}$$

What are the fields for $r = 0$ and $r = \sigma$?

Problem 1.16. Use the results of problem 1.14 and consider a parallel beam at the beginning of a long magnet free drift space. Follow a particle under the influence of the beam self fields starting at a distance from the axis of $r_0 = \sigma$. Derive the radial particle distance from the axis as a function of s.

Problem 1.17. Particles undergo elastic collisions with gas atoms. The rms multiple scattering angle is given by [1.37]

$$\sigma_\theta \approx Z \frac{20(\text{MeV}/c)}{\beta p} \sqrt{\frac{s}{\ell_r}}, \tag{1.48}$$

where Z is the charge multiplicity of the beam particles, s the distance travelled and ℓ_r the radiation length of the scattering material (for air the radiation length at atmospheric pressure is $\ell_r = 500$ m or 60.2 g/cm^2). Derive an expression for the beam radius as a function of s due to scattering. What is the approximate tolerable gas pressure in a proton storage ring if a particle beam is supposed to orbit for 20 hr and the elastic gas scattering shall not increase the beam size by more than a factor of two during that time.

2 Linear Accelerators

Before we address the physics of beam dynamics in accelerators it seems appropriate to discuss briefly various methods of particle acceleration as they have been developed over the years. It would, however, exceed the purpose of this text to discuss all variations of particle accelerators in detail. Fortunately extensive literature is available on a large variety of different accelerators and therefore only fundamental principles of particle acceleration shall be discussed here. A valuable source of information for more detailed information on the historical development of particle accelerators is Livingston's collection of early publications on accelerator developments [2.1].

The development of charged-particle accelerators has progressed along double paths which by the appearance of particle trajectories are distinguished as *linear accelerators* and *circular accelerators*. Particles travel in linear accelerators only once through the accelerator structure while in a circular accelerator they follow a closed orbit periodically for many revolutions accumulating energy at every traversal of the accelerating structure.

2.1 Principles of Linear Accelerators

No fundamental advantage or disadvantage can be claimed for one or the other class of accelerators. It is mostly the particular application and sometimes the available technology that determines the choice between both classes. Both types have been invented and developed early in this century, and continue to be improved and optimized as associated technologies advance. In this chapter we will concentrate on linear accelerators and postpone the discussion on circular accelerators to the next chapter. In linear accelerators the particles are accelerated by definition along a straight path by either *electrostatic fields* or oscillating *rf fields*.

2.1.1 Charged Particles in Electric Fields

In accelerator physics all forces on particles originate from electromagnetic fields. The interaction of such fields with charged particles will be discussed in a very general way to point out the basic process of particle acceleration. For particle acceleration we consider only the electric-field term of the *Lorentz force*. The nature of the electric field can be static, pulsed, generated by a time varying magnetic field or a rf field. Both the electric and magnetic fields are connected by *Maxwell's equation*

$$\nabla \times \mathbf{E} = -\frac{1}{c}\frac{d\mathbf{B}}{dt}.$$ (2.1)

Using *Stokes' integral theorem*

$$\int \nabla \times \mathbf{E} \cdot d\mathbf{a} = \oint \mathbf{E} \cdot d\mathbf{s} = -\frac{1}{c}\frac{d}{dt}\int \mathbf{B} \cdot d\mathbf{a},$$ (2.2)

where d**a** is the differential surface vector directed normal to the surface.

From this equation it becomes apparent that a changing magnetic flux generates an electric field surrounding this flux. Electromagnetic fields at all frequencies can be used for particle acceleration. On one end of the spectrum we have static electric fields which play a significant role for low-energy particle acceleration. Slowly varying magnetic fields are the basis for electric transformers and can be applied as well for the acceleration of particles in *induction linear accelerators* or *betatrons*. The most extensive use of electromagnetic waves utilizes *rf fields* in specially designed *resonant cavities*. At the other end of the frequency spectrum of electromagnetic waves are intense light beams, for example, in the form of laser beams which can provide very high electric fields. At this time, however, the technical problem of turning the purely transverse *laser field* into longitudinal accelerating fields has not been solved yet.

In this chapter we concentrate on the use of static electric fields and rf waves for particle acceleration. Such fields are generated by appropriate sources hooked up to an *accelerating section* which, in the case of electrostatic fields, consists of just two electrodes with the particle source at the potential of one electrode and a hole in the center of the other electrode to let the accelerated particles pass through. Special resonant cavities are used as accelerating sections with two holes on the axis of the cavity to let the beam pass through. Either field can be represented by the *plane wave* equation

$$\mathbf{E}(\psi) = \mathbf{E}_o \cdot e^{i\,(\omega t - ks)} = \mathbf{E}_o \cdot e^{i\psi},$$ (2.3)

where ω is the frequency and k the wave number including the case of static fields with $\omega = 0$ and $k = 0$. The Lorentz force acting on an electric charge is

$$\mathbf{F} = \frac{d}{dt}mc\gamma\beta = e\,\mathbf{E}(\psi)$$ (2.4)

and the *equation of motion* for particles under this force is

$$\frac{d}{dt} mc\gamma\beta = e\,\mathbf{E}(\psi). \tag{2.5}$$

Integration of (2.5) results in an expression for the *momentum gain* of the particle

$$\Delta p = mc\,(\gamma\beta - \gamma_0\beta_0) = e\int \mathbf{E}(\psi)\cdot dt, \tag{2.6}$$

where $mc\gamma_0\beta_0$ is the initial momentum of the particle. Generally it is more complicated to perform a time integration which requires the tracking of particles though the accelerating cavity. To simplify the calculation we look for the gain in kinetic energy which reduces (2.6) to a spatial integration of the electric field in the accelerating cavity. This integral is a property of the cavity and is independent of particle motion. With $\beta\,\Delta cp = \Delta E_{\text{kin}}$ the energy gain for particles passing through the accelerating section is

$$\Delta E_{\text{kin}} = e\int_{L_{\text{cy}}} \mathbf{E}(\psi)\cdot ds, \tag{2.6}$$

where L_{cy} is the length of the accelerating section.

The effectiveness of acceleration in an rf field depends greatly on the phase relationship of the field with the particle motion. For successful particle acceleration we expect therefore the need to meet specific synchronicity conditions to ensure continuous acceleration.

2.1.2 Electrostatic Accelerators

In *electrostatic accelerators* the potential difference between two electrodes is used for particle acceleration as shown in Fig. 2.1.

The most simple such arrangement has been used now for almost two centuries in *glow discharge* tubes for fundamental research on the nature of electrons and plasmas, as light sources or as objects of aesthetic interest due to colorful phenomena in such tubes. In another, more modern application

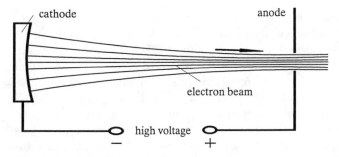

Fig. 2.1. Principle of electrostatic accelerators

electrons are accelerated in an *x-ray tube* by high electrostatic fields and produce after striking a metal target intense x-rays used in medicine and industry.

The voltages that can be achieved by straight voltage transformation and rectification are quite limited by electrical breakdown effects to a few ten thousand volts. More sophisticated methods of producing high voltages therefore have been developed to reach potential differences of up to several million volts.

Tesla Transformer: After the discovery of rf waves by Hertz in 1887 much time and effort went toward the development of sources for rf fields of substantial power. In one such development Tesla built an ironless high frequency transformer consisting of a long secondary coil with many turns placed along the axis of a primary coil with few windings as shown schematically in Fig. 2.2.

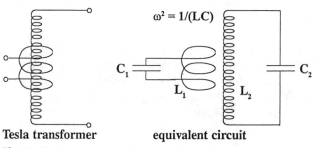

Tesla transformer   equivalent circuit

Fig. 2.2. Tesla transformer (schematic)

The secondary coil in connection with stray capacitance between the coil windings forms a high frequency resonance circuit. Maximum coupling and transformation efficiency is reached when the primary coil is part of a resonant circuit tuned to the resonant frequency of the secondary coil. The voltage on the secondary coil can be used to accelerate a pulse of particles during one half period of the voltage oscillation.

Cascade Generators: The basic method implemented in the *cascade generator* is that of a voltage multiplier circuit which has been proposed by Greinacher [2.2] in 1914 and Schenkel [2.3] in 1919 which allows to achieve a multiplication of the voltage across the plates of a capacitor. A set of capacitors are charged through appropriately placed diodes from an alternating current source (Fig. 2.3) in such a way that during the positive half wave, half the capacitors are charged to a positive voltage and during the negative half wave, the other half of the capacitors are charged to a negative voltage thus providing twice the maximum ac voltage. By arranging $2N$ capacitors in this way the charging voltage can be multiplied by the factor N.

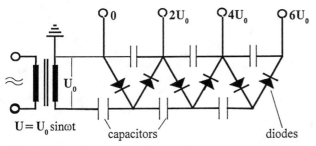

Fig. 2.3. Cascade generator (schematic)

While there is no fundamental limit to the total voltage, high voltage break down will impose a technical limit on the maximum achievable voltage.

Based on this method Cockcroft and Walton [2.4] developed appropriate high-voltage techniques and built the first high energy particle accelerator reaching voltages as high as several million Volt. Applying the high voltage to a beam of protons they were able for the first time to initiate through artificially accelerated protons an atomic reaction, in this case the conversion of a Lithium nucleus into two helium nuclei, in the reaction

$$p + Li \rightarrow 2\,He. \tag{2.8}$$

Such *Cockcroft-Walton accelerators* turned out to be very efficient and are still used as the first step in modern proton accelerator systems. Obviously with this kind of voltage generation it is not possible anymore to produce a continuous stream of particles. Because of the switching process, there is a time to charge the capacitors followed by a time to apply the multiplied voltage to particle acceleration. As a consequence, we observe a pulsed particle beam from a Cockroft-Walton accelerator.

The Van de Graaff Accelerator: Much higher voltages can be reached with a *Van de Graaff accelerator* [2.5]. Here electric charge is extracted by field emission from a pointed metal electrode and sprayed onto an isolated endless belt. This belt is moved by motor action to carry the charge to the inside of a hollow sphere, where the charge is stripped off again by reversed field emission onto a pointed metal electrode which is connected to the sphere. The principle of this *electrostatic generator* is shown in Fig. 2.4.

Electrical charges in a metallic conductor collect on the outside and it is therefore possible to continuously accumulate electrical charge by deposition to the inside surface of a hollow metallic sphere. If the whole system is placed into a high pressure vessel filled with an electrically inert gas like Freon or SF_6 voltages as high as 20 million Volts can be reached.

The high voltage can be used to accelerate electrons as well as protons or ions. In the latter two cases more than double the accelerating voltage can be achieved in a *Tandem Van de Graaff* accelerator. If a proton beam must

29

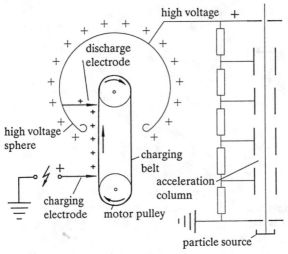

Fig. 2.4. Van de Graaff accelerator (schematic)

be accelerated, the accelerating process is started with negatively charged hydrogen ions H$^-$ from a plasma discharge tube which are then accelerated say from ground potential to the full Van de Graaff voltage $+V$. At that point the two electrons of the negative hydrogen ion are stripped away by a thin foil or gas curtain resulting by *charge exchange* in positively charged protons which can be further accelerated between the potential $+V$ and ground potential to a total kinetic energy $E_{kin} = e\,2V$.

High electrostatic voltages from a Van de Graaff generator cannot be applied directly to just two electrodes as shown in Fig. 2.1. Because of the great distance between the electrodes necessary to avoid voltage break down the fields would not be distributed uniformly along the axis of the *acceleration column*. Therefore, the voltage is applied to a series of resistors connected to electrodes which allow a uniform distribution of the electrical field along the acceleration column as shown in Fig. 2.4. A more detailed review of the development of electrostatic high voltage generators can be found in [2.6].

2.1.3 Induction Linear Accelerator

Very large particle beam intensities can be accelerated through the repeated application of the transformer principle to generate pulsed electric fields across a gap, where the beam serves as the secondary coil [2.7]. In a betatron an azimuthal electric acceleration field is generated around a varying magnetic dipole field. In the principle of the induction linac the field patterns are just reversed. A time varying azimuthal magnetic field is generated resulting in a high electric dipole field across a gap along the particle path.

The first such accelerator was successfully operated in 1964 [2.8]. In a practical application many such transformer units are lined up along a straight path and are triggered in synchronism with the particles. Induction linear accelerators can be designed to accelerate a beam current of up to 1 kA to moderate energies of a few MeV [2.9, 10].

2.2 Acceleration by rf Fields

The most successful acceleration of particles is based on the use of *rf fields* for which by now powerful sources exist. Very high accelerating voltages can be achieved in *resonant rf cavities* far exceeding those obtainable in electrostatic accelerators of similar dimensions. Particle acceleration in *linear accelerators* as well as in circular accelerators are based on the use of rf fields and we will in the following sections and in the next chapter discuss the principles of the more important types of particle accelerators.

2.2.1 Basic Principle of Linear Accelerators

The principle of the *linear accelerator* based on alternating fields and *drift tubes* was proposed by Ising [2.11] and Wideroe [2.12]. In this method particles are accelerated by repeated application of rf fields. Wideroe constructed such an accelerator and was able to accelerate potassium ions up to 50 keV.

While the principle is simple, the realization requires specific conditions to ensure that the particles are exposed to only accelerating rf fields. During the half period when the fields reverse sign the particles must be shielded from the fields in order not to be decelerated again.

Technically this requirement is realized by surrounding the beam path with metallic *drift tubes* as shown in Fig. 2.5. The tubes shield the particles from external rf fields and the length of the tube segments are chosen such that the particles reach the gap between two successive tubes at the moment the rf field is accelerating.

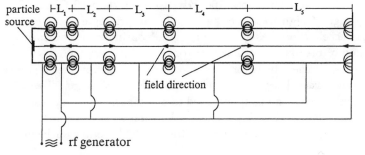

Fig. 2.5. Wideroe linac structure (schematic)

Synchronicity Condition: For efficient acceleration the motion of the particles must be synchronized with the rf fields in the accelerating sections. The drift tubes are dimensioned such that the particles travel for most of the rf period in the field free interior and emerge in a gap to the next drift tube at the moment the accelerating field reaches a maximum. The length of the shielding tubes is therefore almost as long as it take the particles to travel for half an rf period. In this case we have synchronism between particle motion and rf field and the length of the ith drift tube is

$$L_i \lesssim \tfrac{1}{2} v_i T_{\mathrm{rf}} \,, \tag{2.9}$$

where v_i is the velocity of the particles in the ith drift tube and T_{rf} the rf period.

Stimulated by the successful acceleration of potassium ions by Wideroe, a group led by Sloan and Lawrence at Berkeley were able to build a 50 kW rf generator oscillating at 10 MHz and delivering a gap voltage of 42 kV. Applying this to 30 acceleration tubes they were able to accelerate mercury ions to a total kinetic energy of 1.26 MeV [2.13].

In the twenties when this principle was developed it was difficult to build high frequency generators at significant power. In 1928 rf generators were available only up to about 7 MHz and numerical evaluation of (2.9) shows that this principle was useful only for rather slow particles like low energy protons and ions. The drift tubes can become very long for low rf frequencies. Particles traveling with, for example, half the speed of light would require a drift tube length of 10.7 m at 7 MHz. Such long drift tubes add up quickly to a very long accelerators before the particles approach the speed of light. To reduce the length of the tubes higher frequencies are required.

Further progress in the development of rf linear accelerators therefore depended greatly on the development of rf equipment at high frequency which happened during World War II in connection with the development of radar systems. In 1937, Hansen and the Varian brothers invented the *klystron* at Stanford. Soon the feasibility of high power klystrons had been established [2.14] which to this date is one of the most efficient rf amplifiers available. The first klystron was developed for 3000 MHz which is still the preferred frequency for high energy electron linear accelerators. The klystron principle is economically feasible from about 100 MHz to more than 10 GHz. With such a wide range of high frequencies available the principle of rf acceleration in linear accelerators has gained quick and continued prominence for the acceleration of protons as well as electrons.

At higher frequencies, however, the capacitive nature of the *Wideroe structure* becomes very lossy due to electromagnetic radiation. To overcome this difficulty, Alvarez [2.15] proposed to enclose the gaps between the tubes by metallic cavities (Fig. 2.6). The acceleration section would now be composed of a series of tubes forming together with the outer enclosure a resonant cavity.

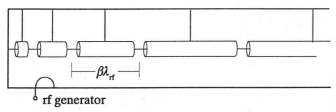

rf generator

Fig. 2.6. Alvarez linac structure (schematic)

This *Alvarez structure* is still the preferred preaccelerator to accelerate protons and ions from a few hundred keV out of a Cockroft Walton electrostatic generator to a few hundred MeV for injection into a booster synchrotron. Because of the lower velocity of protons and ions at up to a few hundred MeV the operating frequency for proton linacs is generally around 200 MHz.

Much higher radio frequencies of 3000 MHz and higher are desired for electron acceleration. We will discuss in some more detail the basic features and scaling of high frequency accelerating structures without going too much into the details of rf technology. The objective of this discussion is to give the interested reader the tools to understand the scaling and limitations of basic accelerator physics. For more detailed discussion of rf aspects in linear electron accelerators the reader is referred to the literature [2.16– 18].

2.2.2 Waveguides for High Frequency EM Waves

The interaction between an electromagnetic wave and charged particles was discussed in Sect. 2.1.1 in a very general way. We ignored the fact that free electromagnetic waves do not have field components in the direction of propagation and are therefore not suited for particle acceleration without further modification.

To make electromagnetic fields useful for particle acceleration, boundary conditions must be introduced such that plane EM waves become modified to exhibit longitudinal field components. For this purpose we study the field patterns and propagation characteristics of electromagnetic waves in a cylindrical or rectangular tube. The wave equation for the accelerating field component E_s is given by the *Laplace equation*

$$\nabla^2 E_s + \frac{\omega^2}{c^2} E_s = 0 \tag{2.10}$$

with the solution

$$E_s = E_{os} e^{i(\omega t - ks)} . \tag{2.11}$$

A similar equation is valid for the azimuthal magnetic field.

We separate the Laplace operator into a longitudinal and transverse part and get with $\partial^2/\partial s^2 = -k^2$ from (2.10)

$$\nabla_\perp^2 E_s + \left(\frac{\omega^2}{c^2} - k^2\right) E_s = 0. \tag{2.12}$$

Cylindrical coordinates (r, θ, s) are more appropriate to express field patterns in a *cylindrical waveguide*. From symmetry considerations we expect the azimuthal field component to vary periodically with the azimuth θ and therefore $\partial^2/\partial\theta^2 = -n^2$, where n is the periodicity. With this (2.12) becomes

$$\frac{\partial^2 E_s}{\partial r^2} + \frac{1}{r}\frac{\partial E_s}{\partial r} + \left(k_c^2 - \frac{n^2}{r^2}\right) E_s = 0, \tag{2.13}$$

where the *cut off wave number* k_c is defined by

$$k_c^2 = \frac{\omega^2}{c^2} - k^2. \tag{2.14}$$

This differential equation can be readily solved with *Bessel functions* of the form

$$E_s = A\,J_n(k_c r) + B\,Y_n(k_c r), \tag{2.15}$$

where the coefficient B must vanish to avoid a singularity for $r = 0$. The remaining solution represent many modes n for the field component E_s consistent with the boundary conditions. Transverse field components $(E_r, E_\theta, H_r, H_\theta)$ can be derived from (2.15) using Maxwell's curl equations and exhibit a similar mode structure.

The multitude of possible modes can be classified into two main groups, the TE *modes*, where all electric field components are transverse, and the TM *modes*, with only transverse magnetic field components. We are interested only in the TM modes which are the only ones to exhibit longitudinal electric fields. Individual TM modes are characterized by three indices, TM_{npq}, where n is the azimuthal periodicity and p and q the radial and longitudinal periodicity, respectively. A TM_{010}-mode therefore has no azimuthal and longitudinal periodicity, e.g., the field $E_s(r, \theta, s)$ is independent of θ and s but has one node in radial direction at the wall, where the electric field must be zero.

In discussing electromagnetic fields in metallic tubes we must consider proper boundary conditions on wall surfaces. All electric field components parallel to any metallic surface must vanish and we have, for example, for a *pill box cavity* of axial length ℓ and radius a the boundary conditions:

$$\begin{aligned} E_s = E_\theta = 0 \quad &\text{for} \quad r = a, \quad \text{and} \\ E_r = E_\theta = 0 \quad &\text{for} \quad s = 0 \quad \text{and} \quad s = \ell. \end{aligned} \tag{2.16}$$

The solutions of (2.13) for $n = 0$, satisfying these boundary conditions, are for the electric field components

$$E_s = E_o J_o(k_c r) \cdot e^{i(\omega t - ks)},$$
$$E_\theta = 0,$$
$$E_r = -i \frac{k}{k_c} E_o J_o'(k_c r) e^{i(\omega t - ks)}, \qquad (2.17)$$

and for the magnetic field components

$$H_r = 0,$$
$$H_\theta = -i \frac{\omega}{ck_c} E_o J_o'(k_c r) e^{i(\omega t - ks)}, \qquad (2.18)$$
$$H_s = 0,$$

where $J_o'(x) = \partial J_o(x)/\partial x$ and $J_o' = -J_1$. For (2.17) to be true the longitudinal electric field must vanish on the wall surface of the cavity which occurs for $J_o(k_c a) = 0$, or at a radius $r = a$ defined by the location of the first root of the Bessel function J_o

$$k_c \cdot a = 2.405. \qquad (2.19)$$

The field pattern for the TM_{o1o}-*mode* is shown in Fig. 2.7. We note from (2.17) and Fig. 2.7 that there exists a longitudinal electric field which can be useful for particle acceleration and this field is maximum on the *waveguide* axis falling off toward the wall boundaries. The fields are independent of θ because we chose $n = 0$ and independent of s.

From (2.12) we conclude that the *cutoff wave number* must be positive

$$k_c^2 = \frac{\omega^2}{c^2} - k^2 > 0, \qquad (2.20)$$

in order to obtain a *travelling wave* rather than a wave decaying exponentially along the waveguide for $k_c^2 < 0$. On the other hand, we require from (2.11) that $k > 0$ or $\omega/c > k_c$ and define a *cutoff frequency*,

$$\omega_c = c k_c = c \frac{2.405}{a}, \qquad (2.21)$$

below which no wave can propagate in the waveguide. The *propagation factor* k is then

$$k^2 = \left(\frac{\omega}{c}\right)^2 \left(1 - \frac{\omega_c^2}{\omega^2}\right). \qquad (2.22)$$

The cutoff frequency is determined by the diameter of the waveguide and limits the propagation of electromagnetic waves in circular waveguides to wavelengths which are less than the diameter of the pipe. Waves with longer wavelengths have an imaginary propagation factor k and decay exponen-

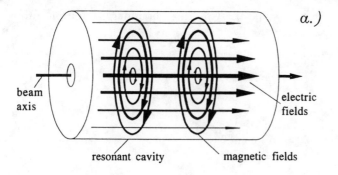

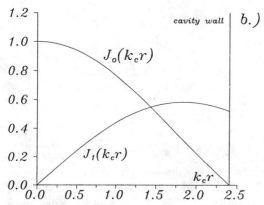

Fig. 2.7. Electromagnetic field pattern for a TM_{010} mode in a circular waveguide a.) three dimensional field configuration, b.) radial dependence of fields

tially along the waveguide. To determine the phase velocity of the wave we set $\psi = \omega t - ks = $ const and get from the derivative $\dot\psi = \omega - k\dot s = 0$ the *phase velocity*

$$v_{\mathrm{ph}} = \dot s = \frac{\omega}{k}.$$

(2.23)

Inserting (2.22) into (2.23) we get a *phase velocity* which exceeds the velocity of light and therefore any velocity a material particle can reach,

$$v_{\mathrm{ph}} = \frac{c}{\sqrt{1 - \omega_c^2/\omega^2}} > c.$$

(2.24)

The phase velocity of the electromagnetic wave in a circular metallic tube is larger than the velocity of light. We were able to modify plane electromagnetic waves in such a way as to produce the desired longitudinal electric field component but note that these fields are not yet suitable for particle acceleration because the phase rolls over the particles and the net acceleration is zero. To make such electromagnetic waves useful for particle acceleration further modifications of the waveguide are necessary to slow down the phase velocity.

To complete our discussion we determine also the group velocity which is the velocity of electromagnetic energy transport along the waveguide. The *group velocity* v_g is defined by

$$v_g = \frac{d\omega}{dk}. \tag{2.25}$$

Differentiating (2.20) with respect to k we get

$$k = \frac{\omega}{c^2} \frac{d\omega}{dk}$$

or with $k = \omega/v_{ph}$ the group velocity

$$v_g = \frac{d\omega}{dk} = \frac{c^2 k}{\omega} = c \cdot \frac{c}{v_{ph}} < c, \tag{2.26}$$

since $v_{ph} > c$.

Disked Loaded Waveguides: The phase velocity v_{ph} must be equal to the particle velocity v_p for efficient acceleration and we need to modify or "load" the wave guide structure to reduce the phase velocity. This can be done by inserting metallic structures into the aperture of the circular wave guide. Many different ways are possible, but we will consider only the *disk loaded waveguide* which is the most common accelerating structure for electron linear accelerators.

In a disk loaded waveguide metallic plates are inserted normal to the waveguide axis at periodic intervals with small holes in the center to allow for the passage of the particle beam as shown in Fig. 2.8.

The boundary conditions and therefore the electromagnetic fields in such a structure are significantly more complicated than those in a simple circular tube. It would exceed the goal of this text to derive the theory of disk loaded

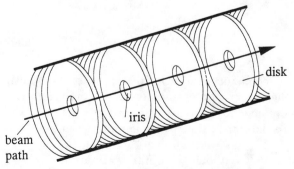

Fig. 2.8. Disk loaded accelerating structure for an electron linear accelerator (schematic)

waveguides and the interested reader is referred to the review article by Slater [2.19].

Insertion of disks in periodic intervals into a uniform waveguide essentially creates a sequence of cavities with electromagnetic coupling through the central hole, holes at some radius on the disks or external coupling cavities. The whole arrangement of cells acts like a band pass filter allowing electromagnetic fields of certain frequencies to propagate. By proper choice of the geometric dimensions the pass band can be adjusted to the desired frequency and the phase velocity can be designed to be equal to the velocity of the particles. For electron linear accelerators the phase velocity is commonly adjusted to the velocity of light since electrons quickly reach such a velocity.

Energy Gain: To calculate the energy gain properly the variation of the field must be taken into account while the particle is moving through the cavity. We assume the particle is in the center of the accelerating gap at $t = 0$ and its velocity is v. Since the cavity fields oscillate, the potential *energy gain* oscillates as well

$$eV(t) = e\widehat{V} \cos \omega t = e\widehat{V} \cos \left(\omega \frac{s}{v} \right) , \qquad (2.27)$$

where $s = vt$, assuming the velocity v to change little during the acceleration in one short accelerating gap. The energy gain in a cavity of length ℓ_c is

$$\Delta E_{\mathrm{kin}} = \int_{-\frac{1}{2}\ell_c}^{\frac{1}{2}\ell_c} \frac{e\widehat{V}_{\mathrm{o}}}{\ell_c} \cos \left(\omega \frac{s}{v} \right) \cdot \mathrm{d}s . \qquad (2.28)$$

Performing the integration, we have finally for the energy gain in a resonant cavity of length $\ell_c = \lambda_{\mathrm{rf}}/2$

$$\Delta E_{\mathrm{kin}} = e\widehat{V}_{\mathrm{o}} \frac{\sin \left(\frac{\omega \lambda_{\mathrm{rf}}}{4v} \right)}{\frac{\omega \lambda_{\mathrm{rf}}}{4v}} , \qquad (2.29)$$

where the factor

$$T_t = \frac{\sin \left(\frac{\omega \lambda_{\mathrm{rf}}}{4v} \right)}{\frac{\omega \lambda_{\mathrm{rf}}}{4v}} \qquad (2.30)$$

is called the *transit time factor*. For small velocities, $v \ll c$, the transit time factor and thereby the energy gain is small or maybe even negative. Maximum energy gain is obtained for particles travelling at or close to the speed of light in which case the transit time factor approaches unity.

Basic Waveguide Parameters: Without going into structure design and detailed determination of geometric parameters we can derive parameters relating to the acceleration capability of such structures. Conservation of energy requires that

$$\frac{\partial W}{\partial t} + \frac{\partial P}{\partial s} + P_{\mathrm{w}} + n\,e\,v\,E_s = 0 \,, \tag{2.31}$$

where W is the stored energy per unit length, P the energy flux along s, P_{w} wall losses per unit length and $nev\,E_s$ the energy transferred to n particles with charge e each moving with the velocity v in the electric field E_s. The *wall losses* are related to the *quality factor Q* of the structure

$$Q = \frac{\omega\,W}{P_{\mathrm{w}}} \,, \tag{2.32}$$

where P_{w}/ω is the energy loss per unit length in the walls per radian of the field oscillation. The energy flux P is

$$P = v_{\mathrm{g}}\,W \,. \tag{2.33}$$

In case of equilibrium the stored energy in the accelerating structure does not change with time, $\partial W/\partial t = 0$, and

$$\frac{\partial P}{\partial s} = -P_{\mathrm{w}} - i_{\mathrm{b}}\,E_s = -\frac{\omega\cdot P}{v_{\mathrm{g}}Q} - i_{\mathrm{b}}E_s \,, \tag{2.34}$$

where $i_{\mathrm{b}} = n\,e\,v$ is the beam current. Considering the case of small beam loading $i_{\mathrm{b}}\,E_s \ll \omega P/(v_{\mathrm{g}}Q)$ we may integrate (2.34) to get

$$P = P_{\mathrm{o}}\cdot\exp\left(-\frac{\omega}{v_{\mathrm{g}}Q}s\right) = P_{\mathrm{o}}\cdot\mathrm{e}^{-2\alpha s} \,, \tag{2.35}$$

where we have defined the *attenuation coefficient*

$$2\alpha = \frac{\omega}{v_{\mathrm{g}}Q} \,. \tag{2.36}$$

Equation (2.35) shows an exponential decay of the energy flux along the accelerating structure with the attenuation coefficient 2α. The wall losses are often expressed in terms of the total voltage or the electrical field defined by

$$P_{\mathrm{w}} = \frac{\widehat{V}_{\mathrm{o}}^2}{Z_{\mathrm{s}}\cdot L} = \frac{\widehat{E}^2}{r_{\mathrm{s}}} \,, \tag{2.37}$$

where Z_{s} is the shunt impedance for the whole section, $\widehat{E}$ the maximum value of the accelerating field, $E_s = \widehat{E}\cdot\cos\psi_{\mathrm{s}}$, ψ_{s} the synchronous phase at which the particle interacts with the wave, r_{s} the *shunt impedance* per unit length, and L the length of the cavity. From (2.37) we get with (2.34,36) for negligible beam current the accelerating field

$$\widehat{E}^2 = \frac{\omega}{v_g} \frac{r_s}{Q} P = 2\alpha r_s P. \tag{2.38}$$

The total *accelerating voltage* along a structure of length L is

$$V_o = \int_0^L E_s ds = \widehat{E} \cos \psi_s \cdot \int_0^L e^{-\alpha s} ds \tag{2.39}$$

or after integration

$$V_o = \frac{1 - e^{-\alpha L}}{\alpha} \cdot \widehat{E} \cdot \cos \psi_s. \tag{2.40}$$

Defining an *attenuation factor* τ by

$$\tau = \alpha \cdot L \tag{2.41}$$

we get with (2.38) for the total accelerating voltage per section of length L

$$V_o = \sqrt{r_s L P_o} \sqrt{2\tau} \frac{1 - e^{-\tau}}{\tau} \cos \psi_s. \tag{2.42}$$

The maximum energy is obtained if the particles are accelerated at the crest of the wave, where $\psi_s = 0$. In praxis it is customary to include the *transit time factor* derived in the previous subsection in the definition of the shunt impedance which then is called the effective shunt impedance. For the effective accelerating voltage we have then $V_o = \widehat{V} \cdot T_t$.

Since the attenuation factor is $\tau = \omega L/(2 v_g Q)$, we get for the *filling time* of the structure

$$t_F = \frac{L}{v_g} = \frac{2Q}{\omega} \tau. \tag{2.43}$$

As an example for an electron linear accelerator, the *SLAC linac structure* has the following parameters [2.17]

$$\begin{aligned}
f_{rf} &= 2856 \text{ MHz}, & L &= 10 \text{ ft} = 3.048 \text{ m}, \\
r_s &= 53 \text{ M}\Omega/\text{m}, & P_o &= 8 \text{ MW}, \\
Q &= 15000, & \tau &= 0.57.
\end{aligned} \tag{2.44}$$

Tacitly it has been assumed that the shunt impedance r_s in the structure is constant leading to a variation of the electrical field strength along the accelerating section. Such a structure is called a *constant impedance structure* and is characterized physically by equal geometric dimensions for all cells.

In a constant impedance structure the electric field is maximum at the beginning of the section and drops off toward the end of the section. A more efficient use of accelerating sections would keep the electric field at the maximum possible value just below field break down throughout the whole section. A structure with such characteristics is called a *constant gradient structure* because the field is now constant along the structure.

A constant gradient structure can be realized by varying the iris holes in the disks to smaller and smaller apertures along the section. This kind of structure is actually used in the SLAC accelerator as well as in most modern linear electron accelerators. The field $\widehat{E} = \text{const}$ and therefore from (2.34, 37) $\partial P/\partial s = \text{const}$, where the constant is with $P(L)/P_o = e^{-2\tau}$

$$\frac{\partial P}{\partial s} = \frac{P(L) - P_o}{L} = -(1 - e^{-2\tau})\frac{P_o}{L} = \text{const}. \tag{2.45}$$

On the other hand from (2.34) we have

$$\frac{\partial P}{\partial s} = -\frac{\omega \cdot P_o}{v_g \cdot Q} = \text{const}. \tag{2.46}$$

From (2.46) we note that the *group velocity* must vary linearly with the local rf power like

$$v_g \sim P(s) = P_o + \frac{\partial P}{\partial s}s. \tag{2.47}$$

Since $\partial P/\partial s < 0$ the group velocity is made to decrease along the section by reducing gradually the iris radii. From (2.46)

$$v_g(s) = -\frac{\omega}{Q}\frac{P(s)}{\partial P/\partial s} \tag{2.48}$$

or with (2.45)

$$v_g(s) = -\frac{\omega}{Q}\frac{P_o + \frac{\partial P}{\partial s}s}{\partial P/\partial s} = +\frac{\omega}{Q}\frac{L - (1 - e^{-2\tau})s}{1 - e^{-2\tau}} \tag{2.49}$$

and the filling time is after integration of (2.49)

$$t_F = \int_o^L \frac{ds}{v_g} = 2\tau\frac{Q}{\omega}. \tag{2.50}$$

The electric field in the accelerating section is from (2.37) with (2.34)

$$\widehat{E} = \sqrt{r_s\left|\frac{\partial P}{\partial s}\right|} \tag{2.51}$$

and the total accelerating voltage V_o or gain in kinetic energy per section is

$$\Delta E_{\text{kin}} = eV_o = e\int_o^L E_s\,ds = e\sqrt{r_sLP_o}\sqrt{1 - e^{-2\tau}}\cos\psi_s, \tag{2.52}$$

where ψ_s is the synchronous phase at which the particles travel with the electromagnetic wave. The energy gain scales with the square root of the accelerating section length and rf power delivered.

As a numerical example, we find for the SLAC structure from (2.52) the gain of kinetic energy per 10 ft section as

$$\Delta E_{\text{kin}}[\text{MeV}] = 10.48 \sqrt{P_{\text{o}}[\text{MW}]}, \tag{2.53}$$

where P_{o} is the rf power delivered to the section. The energy gain (2.53) is the maximum value possible ignoring *beam loading* or energy extraction from the fields by the beam. The total accelerating voltage is reduced when we include beam loading due to a beam current i_b. Refering the interested reader to reference [2.20] we only quote the result for the energy gain in a linear accelerator with constant gradient sections including beam loading

$$V_i = \sqrt{r_s L P_{\text{o}}} \sqrt{1 - e^{-2\tau}} - \tfrac{1}{2} i_b r_s L \left(1 - \frac{2\tau e^{-2\tau}}{1 - e^{-2\tau}} \right). \tag{2.54}$$

For the SLAC linac structure this equation becomes with $\tau = 0.57$

$$E_{\text{kin}} = 10.48 \sqrt{P_{\text{o}}[\text{MW}]} - 37.47\, i_b[\text{A}]. \tag{2.55}$$

The beam loading depends greatly on the choice of the *attenuation factor* τ as is shown in Figs. 2.9,10, where the coefficients $f_v = \sqrt{1 - e^{2\tau}}$ and $f_i = \tfrac{1}{2} \left(1 - \frac{2\tau e^{-2\tau}}{1 - e^{-2\tau}} \right)$ are plotted as functions of τ.

Both coefficients increase as the attenuation factor is increased and reach asymptotic limits. The ratio f_v/f_i, however, decreases from infinity to a factor two which means that beam loading occurs much stronger for large values of the attenuation factor compared to low values. During the design of the linac structure it is therefore useful to know the intended use requiring different optimization for high-energy or high-current acceleration.

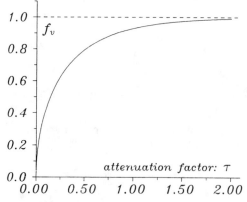

Fig. 2.9. Energy coefficient f_v as a function of τ.

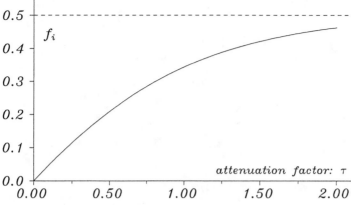

Fig. 2.10. Beam loading coefficient f_i as a function of τ.

We may also ask for the *efficiency* of transferring rf power into beam power which is defined by

$$\eta = \frac{i_b V_i}{P_o} = i_b \sqrt{\frac{r_s L}{P_o}} \sqrt{1 - e^{-2\tau}} - \frac{1}{2} i_b^2 r_s \frac{L}{P_o} \left(1 - \frac{2\tau e^{-2\tau}}{1 - e^{-2\tau}} \right) . \quad (2.56)$$

The *linac efficiency* has clearly a maximum and the optimum beam current is

$$i_{b,opt} = \sqrt{\frac{P_o}{r_s L}} \frac{(1 - e^{-2\tau})^{3/2}}{1 - (1 + 2\tau)e^{-2\tau}} . \quad (2.57)$$

The optimum beam current is plotted in Fig. 2.11 as a function of the attenuation coefficient τ and the linac efficiency is shown in Fig. 2.12 as a function of beam current in units of the optimum current with the attenuation factor as a parameter.

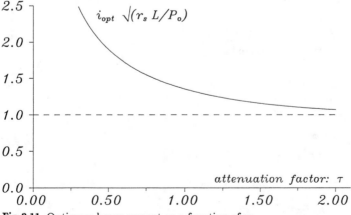

Fig. 2.11. Optimum beam current as a function of τ.

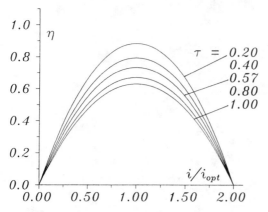

Fig. 2.12. Linac efficiency as a function of beam current.

The optimum beam current increases as the attenuation factor is reduced while the linac efficiency reaches a maximum for $\tau = 1$.

Particle Capture in a Linear Accelerator Field: The *capture of particles* and the resulting particle energy at the end of the accelerating section depends greatly on the relative synchronism of the particle and wave motion. If particles are injected at low energy, $v_p \ll c$, into an accelerator section designed for a phase velocity $v_{ph} \geq v_p$ the electromagnetic wave would roll over the particles with reduced acceleration. The particle velocity and phase velocity must be equal or at least close to each other. Because small mismatches are quite common, we will discuss particle dynamics under those circumstances and note that there is no fundamental difference between electron and proton linear accelerators. The following discussion is therefore applicable to any particle type being accelerated by traveling electromagnetic fields in a linear accelerator.

We observe the relative motion of both the particle and the wave from the laboratory system. During the time Δt particles move a distance $\Delta s_p = v_p \Delta t$ and the wave a distance $\Delta s_{ph} = v_{ph} \Delta t$. The difference in the distance traveled can be expressed in terms of a phase shift

$$\Delta \psi = -k\left(\Delta s_{ph} - \Delta s_p\right) = -k\left(v_{ph} - v_p\right)\frac{\Delta s_p}{v_p}. \tag{2.58}$$

The wave number k is

$$k = \frac{\omega}{v_{ph}} = \frac{2\pi c}{\lambda_{rf}\, v_{ph}} \tag{2.59}$$

and inserted into (2.58) the relative phase shift over a distance Δs_p becomes

$$\Delta \psi = -\frac{2\pi c}{\lambda_{rf}}\frac{v_{ph} - v_p}{v_{ph}\, v_p}\,\Delta s_p. \tag{2.60}$$

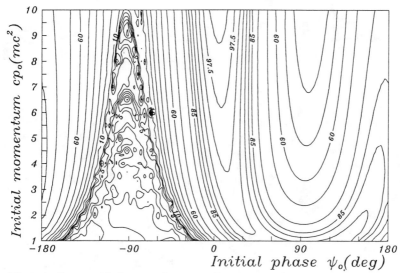

Fig. 2.13. Capture of electrons in a 3 m linac section for initial phase ψ_0 and initial momentum p_0. Contour lines are lines of constant particle momentum (in units of mc^2) at the end of the section and the phase $\psi_0 = 0$ corresponds to the crest of the accelerating wave.

To complete the equation of motion we consider the energy gain of the particles along the same distance Δs_p which is

$$\Delta E_\mathrm{kin} = -eE_s(\psi)\,\Delta s_\mathrm{p}\,. \tag{2.61}$$

Equations (2.60,61) form the equations of motion for particles in phase space. Both equations are written as difference equations for numerical integration since no analytic solution exists but for the most trivial case, where $\beta_\mathrm{ph} = \beta_\mathrm{p}$ in which case $\psi = \mathrm{const}$ and (2.61) can be integrated readily. This trivial case becomes the overwhelming common case for electrons which reach a velocity very close to the speed of light. Consistent with this, most accelerating sections are dimensioned for a phase velocity also equal to the speed of light.

As an illustrative example, we integrate (2.60, 61) numerically to determine the beam parameters at the end of a single 3 m long accelerating section with $v_\mathrm{ph} = c$ for an initial particle distribution in phase and momentum at the entrance to the accelerating section. This situation is demonstrated in Figs. 2.13, 14 for a constant field gradient of $\widehat{E} = 12.0$ MeV/m [2.21]. The momentum and phase at the end of the accelerating section are shown as functions of the initial momentum and phase. We note from Figs. 2.13, 14 that particles can be captured in the accelerating field at almost any initial phase and momentum. Only in the vicinity of $\psi_0 = -90°$ particles become trapped in the negative field performing random motion. This situation is particularly severe for low momentum particles because they do

45

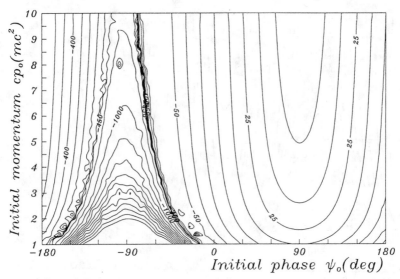

Fig. 2.14. Capture of electrons in a 3 m linac section for initial phase ψ_o and initial momentum cp_o. Contour lines are lines of constant phase at the end of the section and the initial phase $\psi_\mathrm{o} = 0$ corresponds to the crest of the accelerating wave.

not travel with the phase velocity of the accelerating wave yet. On the other hand, particles which enter the accelerating section ahead of the crest gain maximum momentum while the wave's crest just sweeps over them.

Such diagrams calculated for particular parameters under consideration provide valuable information needed to prepare the beam for optimum acceleration. The most forgiving operating parameters are, where the contour lines are far apart. In those areas a spread in initial phase or momentum has little effect on the final phase or energy. If a beam with a small *momentum spread* at the end of acceleration is desired, the initial phase should be chosen to be at small positive values or just ahead of the wave crest as shown in Fig. 2.13. Even for a long bunch the final momentum spread is small while reaching the highest total energy.

On the other hand, if a short bunch length at the end of acceleration is of biggest importance, an initial phase of around $\psi_\mathrm{o} \approx 100°$ seems to be more appropriate. In this case, however, the final momentum is lower than the maximum possible momentum and the momentum spread is large.

Once the particular particle distribution delivered to the linear accelerator and the desired beam quality at the end is known one can use Figs. 2.13, 14 for optimization. Conversely such diagrams can be used to judge the feasibility of a particular design to reach the desired beam characteristics.

2.3 Preinjector Beam Preparation

Although the proper choice of the initial phase with respect to the particle beam greatly determines the final beam quality, the flexibility of such adjustments are limited. Special attention must be given to the preparation of the beam before acceleration. Depending on the particle source special devices are used for initial acceleration and *bunching* of the beam. We will discuss basic principles of beam preparation.

2.3.1 Prebuncher

Many particle sources, be it for electrons, protons or ions, produce a continuous stream of particles at modest energies limited by electrostatic acceleration between two electrodes. Not all particles of such a beam will be accelerated because of the oscillating nature of the accelerating field. For this reason and also in case short bunches or a small energy spread at the end of the linac is desired, the particles should be concentrated at a particular phase. This concentration of particles in the vicinity of an optimum phase maximizes the particle intensity in the bunch in contrast to a mechanical *chopping* of a continuous beam. To bunch particles requires specific beam manipulation which we will discuss here in more detail.

A bunched beam can be obtained from a continuous stream of nonrelativistic particles by the use of a *prebuncher*. The basic components of a prebuncher is an rf cavity followed by a drift space. The cavity is excited to a TM_{010} mode to produce an accelerating field. As a continuous stream of particles passes through the prebuncher, some particles get accelerated and some are decelerated. The manipulation of the continuous beam into a bunched beam is best illustrated in the phase space diagrams of Fig. 2.15.

Figure 2.15a shows the continuous particle distribution in energy and phase at the entrance of the prebuncher. Depending on the phase of the electric field in the prebuncher at the time of passage, a particle becomes accelerated or decelerated and the particle distribution at the exit of the prebuncher is shown in Fig. 2.15b. The particle distribution has been distorted into a sinusoidal energy variation. Since the particles are nonrelativistic the energy modulation reflects also a velocity modulation. We concentrate on the origin of the distribution at $\varphi = 0$ and $\Delta E_{kin} = 0$ as the reference phase and note that particles ahead of this reference phase have been decelerated and particles behind the reference phase have been accelerated. Following this modulated beam through the drift space we observe due to the velocity modulation a bunching of the particle distribution which reaches an optimum at some distance L as shown in Fig. 2.15c.

A significant beam intensity has been concentrated close to the reference phase of the prebuncher. At a distance $s \lesssim L$ the main accelerator starts and accelerates the particles quickly to relativistic energies to preserve the short bunch. In case of proton and ion beams this is not possible but here the

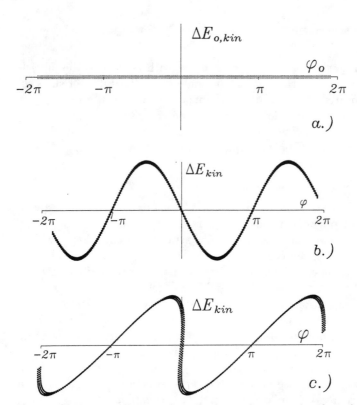

Fig. 2.15. Phase space diagrams for a continuous beam passing through a prebuncher, a.) before acceleration, b.) right after acceleration, c.) at end of drift space L

interaction with the accelerating fields provides more efficient phase focusing as discussed in the previous section in connection with beam capture.

The frequency used in the prebuncher depends on the desired bunch distribution. For straight acceleration in a linear accelerator one would choose the same frequency for both systems. Often, however, the linear accelerator is only an injector into a bigger circular accelerator which generally operates at a lower frequency. For optimum matching to the final circular accelerator the appropriate prebuncher frequency would be the same as the cavity frequency in the ring.

The effect of the prebuncher can be formulated analytically in the vicinity of the reference phase. At the exit of the prebuncher operating at a voltage $V = V_0 \sin\varphi$ the energy spread is

$$\Delta E_{\text{kin}} = eV_0 \sin\varphi = mc^2\, \beta\gamma^3\, \Delta\beta, \qquad (2.62)$$

which is related to a velocity spread $\Delta\beta$. Perfect bunching occurs a time Δt later when

$$\Delta v\, \Delta t = \frac{\varphi}{2\pi}\, \lambda_{\text{rf}}. \qquad (2.63)$$

Solving for Δt we get for nonrelativistic particles with $\gamma = 1$ and $\beta \ll 1$

$$\Delta t = \frac{\lambda_{\mathrm{rf}}}{2\pi} \frac{mv}{eV_{\mathrm{o}}} \tag{2.64}$$

and the distance L for optimum bunching is

$$L = v\,\Delta t = \frac{2\,E_{\mathrm{kin}}}{k_{\mathrm{rf}}\,eV_{\mathrm{o}}}, \tag{2.65}$$

where $k_{\mathrm{rf}} = 2\pi/\lambda_{\mathrm{rf}}$ and λ_{rf} is the rf wavelength in the prebuncher. The minimum bunch length in this case is then

$$\delta L = \frac{\delta E_{\mathrm{kin}}}{k_{\mathrm{rf}}\,eV_{\mathrm{o}}}, \tag{2.66}$$

where δE_{kin} is the total energy spread in the beam before the prebuncher.

In this derivation we have greatly idealized the field variation being linear instead of sinusoidal. The real *bunching* is therefore less efficient than the above result and shows some wings as is obvious from Fig. 2.15c. In a compromise between beam intensity and bunch length one might let the bunching go somewhat beyond the optimum and thereby pull in more of the particle intensity in the wings.

There are still particles between the bunches which could either be eliminated by an *rf chopper* or let go to be lost in the linear accelerator because they are mainly distributed over the decelerating field period in the linac.

2.3.2 Beam Chopper

A conceptually simple way to produce a bunched beam is to pass a continuous beam from the source through a *chopper* system, where the beam is deflected across a narrow slit resulting in a pulsed beam behind the slit. The principle components of such a chopper system are shown in Fig. 2.16.

As was mentioned in the previous section this mode of bunching is rather wasteful and therefore an rf prebuncher which concentrates particles from a large range of phases towards a particular phase is more efficient for particle bunching. However, we still might want to add a beam chopper for specific reasons.

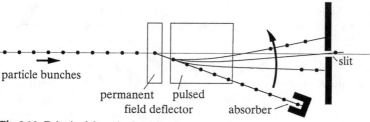

Fig. 2.16. Principal functioning of a chopper system

One such reason is to eliminate most of the remaining particles between the bunches from the prebuncher. Although these particles most likely get lost during the acceleration process a significant fraction will still reach the end of the linac with an energy spread between zero and maximum energy. Because of their large energy deviation from the energy of the main bunches, such particles will be lost in a subsequent beam transport system and therefore create an unnecessary increase in the radiation level. It is therefore prudent to eliminate such particles at low energies. A suitable device for that is a chopper which consists of an rf cavity excited similar to the prebuncher cavity but with the beam port offset by a distance r from the cavity axis. In this case the same rf source as for the prebuncher or main accelerator can be used and the deflection of particles is effected by the azimuthal magnetic field in the cavity which from (2.18) reaches a maximum at a radius $r = 1.825/k_c$.

The prebuncher produces a string of bunches at the prebuncher frequency. For many applications, however, a different bunch structure is desired. Specifically it often occurs that only one single bunch is desired. To produce a single pulse, the chopper system may consist of a permanent magnet and a fast pulsed magnet. The permanent magnet deflects the beam into an absorber while the pulsed magnet deflects the beam away from the absorber across a small slit. The distance between the center of the pulsed magnet and the slit be D (Fig. 2.16), the slit opening s and the rate of change of the magnetic field $\dot{B}$. For an infinitely thin beam the pulse length behind the slit is then

$$\tau_b = \frac{s}{D\dot{\varphi}} = \frac{s}{D}\frac{cp}{e\dot{B}\ell}, \tag{2.67}$$

where φ is the deflection angle, ℓ the effective magnetic length of the pulsed magnet and cp the momentum of the particles. In order to clean the beam between bunches or to select a single bunch from a train of bunches the chopper parameters must be chosen such that only the desired part of the beam passes through.

Problems

Problem 2.1. Calculate the minimum power rating for the motor driving the charging belt of a Van de Graaff accelerator producing a charge current of 1 A at 5 MV.

Problem 2.2. Calculate the length for the first four drift tubes of a Wideroe linac for the following parameters: starting energy is 100 keV, the energy gain per gap 1 MeV, and the field frequency 7 MHz. Perform the calculations for both electrons and protons and compare the results.

Problem 2.3. Verify the validity of the solutions (2.17,18). Derive a vector relation between the electric and magnetic fields.

Problem 2.4. The electromagnetic field for a cylindrical waveguide have been derived in Sect. 2.2.2. Derive in a similar way expressions for resonant field modes in a rectangular waveguide.

Problem 2.5. Consider a pill box cavity made from copper with dimensions suitable for 3 GHz. Surface currents cause heating losses. Apply Maxwell's equations to the surfaces of the pill box cavity to determine these surface currents and calculate the power loss due to surface resistance. Calculate the value of the quality factor Q as defined in (2.32). Determine the resonance width of the cavity and the temperature tolerance to keep the driving frequency within the resonance.

Problem 2.6. Calculate for a SLAC type linac section the noload energy gain, the optimum beam current, and the energy gain with optimum beam loading for an rf power of $P_o = 25$ MW and a pulse length of 2.5 μsec. What is the linac efficiency for this current? Assume the acceleration of only one bunch of 1×10^{10} electrons. What is the energy gain and the linac efficiency in this case?

Problem 2.7. Consider a SLAC type linear accelerator section connected to an rf source of 25 MW. The stored field energy is depleted by the particle beam. If there are 10^9 electrons per micro bunch, how many such bunches can be accelerated before the energy gain is reduced by 1%? The field energy is replenished from the rf source at a rate given by the rf power and the filling time. What is the average beam current for which the energy gain is constant for all bunches?

Problem 2.8. Design a 500 MHz prebuncher system for a 3 GHz linear accelerator. Particles in a continuous beam from the source have a kinetic energy of 100 keV with a spread of $\pm0.01\%$. Specify the optimum prebuncher voltage and drift length to compress the maximum intensity into a bunch with a phase distribution of $\pm10°$ at 3 GHz.

Problem 2.9. Consider a rectangular box cavity with copper walls and dimensioned for an rf wavelength of $\lambda_{rf} = 10.5$ cm. Calculate the wall losses due to the fundamental field only and determine the shunt impedance per unit length r and the quality factor Q for this cavity. These losses are due to surface currents within a skin depth generated by the EM fields on the cavity surface due to Maxwell's equations. Compare these parameters with those of (2.44). Is the shape of the cavity very important? Determine the resonance width and temperature tolerance for the cavity.

Problem 2.10. Discuss the graphs in Figs. 2.13, 14. Specifically explain in words the particle dynamics behind nonuniform features. How come particles get accelerated even though they enter the linac while the accelerating field is negative? Use a computer to reproduce these graphs and then produce such graphs for particles with an initial kinetic energy of between 50 keV and 150 keV which is case in conventional thermionic electron guns.

3 Circular Accelerators

Parallel with the development of electrostatic and linear rf accelerators the potential of *circular accelerators* was recognized and a number of ideas for such accelerators have been developed over the years. Technical limitations for linear accelerators encountered in the early twenties to produce high-power rf waves stimulated the search for alternative accelerating methods or ideas for accelerators that would use whatever little rf fields could be produced as efficiently as possible.

Interest in circular accelerators quickly moved up to the forefront of accelerator design and during the thirties made it possible to accelerate charged particles to many million electron volts. Only the invention of the *rf klystron* by the Varian brothers at Stanford in 1937 gave the development of *linear accelerators* the necessary boost to reach par with circular accelerators again. Since then both types of accelerators have been developed further and neither type has yet outperformed the other. In fact, both types have very specific advantages and disadvantages and it is mainly the application that dictates the use of one or the other.

Circular accelerators are based on the use of magnetic fields to guide the charged particles along a closed orbit. The acceleration in all circular accelerators but the betatron is effected in one or few accelerating cavities which are traversed by the particle beam many times during their orbiting motion. This greatly simplifies the rf system compared to the large number of energy sources and accelerating sections required in a linear accelerator. While this approach seemed at first like the perfect solution to produce high energy particle beams, its progress soon became limited for the acceleration of electrons by copious production of *synchrotron radiation*.

The simplicity of circular accelerators and the absence of significant synchrotron radiation for protons and heavier particles like ions has made circular accelerators the most successful and affordable principle to reach the highest possible proton energies for fundamental research in high energy physics. At present protons are being accelerated up to 1000 GeV in the presently highest energy proton synchrotron at the *Fermi National Accelerator Laboratory* FNAL [3.1]. Higher energy *proton accelerators* up to 20 TeV are under construction or in planning [3.2].

For electrons the principle of circular accelerators has reached a technical and economic limit at about 100 GeV [3.3] due to synchrotron radiation losses, which make it increasingly harder to accelerate electrons to higher energies [3.4]. Further progress in the attempt to reach higher electron energies is being pursued though the principle of *linear colliders* [3.5, 6], where synchrotron radiation is avoided.

Many applications for accelerated particle beams, however, exist at significantly lower energies and a multitude of well developed principles of particle acceleration are available to satisfy those needs. We will discuss only the basic principles behind most of these low- and medium-energy accelerators in this text and concentrate in more detail on the beam physics in synchrotrons and storage rings. Well documented literature exists for smaller accelerators and the interested reader is referred to the bibliography at the end of this volume.

3.1 Betatron

As the first "circular electron accelerator" we may consider what has been invented and developed a hundred years ago in the form of the electrical current transformer. Here we find the electrons in the wire of a secondary coil accelerated by an electro motive force generated by a time varying magnetic flux penetrating the area enclosed by the secondary coil.

This idea was picked up independently by several researchers [3.7, 8]. Wideroe finally recognized the importance of a fixed orbit radius and formulated the *Wideroe* $1/2$-*condition*, which is a necessary although not sufficient condition for the successful operation of a *beam transformer* or *betatron* as it was later called, because it functions optimally only for the acceleration of beta rays or electrons [3.9].

The betatron makes use of the transformer principle, where the secondary coil is replaced by an electron beam circulating in a closed doughnut shaped vacuum chamber. A time-varying magnetic field is enclosed by the electron orbit and the electrons gain an energy in each turn which is equal to the electro-motive force generated by the varying magnetic field. The principle arrangement of the basic components of a betatron are shown in Fig. 3.1.

The accelerating field is determined by integrating Maxwell's equation

$$\nabla \times \mathbf{E} = -\frac{1}{c}\frac{\mathrm{d}}{\mathrm{d}t}\mathbf{B} \tag{3.1}$$

and utilizing Stokes's theorem, we obtain the energy gain per turn

$$\oint \mathbf{E}\,\mathrm{d}s = -\frac{1}{c}\frac{\mathrm{d}\Phi}{\mathrm{d}t}, \tag{3.2}$$

where Φ is the magnetic flux enclosed by the integration path, which is identical to the *design orbit* of the beam.

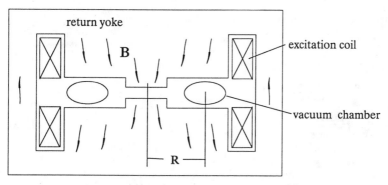

Fig. 3.1. Principle of acceleration in a betatron (schematic)

The particles follow a circular path under the influence of the *Lorentz force* from a uniform magnetic field. We use a *cylindrical coordinate system* (r, φ, y), where the particles move with the coordinate φ clockwise along the orbit. The Lorentz force is compensated by the *centrifugal force* and we have in equilibrium the *equation of motion*

$$\frac{\gamma m v^2}{r} - \frac{e}{c} v B_\perp = 0, \tag{3.3}$$

where γ is the particle energy in units of the rest energy and the magnetic field direction is normal to the plane of the circular orbit. From (3.3) we get for the particle momentum

$$p = \gamma m v = \frac{e}{c} r B_\perp. \tag{3.4}$$

The accelerating force is equal to the rate of change of the particle momentum and can be obtained from the time derivative of (3.4). This force must be proportional to the azimuthal electric field component E_φ on the orbit

$$\frac{dp}{dt} = -\frac{e}{c} \left(\frac{dr}{dt} B_\perp + r \frac{dB_\perp}{dt} \right) = e E_\varphi. \tag{3.5}$$

Following Wideroe's requirement for a constant orbit $dr/dt = 0$ allows the containment of the particle beam in a doughnut shaped vacuum chamber surrounding the magnetic field. The induced electric field has only an angular component E_φ since we have assumed that the magnetic field enclosed by the circular beam is uniform or at least rotationally symmetric. The left hand side of (3.2), while noting that for a positive field the induced azimuthal electric field is negative, then becomes simply

$$\oint \mathbf{E} \, ds = -\int E_\varphi R \, d\varphi = -2\pi R E_\varphi. \tag{3.6}$$

On the other hand, we have from (3.5)

$$e\,E_\varphi = \frac{e}{c}\,R\,\frac{\mathrm{d}B(R)}{\mathrm{d}t} \tag{3.7}$$

and using (3.6,7) in (3.2) we get

$$\frac{\mathrm{d}\Phi}{\mathrm{d}t} = 2\pi\,R^2\,\frac{\mathrm{d}B(R)}{\mathrm{d}t}\,. \tag{3.8}$$

Noting that the complete magnetic flux enclosed by the particle orbit can also be expressed by an average field enclosed by the particle orbit, we have $\Phi = \pi\,R^2\,\overline{B}(R)$, where $\overline{B}(R)$ is the average magnetic induction within the orbit of radius R. The rate of change of the magnetic flux becomes

$$\frac{\mathrm{d}\Phi}{\mathrm{d}t} = \pi\,R^2\,\frac{\mathrm{d}\overline{B}(R)}{\mathrm{d}t} \tag{3.9}$$

and comparing this with (3.8) we obtain the *Wideroe* 1/2-*condition* [3.9]

$$B(R) = \tfrac{1}{2}\,\overline{B}(R)\,, \tag{3.10}$$

which requires for orbit stability that the field at the orbit be half the average flux density through the orbit. This condition must be met in order to obtain orbital-beam stability in a betatron accelerator. By adjusting the total magnetic flux through the particle orbit such that the average magnetic field within the orbit circle is twice the field strength at the orbit we are in a position to accelerate particles on a circle with a constant radius, R, within a doughnut shaped vacuum chamber.

The basic components of a betatron, shown in Fig. 3.1, have rotational symmetry. In the center of the magnet we recognize two magnetic gaps of different aperture. One gap at $r = R$ provides the *bending field* for the particles along the orbit. The other gap in the midplane of the central return yoke is adjustable and is being used to tune the magnet such as to meet the Wideroe 1/2-condition. The magnetic field is generally excited by a resonance circuit cycling at the ac frequency of the main electricity supply. In this configuration the *magnet coils* serve as the inductance and are connected in parallel with a capacitor bank tuned to the ac frequency of 50 or 60 Hz.

The rate of *momentum gain* is derived by integration of (3.5) with respect to time and we find that the change in momentum is proportional to the change in the magnetic field

$$\Delta p = \frac{e}{c}\,R\int\frac{\mathrm{d}B_\perp}{\mathrm{d}t}\,\mathrm{d}t = \frac{e}{c}\,R\,\Delta\,B_\perp\,. \tag{3.11}$$

The particle momentum depends only on the momentary magnetic field and not on the rate of change of the field. For slowly varying magnetic fields the electric field is smaller but the particles will make up the reduced acceleration by travelling a longer time in this lower accelerating field. While the magnet cycling rate does not affect the particle energy it certainly determines the available flux of accelerated particles per unit time. The maximum particle momentum is determined only by the orbit radius and the maximum magnetic field at the orbit during the acceleration cycle

$$p_{max} = \frac{e}{c} R B_{max}(R).$$ (3.12)

The betatron principle works for any charged particle and for all energies since the stability condition (3.10) does not depend on particle parameters. In praxis, however, we find that the betatron principle is unsuitable to the acceleration of heavy particles like protons. The magnetic fields in a betatron as well as the size of the betatron magnet set practical limits to the maximum momentum achievable. Kerst built the largest betatron ever with an orbit radius of $R = 1.23$ m, a maximum magnetic field at the orbit of 8.1 kG and a total magnet weight of 350 tons reaching the maximum expected particle momentum of 300 MeV/c.

For experimental applications we are interested in the kinetic energy $E_{kin} = \sqrt{(cp)^2 + (mc^2)^2} - mc^2$ of the accelerated particles. In case of electrons the rest energy $m_e c^2$ is small compared to the maximum momentum of $cp = 300$ MeV and therefore the kinetic electron energy from this betatron is

$$E_{kin} \approx c p = 300 \, \text{MeV}.$$ (3.13)

In contrast to this result, we find the achievable kinetic energy for a proton to be much smaller

$$E_{kin} \approx \frac{1}{2} \frac{(cp)^2}{m_p c^2} = 48 \, \text{MeV},$$ (3.14)

because of the large mass of protons, $m_p c^2 = 938.28$ MeV.

Different, more efficient accelerating methods have been developed for protons and betatrons are therefore used exclusively for the acceleration of electrons as indicated by it's name. Most betatrons are designed for modest energies of up to 45 MeV and are used to produce electron and hard *x-ray beams* for medical applications or in technical applications to, for example, examine the integrity of full penetration welding seams in heavy steel containers.

3.2 Weak Focusing

The *Wideroe 1/2-condition* is a necessary condition to obtain a stable particle orbit at a fixed radius R. This stability condition, however, is not sufficient for particles to survive the accelerating process. Any particle starting out with, for example, a slight vertical slope would, during the acceleration process, follow a continuously spiraling path until it hits the top or bottom wall of the vacuum chamber and gets lost.

Constructing and testing the first, although unsuccessful, beam transformer, Wideroe recognized [3.9] the need for *beam focusing*, a need which has become a fundamental part of all future particle accelerator designs. First theories on beam stability and focusing have been pursued by Walton [3.10] and later by Steenbeck, who formulated a stability condition for weak focusing and applied it to the design of the first successful construction and operation of a betatron in 1935 at the Siemens-Schuckert Company in Berlin reaching an energy of 1.9 MeV [3.11] although at a very low intensity measurable only with a Geiger counter. The focusing problems in a betatron were finally solved in a detailed orbit analysis by Kerst and Serber [3.12].

To derive the beam *stability conditions* we note that (3.3) is true only at the ideal orbit R. For any other orbit radius r the restoring force is

$$F_x = \frac{\gamma m v^2}{r} - \frac{e}{c} v B_y . \tag{3.15}$$

Here we use a cartesian coordinate system moving with the particle along the *circular orbit* with **x** pointing in the radial and **y** in the axial direction.

In a uniform magnetic field the restoring force would be zero for any orbit. To include focusing we assume that the magnetic field at the orbit includes a gradient such that for a small deviation x from the ideal orbit, $r = R + x = R(1 + x/R)$, the magnetic guide field becomes

$$B_y = B_{oy} + \frac{\partial B_y}{\partial x} x = B_{oy} \left(1 + \frac{R}{B_{oy}} \frac{\partial B_y}{\partial x} \frac{x}{R} \right). \tag{3.16}$$

After insertion of (3.16) into (3.15) the *restoring force* is

$$F_x \approx \frac{\gamma m v^2}{R} \left(1 - \frac{x}{R} \right) - \frac{e}{c} v B_{oy} \left(1 - n \frac{x}{R} \right), \tag{3.17}$$

where we assumed $x \ll R$ and defined the *field index*

$$n = -\frac{R}{B_{oy}} \frac{\partial B_y}{\partial x} . \tag{3.18}$$

With (3.3) we get for the horizontal restoring force

$$F_x = -\frac{\gamma m v^2}{R} \frac{x}{R} (1 - n) . \tag{3.19}$$

The equation of motion under the influence of the restoring force in the deflecting or horizontal plane is with $F_x = \gamma m \ddot{x}$

$$\ddot{x} + \omega_x^2 x = 0, \tag{3.20}$$

which has the exact form of a harmonic oscillator with the frequency

$$\omega_x = \frac{v}{R} \sqrt{1-n} = \omega_o \sqrt{1-n}, \tag{3.21}$$

where ω_o is the orbital revolution frequency. The particle performs oscillation about the ideal or reference orbit with the amplitude $x(s)$ and the frequency ω_x. Because this focusing feature was discovered in connection with the development of the betatron we refer to this particle motion as *betatron oscillations* with the *betatron frequency* ω. From (3.21) we note a *stability criterion*, which requires that the field index not exceed unity to prevent the betatron oscillation amplitude to grow exponentially,

$$n < 1. \tag{3.22}$$

The particle beam stability discussion is complete only if we also can show that there is stability in the vertical plane. A vertical restoring force requires a finite horizontal field component B_x and the equation of motion becomes

$$\gamma m \ddot{y} = \frac{e}{c} v B_x. \tag{3.23}$$

Maxwell's curl equation

$$\frac{\partial B_x}{\partial y} - \frac{\partial B_y}{\partial x} = 0$$

can be integrated and the horizontal field component is with (3.16,18)

$$B_x = \int \frac{\partial B_y}{\partial x} \, dy = - \int n \frac{B_{oy}}{R} \, dy = -n \frac{B_{oy}}{R} y. \tag{3.24}$$

Insertion of (3.24) into (3.23) results after some manipulations with (3.3) in the equation of motion for the vertical plane

$$\ddot{y} + \omega_y^2 y = 0. \tag{3.25}$$

The particles perform stable betatron oscillations about the horizontal mid plane with the vertical betatron frequency

$$\omega_y = \omega_o \sqrt{n} \tag{3.26}$$

as long as the field index is positive

$$n > 0. \tag{3.27}$$

In summary, we have found that a field gradient in the magnetic guide field can provide beam stability in both the horizontal and vertical plane provided that the field index meets the criterion

$$0 < n < 1,$$
(3.28)

which has been first formulated and applied by Steenbeck [3.11] and is therefore also called *Steenbeck's stability criterion*.

A closer look at the stability criterion shows that the field gradient provides only focusing in the vertical plane but is defocusing in the horizontal plane. The reason why we get focusing in both planes is that there is a strong natural focusing from the sector type of magnet which is larger than the defocusing from the field gradient. This focusing is of geometric nature. A particle travelling, for example, parallel to and outside the ideal orbit is deflected more by a uniform magnetic field than a particle following the ideal orbit leading to effective focusing toward the ideal orbit. Conversely a particle traveling parallel to and inside the ideal orbit is deflected less and therefore again is deflected toward the ideal orbit.

The stability condition actually stipulates that the defocusing in the horizontal plane be less than the focusing of the sector magnet allowing to provide focusing in the vertical plane. Basically the field gradient provides a means to distribute the strong sector magnet focusing into both planes. This method of beam focusing is known as *weak focusing* in contrast to the principle of strong focusing, which will be discussed extensively in the remainder of this text.

3.3 Adiabatic Damping

During the discussion of *transverse focusing* we have neglected the effect of acceleration. To include the effect of acceleration into our discussion on beam dynamics, we use as an example the *Lorentz force equation* for the vertical motion. The equation of motion is

$$\frac{d}{dt}(\gamma m \dot{y}) = \frac{e}{c} v_s B_x,$$
(3.30)

where we used the fields $\mathbf{E} = (0, 0, E_z)$ and $\mathbf{B} = (B_x, B_y, 0)$ in a cartesian coordinate system (x, y, z). Multiplying (3.30) with c^2 and evaluating the differentiation we get the equation of motion at the equilibrium orbit

$$\gamma m c^2 \ddot{y} + \dot{\gamma} m c^2 \dot{y} = e c \omega_o R B_x(R).$$
(3.31)

Inserting (3.24) into (3.31) we get the equation of motion in the vertical plane under the influence of accelerating electrical and focusing magnetic fields

$$\ddot{y} + \frac{\dot{E}}{E}\dot{y} + n\,\omega_o^2\,y = 0\,, \tag{3.32}$$

where $\dot{E}$ is the energy gain per unit time. This is the differential equation of a *damped harmonic oscillator* with the solution

$$y = y_o\,e^{-\alpha_y t}\,\cos\omega t\,, \tag{3.33}$$

where $\omega \approx \omega_o\,\sqrt{n}$ and the *damping decrement*

$$\alpha_y = \tfrac{1}{2}\frac{\dot{E}}{E}\,. \tag{3.34}$$

For technically feasible acceleration $\dot{E}$ the *damping time* $\tau_y = \alpha_y^{-1}$ is very long compared to the oscillation period and we therefore may consider for the moment the damping as a constant. The envelope $y_{\max} = y_o\,e^{-\alpha_y t}$ of the oscillation (3.33) decays like

$$\mathrm{d}y_{\max} = -\tfrac{1}{2}\frac{\dot{E}}{E}\,y_{\max}\,\mathrm{d}t\,, \tag{3.35}$$

which after integration becomes

$$\frac{y_{\max}}{y_{o,\max}} = \sqrt{\frac{E_o}{E}}\,. \tag{3.36}$$

The betatron oscillation amplitude is reduced as the particle energy increases. This type of damping is called *adiabatic damping*. Similarly the slope y' as well as the horizontal oscillation parameters experience the same effect of adiabatic damping during acceleration. We define a beam emittance in both planes by the product $\epsilon_u = u_{\max}\,u'_{\max}$, where u stands for x or y. Due to adiabatic damping this beam emittance is reduced inversely proportional to the energy like

$$\epsilon \sim \frac{1}{E}\,. \tag{3.37}$$

No specific use has been made of the principle of betatron acceleration to derive the effect of adiabatic damping. We therefore expect this effect to be generally valid for any kind of particle acceleration.

The development of the betatron was important for accelerator physics for several reasons. It demonstrated the need for particle focusing, the phenomenon of adiabatic damping and stimulated Schwinger to formulate the theory of *synchrotron radiation* [3.4]. He realized that the maximum achievable electron energy in a betatron must be limited by the energy loss due to synchrotron radiation. Postponing a more detailed discussion of synchrotron radiation to Chap. 9 we note that the instantaneous *synchrotron radiation power* is given by

$$P_\gamma = \tfrac{2}{3} \frac{r_e\,c}{(mc^2)^3}\, E^2\, F_\perp^2, \tag{3.38}$$

where the transverse force is

$$F_\perp = \frac{e}{c}\, vB \approx eB = \frac{cp}{R} \approx \frac{E}{R}. \tag{3.39}$$

The energy loss per turn in a circular electron accelerator is then

$$\Delta E_\gamma = P_\gamma\, T_{\mathrm{rev}} = P_\gamma \frac{2\pi R}{c} = C_\gamma \frac{E^4}{R} \tag{3.40}$$

with

$$C_\gamma = 8.85 \times 10^{-5}\ \frac{\mathrm{m}}{\mathrm{GeV}^3}. \tag{3.41}$$

The *energy loss* to synchrotron radiation increases rapidly with the fourth power of energy and can lead quickly to an energy limitation of the accelerator when the energy loss per turn becomes equal to the energy gain.

3.4 Acceleration by rf Fields

Most types of circular particle accelerators utilize compact accelerating cavities, which are excited by a rf amplifier. Particles traverse this cavity periodically and gain energy from the electromagnetic fields in each passage. The bending magnet field serves only as a beam guidance system to allow the repeated passage of the particle beam through the cavity. Technically this type of accelerators seem to be very different from the principle of the betatron. Fundamentally, however, there is no difference. We still rely on the transformer principle, which in the case of the betatron looks very much like the familiar transformers at low frequencies, while *accelerating cavities* are transformer realizations for very high frequencies. Electric fields are generated in both cases by time varying magnetic fields.

Since the cavity fields are oscillating, acceleration is not possible at all times and for multiple accelerations we must meet specific conditions of synchronization between the motion of particles and the field oscillation. The time it takes the particles to travel along the orbital path must be an integer multiple of the oscillation period for the radio frequency field. This synchronization depends on the particle velocity, path length, magnetic fields employed, and on the rf frequency. Specific control of one or more of these parameters defines the different types of particle accelerators to be discussed in the following subsections.

3.4.1 Microtron

The schematic configuration of a microtron [3.13] is shown in Fig. 3.2. Particles emerging from a source pass through the accelerating cavity and follow then a circular orbit in a uniform magnetic field leading back to the accelerating cavity. After each acceleration the particles follow a circle with a bigger radius till they reach the boundary of the magnet. The bending radius of the orbit can be derived from the Lorentz force equation (3.3)

$$\frac{1}{r} = \frac{eB}{cp} = \frac{eB}{mc^2\gamma\beta},\tag{3.42}$$

and the *revolution time* for a particle traveling with velocity v is

$$\tau = \frac{2\pi r}{v} = \frac{2\pi\,mc}{e}\,\frac{\gamma}{B}.\tag{3.43}$$

The revolution time is therefore proportional to the particle energy and inversely proportional to the magnetic field. It is interesting to note that for subrelativistic particles, where $\gamma \approx 1$, the revolution time is constant even though the particle momentum increases. The longer *path length* for the higher particle momentum is compensated by the higher velocity. As particles reach relativistic energies, however, this synchronism starts to fail. To still achieve continued acceleration, specific conditions must be met.

A particle having completed the nth turn passes through the cavity and it's energy is increased because of acceleration. The change in the revolution time during the $(n + 1)$st turn compared to the nth turn is proportional

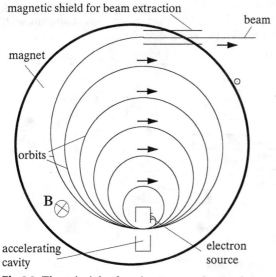

Fig. 3.2. The principle of a microtron accelerator (schematic)

to the energy increase $\Delta\gamma$. The increase in the revolution time must be an integer multiple of the radio frequency period. Assuming that the revolution time along the first most inner circle, when the particle energy is still $\gamma \approx 1$, is equal to one rf period we conclude that synchronism is preserved for all turns if

$$\Delta\gamma = 1 \tag{3.44}$$

or integer multiples. In order to make a microtron functional the energy gain from the accelerating cavity in each passage must be

$$\begin{aligned}
\Delta E_{\mathrm{e}} &= 511 \text{ keV} \qquad \text{for electrons and} \\
\Delta E_{\mathrm{p}} &= 938 \text{ MeV} \qquad \text{for protons.}
\end{aligned} \tag{3.45}$$

While it is possible to meet the condition for electrons it is technically impossible at this time to achieve accelerating voltages of almost 1 GV in an accelerating cavity. The principle of microtrons is therefore specifically suited for the acceleration of electrons.

The size of the magnet imposes a practical limit to the maximum particle energy. A single magnet scales like the third power of the bending radius and therefore the weight of the magnet also scales like the third power of the maximum particle energy. Basically single magnet microtrons are used to accelerated electrons to energies up to about 25 - 30 MeV.

To alleviate the technical and economic limitations as well as to improve control of the *synchronicity condition*, the concept of a *race track microtron* has been developed [3.14, 15]. In this type of microtron the magnet is split in the middle and normal to the orbit plane and pulled apart as shown in Fig. 3.3. The space opened up provides space for a short linear accelerator which allows the acceleration of electrons by several units in γ thus reducing the number of orbits necessary to reach the desired energy. The magnets are flat and scale primarily only like the square of the bending radius.

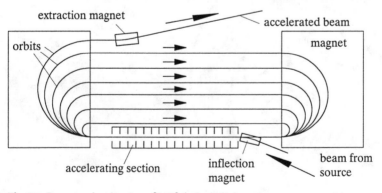

Fig. 3.3. Race track microtron [3.15] (schematic)

3.4.2 Cyclotron

The synchronicity condition of a microtron proved to be too severe for the successful acceleration of heavier particles like protons. In drawing this conclusion, however, we have ignored the trivial synchronicity condition, $\Delta\gamma = 0$. This condition demands that the particle energy be nonrelativistic which limits the maximum achievable energy to values much less than the rest energy. This limitation is of no interest for electrons since electrostatic accelerators would provide much higher energies than that. For protons, however, energies much less than the rest energy of 938 MeV are of great interest. This was recognized in 1930 by Lawrence and Edlefsen [3.16] in the process of inventing the principle of the *cyclotron* and the first such device was built by Lawrence and Livingston in 1932 [3.17].

The cyclotron principle employs a uniform magnetic field and an rf cavity that extends over the whole aperture of the magnet as shown in Fig. 3.4.

The accelerating cavity has basically the form of a pill box cut in two halves, where the accelerating fields are generated between those halves and are placed between the poles of the magnet. Because of the form of the half

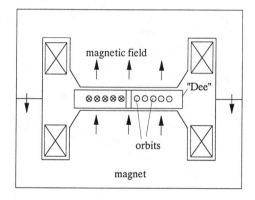

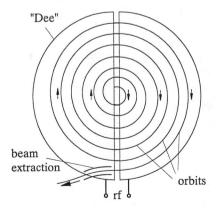

Fig. 3.4. Principle of a cyclotron [3.17] (schematic). Vertical (top) and horizontal (bottom) mid plane cross sections

pill boxes, these cavities are often called the *Dees of a cyclotron*. The particle orbits occur mostly in the field free interior of the Dees and traverse the accelerating gaps between the two Dees twice per revolution. Due to the increasing energy, the particle trajectories spiral to larger and larger radii. The travel time within the Dees is adjusted by the choice of the magnetic field such that it is equal to half the radio frequency period. The principle of the cyclotron is basically the application of the Wideroe linac in a coiled up version to save space and rf equipment. Fundamentally, however, the use of field free tubes or Dees with increasing path length between accelerating gaps is the same.

The revolution time in a cyclotron is from (3.43)

$$\tau = \frac{2\pi r}{v} = \frac{2\pi mc}{e} \frac{\gamma}{Z B}, \tag{3.46}$$

where we set $\gamma = 1$ and allow for the acceleration of ions with a charge multiplicity Z. Keeping the magnetic field constant we have a constant revolution frequency and may therefore apply a constant radio frequency

$$B = \text{const.} \Rightarrow f_{\text{rev}} = \frac{Z eB}{2\pi mc\gamma} = \text{const.} = f_{\text{rf}}, \tag{3.47}$$

where f_{rf} is the radio frequency in the accelerating cavity.

The principle of the cyclotron is limited to nonrelativistic particles. Protons are sufficiently nonrelativistic up to kinetic energies of about 20 - 25 MeV or about 2.5 % of the rest energy. As the particles become relativistic, $\gamma > 1$, the revolution frequency becomes smaller and the particles get out of synchronism with the radio frequency f_{rf}.

The radio frequency depends on the charge multiplicity Z of the particles to be accelerated and on the magnetic field B. From (3.47)the following frequencies are required for different types of particles

$$
\begin{aligned}
f_{\text{rf}} \, [\text{MHz}] &= 1.53 \, B[\text{kG}] &\quad &\text{for protons}, \\
&= 0.76 \, B[\text{kG}] &\quad &\text{for deuterons}, \\
&= 0.76 \, B[\text{kG}] &\quad &\text{for He}^{++}.
\end{aligned}
\tag{3.48}
$$

A closer inspection of the synchronicity condition, however, reveals that these are only the lowest permissible rf frequencies. Any odd integer multiple of the frequencies (3.48) would be acceptable too.

As long as particles do not reach relativistic energies the maximum achievable kinetic energy E_{kin} depends on the type of the particle, the magnetic field B, and the maximum orbit radius R possible in the cyclotron and is given by

$$E_{\text{kin}} = \tfrac{1}{2}mv^2 = \frac{(cp)^2}{2\,mc^2} = \frac{Z^2\,e^2\,B^2\,R^2}{2\,mc^2}. \tag{3.49}$$

Examples of numerical relations are

$$E_{\text{kin}}[\text{MeV}] = 0.48\,B^2[\text{kG}^2]\,R^2[\text{m}^2] \qquad \text{for protons},$$
$$= 0.24\,B^2[\text{kG}^2]\,R^2[\text{m}^2] \qquad \text{for deuterons}, \qquad (3.50)$$
$$= 0.48\,B^2[\text{kG}^2]\,R^2[\text{m}^2] \qquad \text{for He}^{++}.$$

The particle flux reflects the time structure of the radio frequency field. For a continuous radio frequency field the particle flux is also "continuous" with micro bunches at distances equal to the oscillation period of the accelerating field. For a pulsed rf system obviously the particle flux reflects this macropulse structure on top of the micropulses.

3.4.3 Synchro Cyclotron

The limitation to nonrelativistic energies of the cyclotron principle is due only to the assumption that the radio frequency be constant. This mode of operation for rf systems is desirable and most efficient but is not a fundamental limitation. Technical means are available to vary the radio frequency in an accelerating cavity.

As the technology for acceleration to higher and higher energies advances, the need for particle beam focusing becomes increasingly important. In the transverse plane this is achieved by the weak focusing discussed earlier. In the *longitudinal phase space stability* criteria have not been discussed yet. Veksler [3.18] and McMillan [3.19] discovered and formulated independently the principle of *phase focusing*, which is a fundamental focusing property for high energy particle accelerators based on accelerating radio frequency fields and was successfully tested only one year later [3.20]. We will discuss this principle of phase focusing in great detail in Chap. 8.

Both the capability of varying the rf frequency and the principle of phase focusing is employed in the *synchro cyclotron*. In this version of the cyclotron, the radio frequency is varied as the relativistic factor γ deviates from unity. Instead of (3.47) we have for the revolution frequency or radio frequency

$$f_{\text{rf}} = \frac{Z\,eB}{2\pi\,\gamma\,mc}. \qquad (3.51)$$

Since $B = \text{const}$ the radio frequency must be adjusted like

$$f_{\text{rf}} \sim 1/\gamma(t) \qquad (3.52)$$

to keep synchronism. The momentary particle energy $\gamma(t)$ can be derived from the equation of motion $1/r = ZeB/(cp)$ which we solve for the kinetic energy

$$\sqrt{E_{\text{kin}}\,(E_{\text{kin}} + 2\,mc^2)} = eZ\,B\,r. \qquad (3.53)$$

67

The largest accelerator ever built, based on this principle, is the 184 inch synchro cyclotron at the *Lawrence Berkeley Laboratory* LBL in 1946 [3.21]. The magnet weighs 4300 tons, produces a maximum magnetic field of 15 kG and has a maximum orbit radius of 2.337 m. From (3.53) we conclude that the maximum kinetic proton energy should be $T_{max} = 471$ MeV while 350 MeV have been achieved. The discrepancy is mostly due to the fact that the maximum field of 15 kG does not extend out to the maximum orbit due to focusing requirements. The principle of the synchro cyclotron allows the acceleration of particles to rather large energies during many turns within the cyclotron magnet. This long path requires the addition of *weak focusing* as discussed in connection with the betatron principle to obtain a significant particle flux at the end of the acceleration period. From the discussion in Sect. 3.2 we know that efficient focusing in both planes requires the vertical magnetic field component to drop off with increasing radius. For equal focusing in both planes the *field index* should be $n = 1/2$ and the magnetic field scales therefore like

$$B_y(r) \sim \frac{1}{\sqrt{r}}. \tag{3.54}$$

The magnetic field is significantly lower at large radii compared to the center of the magnet.

This magnetic field dependence on the radial position leads also to a modification of the frequency tracking condition (3.52). Since both the magnetic field and the particle energy change, synchronism is preserved only if the rf frequency is modulated like

$$f_{rf} \sim \frac{B[r(t)]}{\gamma(t)}. \tag{3.55}$$

Because of the need for frequency modulation, the particle flux has a pulsed macro structure equal to the cycling time of the rf modulation. A detailed analysis of the accelerator physics issues of a synchro cyclotron can be found in [3.22]

3.4.4 Isochron Cyclotron

The frequency modulation in a synchro cyclotron is technically complicated and must be different for different particle species. A significant breakthrough occurred in this respect when Thomas [3.23] realized that the radial dependence of the magnetic field could be modified in such a way as to match the particle energy γ. The condition (3.55) becomes in this case

$$f_{rf} \sim \frac{B[r(t)]}{\gamma(t)} = \text{const}. \tag{3.56}$$

To reconcile (3.56)with the focusing requirement, strong azimuthal variations of the magnetic fields are introduced

$$\frac{\partial B_y(r,\varphi)}{\partial \varphi} \neq 0. \tag{3.57}$$

In essence, the principle of weak focusing is replaced by *strong focusing*, to be
discussed later, with focusing forces established along the particle trajectory
while meeting the synchronicity condition only on average in each turn such
that

$$\frac{1}{2\pi} \oint B_y[r(t),\varphi] \, d\varphi \sim \gamma(t). \tag{3.58}$$

The development of circular accelerators has finally made a full circle. Starting from the use of a constant radio frequency field to accelerate particles
we found the need for frequency modulation to meet the synchronicity condition for particles through the relativistic transition regime. Application
of sophisticated magnetic focusing schemes, which are now known as *strong
focusing*, finally allowed to revert back to the most efficient way of particle
acceleration with constant fixed radio frequency fields.

Isochron cyclotrons produce a continuous beam of micro bunches at the
rf frequency. The high proton flux makes this accelerator a very efficient
source of high energetic protons which are often aimed at a target to create
high fluxes of kaon and pion mesons.

3.5 Synchrotron

The maximum particle energy is limited to a few hundred MeV as long
as one stays with the basic cyclotron principle because the volume and
therefore cost for the magnet becomes prohibitively large. Higher energies
can be achieved and afforded if the orbit radius R is kept constant. In this
case the center of the magnet is not needed anymore and much smaller
magnets can be employed along the constant particle orbit. In Fig. 3.5
the arrangement of magnets is shown for a strong focusing synchrotron
[3.24] based on a FODO structure, which will be discussed in more detail in
Chap. 6.

Equation (3.42) is still applicable but we keep now the orbit radius
constant and have the design condition

$$\frac{1}{R} = \frac{eB}{cp} = \text{const}. \tag{3.59}$$

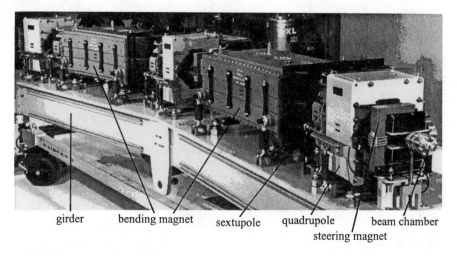

girder   bending magnet   sextupole   quadrupole   beam chamber
steering magnet

Fig. 3.5. Magnet lattice and accelerating cavities in a synchrotron [3.24]

This condition can be met for all particle energies by ramping the magnet fields proportional to the particle momentum. Particles are injected at low momentum and are then accelerated while the bending magnet fields are increased to keep the particles on a constant radius while they gain energy. The particle beam from such a synchrotron is pulsed with a repetition rate determined by the magnetic field cycling. The *synchronicity condition* (3.51),

$$f_{\text{rf}} = \frac{Z\,eB}{2\pi\,\gamma\,mc},$$ (3.60)

is still valid, but because the magnetic field is varied proportional to the particle momentum we expect the frequency to require adjustment as long as there is sufficient difference between particle energy and momentum.

For highly relativistic particles a solution for particle acceleration has been found which does not require a prohibitively large magnet and where the radio frequency fields can be of constant frequency for optimal efficiency. This is the case for electron synchrotrons with an initial energy of at least a few tens of MeV. For this reason electrons are generally injected into a synchrotron at energies of more than about 10 - 20 MeV from a linear accelerator or a microtron.

For heavier particles, however, we are back to the need for some modest frequency modulation during the early phases of acceleration. From (3.60) we expect the revolution frequency to vary like

$$f_{\text{rev}}(t) = \frac{Z\,ecB}{2\pi\,cp}\,\beta(t) \sim \beta(t).$$ (3.61)

To preserve the synchronicity condition, the radio frequency must be an integer multiple of the revolution frequency and must be modulated in proportion of the varying revolution frequency. The ratio of the rf frequency to revolution frequency is called the *harmonic number*

$$f_{rf} = h f_{rev}. \tag{3.62}$$

The maximum energy in a synchrotron is determined by the ring radius R, and the maximum magnetic field B, and is from (3.59)

$$cp_{max} = \sqrt{E_{kin}(E_{kin} + 2mc^2)} = C_p B[\text{kG}] R[\text{m}], \tag{3.63}$$

where

$$C_p = e = 0.02997926 \frac{\text{GeV}}{\text{kG m}}. \tag{3.63}$$

Early synchrotrons have been constructed with weak focusing bending magnets which in addition to the dipole field component also included a field gradient consistent with a field index meeting the focusing condition (3.28). More detailed information about early weak focusing synchrotrons can be obtained from [3.25, 26].

With the discovery of the principle of *strong focusing* by Christofilos [3.27] and independently by Courant et. al. [3.28] in 1952 much more efficient synchrotrons could be designed. The apertures in the magnets could be reduced by up to an order of magnitude thus allowing the design of high-field magnets at greatly reduced overall magnet size and cost. The physics of strong focusing will be discussed in great detail in subsequent chapters.

Synchrotrons are the workhorse in particle acceleration and are applied for electron acceleration as well as proton and ion acceleration to the highest energies. In more modern proton accelerators superconducting magnets are employed to reach energies in excess of 1000 GeV.

3.5.1 Storage Ring

Although not an accelerator in the conventional sense, a particle *storage ring* can be considered as a synchrotron frozen in time. While the basic functioning of a storage ring is that of a synchrotron, particle beams are generally not accelerated but only stored to orbit for long times of several hours. The original application was in high energy physics to bring two counter rotating beams of particle and antiparticles into collision and study high energy elementary particle processes. A newer and more copious application of the storage ring principle arose for the dedicated production of synchrotron radiation for basic and applied research, and technology.

The principles and functioning of synchrotrons and storage rings will occupy most of our discussions in this text.

3.6 Summary of Characteristic Parameters

It is interesting to summarize the basic principles for the different particle accelerators discussed. All are based on two relations, the *Lorentz force equation*

$$\frac{1}{r} = \frac{eB_y}{\gamma\,mc^2\,\beta}\,,$$

and the *synchronicity condition*

$$f_{rf} = \frac{c\,eB_y}{2\pi\,\gamma\,mc^2}\,h\,.$$

Depending on which parameter in these two relations we want to keep constant or let vary, different acceleration principles apply with varying advantages and disadvantages. In Table 3.1 these parameters and their disposition are compiled for the acceleration principles discussed above.

Table 3.1. Parameter disposition for different acceleration principles

principle	energy	velocity	orbit	field	frequency	flux
	γ	v	r	B	f_{rf}	
Cyclotron:	1	var.	$\sim v$	const.	const.	cont.[a]
Synchro cyclotron:	var.	var.	$\sim p$	$B(r)$	$\sim \frac{B(r)}{\gamma(t)}$	pulsed
Isochron cyclotron:	var.	var.	$r = f(p)$	$B(r,\varphi)$	const.	cont.[a]
Proton/Ion-synchrotron:	var.	var.	R	$\sim p(t)$	$\sim v(t)$	pulsed
Electron-synchrotron:	var.	const.	R	$\sim p(t)$	const.	pulsed

[a] continuous beam, but rf modulated

There are two particle parameters and three technical device parameters which define the mode of accelerator operation. Different acceleration principles are defined by choosing one or the other parameter to be variable or fixed. Interestingly enough, most of the discussed acceleration methods have their proper application and are used either as stand alone accelerators for research and technology or are part of an acceleration chain for high energy particle accelerators. Depending on the energy of particular particles, different acceleration principles are optimum.

For example, it makes no sense to construct a proton synchrotron, where the protons must be injected directly from the source at very low energies. The proper way is to first accelerate the protons with electrostatic fields, for example in a Cockcroft-Walton accelerator followed by a medium energy linear accelerator to reach a high enough energy of a few hundred MeV for efficient injection into a synchrotron.

Problems

Problem 3.1. Consider the Kerst betatron cycling at 60 Hz. Electrons are injected at 50 keV into this betatron, calculate the magnet field at injection and the energy gain per turn for the first turn and at a time when the electron has gained 20 MeV. Discuss the reason for the difference in the energy gain per turn.

Problem 3.2. What is the total excitation current in each of the two coils for a betatron with an orbit radius of R = 0.4 m, a maximum electron momentum of cp = 42 MeV and a gap of g = 10 cm between the poles.

Problem 3.3. Calculate the electron beam current in the Kerst betatron that would produce a total synchrotron radiation power of 1 Watt at $cp = 300$ MeV.

Problem 3.4. Try to "design" a microtron for a maximum electron energy of E = 25 MeV. Use a magnetic field of B = 2140 G.
a) What is the diameter of the last circular trajectory at 25 MeV ?
b) Sketch a cross section of the magnet with excitation coils. Magnet poles must extend radially at least by 1.5 gap heights beyond the maximum orbit to obtain good field quality. Use a total coil cross section of 5 cm². Choose your own gap height.
c) What is the electrical power required to operate each coil assuming copper and a copper fill factor of 75 %. That means 75% of the coil cross section is copper and the rest is for insulation and cooling. Do you think your coil needs water cooling?
d) How does the electrical power requirement change if you change the number of turns in the coil thereby changing the electrical current. Keep in either case 75% fill factor.

Problem 3.5. Calculate the frequency variation required to accelerate protons or deuterons in a synchro cyclotron from a kinetic energy of $E_{kin,o} = 100$ keV to an end energy of $E_{kin} = 600$ MeV. Keep the magnetic field constant and ignore weak focussing. Derive formula for the rf frequency as a function of the kinetic energy and generate a graph of f_{rf} vs E_{kin} from just a few points. How big would the frequency swing be for electrons ?

Problem 3.6. Consider the Fermilab 400 GeV synchrotron which has a circumference of 6000 m. Protons are injected at 10 GeV. Calculate the frequency swing necessary for synchronicity during the acceleration cycle. The acceleration from 10 GeV to 400 GeV takes 6 sec. If the rf frequency is not modulated, by how much will the actual beam slip with respect to an ideal beam traveling with the speed of light? What would the maximum energy be? Sketch the energy variation during these 6 sec.

Problem 3.7. Verify the numerical validity of Eqs. (3.41) and (3.63) as well as (3.48) and (3.50).

4 Charged Particles in Electromagnetic Fields

The most obvious components of particle accelerators and beam transport systems are those that provide the beam guidance system. Whatever the application may be, a beam of charged particles is designed to follow closely a predescribed path along a desired beam transport line or along a closed orbit in case of circular accelerators. The forces required to bend and direct the charged particle beam or to hold particles close to the ideal path are known as the *Lorentz forces* and are derived from electric and magnetic fields through the *Lorentz equation*.

4.1 The Lorentz Force

For a particle carrying a single basic unit of electrical charge the Lorentz force is expressed by

$$\mathbf{F} = e\mathbf{E} + \frac{e}{c} \left[\mathbf{v} \times \mathbf{B} \right],$$

(4.1)

where e is the basic unit of electrical charge [4.1],

$$e = 4.8032068 \times 10^{-10} \text{ esu} = 1.60217733 \times 10^{-19} \text{ C},$$

(4.2)

and c the speed of light [4.1],

$$c = 2.99792458 \times 10^{10} \text{ cm/s}.$$

(4.3)

The vectors $\mathbf{E}$ and $\mathbf{B}$ are the electrical and magnetic field vectors, respectively, and $\mathbf{v}$ is the velocity vector of the particle. These Lorentz forces will be applied not only to guide particles along a predefined path but will also be used for beam focusing to confine a beam of particles to within a narrow vicinity of the ideal path. The evolution of particle trajectories under the influence of Lorentz forces is called *beam dynamics* or *beam optics*. The basic formulation of beam dynamics relies only on *linear fields* which are independent of or only linearly dependent on the distance of a particular particle from the ideal trajectory. The mathematical description of particle trajectories in the presence of only such linear fields is called *linear beam dynamics*. We will discuss the theory of linear beam dynamics here and postpone the investigation of nonlinear fields to Vol. II.

The Lorentz force has two components originating from an electrical field **E** and a magnetic field **B**. For relativistic particles, $(v \approx c)$, we find that the force from a magnetic field of 1 G is equivalent to that for an electrical field of 300 V/cm. Since it is technically straight forward to generate magnetic fields of the order of 10,000 G but rather difficult to establish the equivalent electric fields of 3 Million V/cm, it becomes apparent that most beam guidance and focusing elements for relativistic particle beams are based on magnetic fields. At low particle energies, however, this preference is less clear and justified since the effectiveness of magnetic fields to bend particles is reduced proportional to the particle velocity $\beta = v/c$.

4.2 Coordinate System

To develop a useful mathematical formalism for the description of charged particle beam dynamics we must choose the appropriate *coordinate system* to minimize mathematical complexity and maximize physical clarity. Any coordinate system could be used but some are more practical than others. In accelerator physics we use coordinate systems which most clearly exhibit the parameters of interest. To describe particle trajectories we separate the particle position in space or its coordinates into two parts. One part describes the path the particle or beam is supposed to follow and this path can be of any arbitrary form as desired by the application. This path is called the ideal path or reference path . The other part of the coordinates is the deviation of a particular particle from this reference path.

We use a fixed, right-handed cartesian *reference system* $(\mathbf{x}, \mathbf{y}, \mathbf{z})$ to define the ideal path as determined by the *Lorentz force*. In most cases, however, this ideal path is defined once and for all as soon as the actual location of deflecting magnets along a beam transport line is known. Once the ideal path is known and fixed we are interested only in the deviation of individual particle trajectories from this ideal path. Meaningful formulation of particle trajectories in the vicinity of the ideal path becomes very awkward in a fixed cartesian coordinate system because the mathematical formula for the reference path would be superimposed on any deviations from the ideal path a particle may have. This is no problem in a straight beam line, where the reference path coincides with the z coordinate. In curved transport lines, however, the expressions for the particle locations becomes very complicated, if analytical formulas exist at all and require additional manipulations to isolate the quantity of interest which is the deviation from the ideal path.

The natural reference system avoiding this complication is an orthogonal, right-handed coordinate system (x, y, s) that follows an ideal particle traveling along the ideal path. We use the vector **s** for the coordinate along the path in contrast to **z** in the fixed coordinate system. In this case the

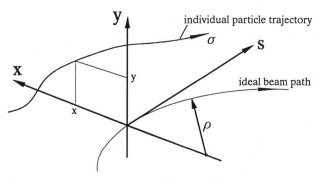

Fig. 4.1. Coordinate system

transverse coordinates will measure only the deviations from the ideal path, like in the linear case.

This moving reference system is shown in Fig. 4.1 and, because of its wide use in accelerator physics, it is often called the standard coordinate system. Its origin is defined by the vector $\mathbf{S}(s)$ following the reference path in an osculatory motion through straight as well as curved sections and is therefore a curvilinear coordinate system .

The coordinate vector $\mathbf{s}$ is positioned tangentially to the reference orbit and the orthogonal coordinate vectors $\mathbf{x}$ and $\mathbf{y}$ may have any arbitrary azimuthal orientation as long as they are orthogonal to $\mathbf{s}$ and to each other. The most convenient orientation is the one in which the plane of the complete curved beam transport system or a segment thereof includes the vector $\mathbf{s}$ and one of the transverse coordinate vectors. Since most accelerators are built in a horizontal plane, it is customary to identify the $(\mathbf{x}, \mathbf{s})$-plane with the horizontal plane and the vector $\mathbf{y}$ with the vertical direction as shown in Fig. 4.1. The $\mathbf{s}$-coordinate axis in such a right-handed reference system points in the direction of the particle reference path while the $\mathbf{x}$-axis points to the left and the $\mathbf{y}$-axis upwards if we look along the $\mathbf{s}$-axis. For *path length* measurements we will use the variable σ as the independent coordinate to describe the trajectory of an individual particle.

The *reference time t* is measured at the origin of the coordinate system $t = s/v_s$ and the "particle time" $\tau = \sigma/v_\sigma$ is the time measured along a particular particle trajectory. Before we consider individual particle trajectories in this curvilinear coordinate system, we must discuss the basic dynamics of charged particles under the influence of the Lorentz force in fixed space in an effort to define the ideal path. The *curvature* vector for any individual particle trajectory is defined by

$$\kappa = -\frac{\mathrm{d}^2\mathbf{S}(s)}{\mathrm{d}s^2} , \qquad (4.4)$$

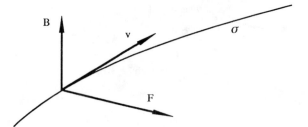

Fig. 4.2. Lorentz force (**B** magnetic field, **v** particle velocity, **F** Lorentz force)

where **S** is the location vector for the particle. In component form the curvature is

$$\boldsymbol{\kappa} = (\kappa_x, \kappa_y) = (-x'', -y'') \tag{4.5}$$

and is determined by the Lorentz force. With τ being the real time defined as $\tau = \sigma/v_\sigma$ we express the Lorentz force **F** by the rate of change in the particle momentum

$$\mathbf{F} = \frac{d\mathbf{p}}{d\tau} = \frac{e}{c}[\mathbf{v} \times \mathbf{B}], \tag{4.6}$$

where the vectors $[\mathbf{v}, \mathbf{B}, \dot{\mathbf{p}}]$ or $[\mathbf{v}, \mathbf{B}, \mathbf{F}]$ form an orthogonal, right-handed coordinate system as shown in Fig. 4.2. A purely vertical magnetic field, for example, generates a force in the horizontal (**x**, **s**)-plane only and we get with $\mathbf{B} = (0, B_y, 0)$ from (4.6)

$$\frac{dp_x}{d\tau} = -\frac{e}{c} v_\sigma B_y. \tag{4.7}$$

On the other hand we have

$$\frac{dp_x}{d\tau} \approx p\frac{dx'}{d\tau} \approx \beta c p\, x'', \tag{4.8}$$

where we have made use of $p_x \approx p x'$, with $x' = dx/ds$ and $ds \approx d\sigma = c\beta\, d\tau$. The particle momentum in the **s**-direction $p_s = p(1 - x'^2 - y'^2)^{1/2} \approx p$ since for most beam transport systems we may assume that the slopes of the particle trajectories x' and y' are small compared to unity. This linear approximation may not be sufficient in some special cases of beam dynamics and a more rigorous derivation will therefore be pursued in Vol. II. The linear solution, however, will be sufficient for most practical applications and the study of particle motion in this approximation is called *linear particle dynamics* or *linear optics*. From (4.6,7) we get with $v_\sigma = \beta c$

$$x'' \approx -\frac{e}{cp}\frac{v_\sigma}{\beta c} B_y = -\frac{e}{cp} B_y. \tag{4.9}$$

A positive magnetic field component $B_y > 0$ causes therefore a positive curvature $\kappa_x > 0$, for a positively charged particle ($e > 0$) traveling in the direction of the coordinate s as shown in Fig. 4.1. The generalization to a trajectory in three-dimensional space with horizontal as well as vertical deflection is straightforward. A positive horizontal field component, $B_x > 0$, would therefore deflect the same particle into the positive vertical direction and the vertical equation of motion becomes

$$y'' \approx \frac{e}{cp} B_x ,\tag{4.10}$$

while the curvature κ_y is negative.

Equations (4.9,10) represent the fundamental *equations of motion* in transverse magnetic fields for charged particles in the horizontal and vertical plane, respectively. As a consequence of the vector product in (4.6) we note that positive field components in the horizontal and vertical planes cause the particle trajectories to be curved in opposite directions. These equations can be derived in a general vector form by noting that

$$\frac{d\mathbf{p}}{d\tau} = m\gamma \frac{d^2\mathbf{S}}{d\tau^2} = m\gamma v_\sigma^2 \frac{d^2\mathbf{S}}{d\sigma^2} = \frac{e}{c}[\mathbf{v} \times \mathbf{B}] ,\tag{4.11}$$

where m is the mass of the particle and γ is the particle energy in units of its rest energy. With (4.4) Eq. (4.11) becomes

$$\boldsymbol{\kappa} = -\frac{d^2\mathbf{S}}{ds^2} \approx -\frac{e}{cp}\left[\frac{\mathbf{v}}{v} \times \mathbf{B}\right] ,\tag{4.12}$$

or in the approximation of paraxial trajectories, where $v_x \ll v$ and $v_y \ll v$,

$$\begin{aligned}
\kappa_x &= -x'' = +\frac{e}{cp} B_y \qquad \text{with} \qquad \left|\frac{e}{cp} B_y\right| = \frac{1}{\rho_x} , \\
\kappa_y &= -y'' = -\frac{e}{cp} B_x \qquad \text{with} \qquad \left|\frac{e}{cp} B_x\right| = \frac{1}{\rho_y} ,
\end{aligned}\tag{4.13}$$

where ρ_x and ρ_y are the local bending radii of the particle trajectory in the horizontal and vertical plane respectively. We select from all particle trajectories a special one, which is defined by taking only the pure dipole fields into account. This trajectory is defined by

$$\boldsymbol{\kappa}_o = -\frac{d^2\mathbf{S}_o}{ds^2} \approx -\frac{e}{cp}\left[\frac{\mathbf{v}}{v} \times \mathbf{B}_o\right]\tag{4.14}$$

and is called the *reference trajectory, reference path* or *ideal path* of the beam line. The geometry of the reference trajectory is determined by the placement and strength of bending magnets with bending radii

$$\frac{1}{\rho_{ox}} = \left|\frac{e}{cp}B_{oy}\right| \quad \text{and} \quad \frac{1}{\rho_{oy}} = \left|\frac{e}{cp}B_{ox}\right|, \tag{4.15}$$

and the equation of motion for the reference path is from (4.13)

$$x_o'' = -\kappa_{ox} \quad \text{and} \quad y_o'' = -\kappa_{oy}. \tag{4.16}$$

Later we will subtract this reference path from the general solution for the equation of motion and obtain equations describing the deviation from the reference path only.

Consistent with the *sign convention* adopted here we will also determine the strength parameters for higher multipoles. In nonmathematical terms, the sign convention used in this book is such that an observer looking into the positive **s**-direction finds a positive charge traveling along the positive **s**-direction to be deflected to the right by a positive vertical magnetic field $B_y > 0$. As a consequence, a positive horizontal field component $B_x > 0$ would deflect this same charge upwards, as illustrated in Fig. 4.1. We have derived the tools to determine the ideal reference beam path of charged particles in the presence of Lorentz forces and in Sect. 4.7 we will derive the equations of motion of individual particles relative to this ideal path.

Before we leave the discussion on coordinate systems used in accelerator physics it is useful to introduce here the metric for the curvilinear coordinate system. The infinitesimal path length element along a trajectory in a curvilinear coordinate system is

$$d\sigma^2 = dx^2 + dy^2 + (1 + \kappa_x\, x + \kappa_y\, y)^2\, ds^2. \tag{4.17}$$

We use this metric to transform commonly used vector relations from the cartesian to the curvilinear coordinate system. With

$$H = (1 + \kappa_x\, x + \kappa_y\, y) \tag{4.18}$$

we have

$$\nabla f = \frac{\partial f}{\partial x}\mathbf{x} + \frac{\partial f}{\partial y}\mathbf{y} + \frac{1}{H}\frac{\partial f}{\partial s}\mathbf{s}, \tag{4.19}$$

$$\nabla \cdot \mathbf{A} = \frac{1}{H}\left[\frac{\partial(H\,A_x)}{\partial x} + \frac{\partial(H\,A_y)}{\partial y} + \frac{\partial A_s}{\partial s}\right], \tag{4.20}$$

and

$$\begin{aligned}
\nabla \times \mathbf{u} = \ & \frac{1}{H}\left[\frac{\partial(H\,u_s)}{\partial y} - \frac{\partial u_y}{\partial s}\right]\mathbf{x} \\
& + \frac{1}{H}\left[\frac{\partial u_x}{\partial s} - \frac{\partial(H\,u_s)}{\partial x}\right]\mathbf{y} + \left[\frac{\partial u_y}{\partial x} - \frac{\partial u_x}{\partial y}\right]\mathbf{s},
\end{aligned} \tag{4.21}$$

where f is a scalar function and $\mathbf{A}$ and $\mathbf{u}$ are vectors. Once the metric (4.17) with the appropriate expression H of a particular coordinate system is known, we may express any vector equation in these new coordinates.

4.3 Fundamentals of Charged Particle Beam Optics

Magnetic as well as electric fields can be produced in many ways and appear in general in arbitrary directions and varying strength at different locations. It is impossible to derive a general mathematical formula for the complete path of charged particles in an arbitrary field distribution. To design particle beam transport systems we therefore include some organizing and simplifying requirements on the characteristics of electromagnetic fields used. In this section first general expressions for the electromagnetic fields will be derived which are then introduced into the equations of motions. At that point it becomes obvious which field components are the most useful to design predictable beam transport systems. By appropriate design of magnets less desirable terms become negligibly small.

The general task in beam optics is to transport charged particles from point A to point B along a desired path. We call the collection of bending and focusing magnets installed along this ideal path the *magnet lattice* and the complete optical system including the bending and focusing parameters a *beam transport system*. Two general cases can be distinguished in beam transport systems, such systems that display neither symmetry nor periodicity and transport systems that include a symmetric or periodic array of magnets. Periodic or symmetric transport systems can be repeated an arbitrary number of times to produce longer transport lines. A specific periodic magnet lattice is obtained if the arrangement of bending magnets forms a closed loop. For a particle beam such a lattice appears like a periodic transport system with a period length equal to the length of the closed path. In our discussions of transverse beam dynamics, we will make no particular distinction between open beam transport lines and circular lattices except in such cases when we find the need to discuss special eigensolutions for closed periodic lattices. We will therefore use the terminology of beam transport systems when we discuss beam optics results applicable to both types of lattices and refer to circular accelerator lattices when we derive eigenfunctions characteristic only to periodic and closed magnet lattices.

4.3.1 Particle Beam Guidance

To guide a charged particle along a predefined path, magnetic fields are used which deflect particles as determined by the equilibrium of the centrifugal force and Lorentz force

$$m\gamma v^2 \, \boldsymbol{\kappa} + \frac{e}{c} \, [\, \mathbf{v} \times \mathbf{B} \,] \; = \; 0 \, , \tag{4.22}$$

where $\boldsymbol{\kappa} = (\kappa_x, \kappa_y, 0)$ is the local curvature vector of the trajectory.

We assume in general that the magnetic field vector $\mathbf{B}$ is oriented normal to the velocity vector $\mathbf{v}$. This means we restrict the treatment of linear beam dynamics to purely *transverse fields*. The restriction to purely transverse field components has no fundamental reason other than to simplify the formulation of particle beam dynamics. The dynamics of particle motion in longitudinal fields will be discussed in Vol. II. As mentioned earlier, the transverse components of the particle velocities for relativistic beams are small compared to the particle velocity v_s, ($v_x \ll v_s, v_y \ll v_s, v_s \approx v_\sigma$). With these assumptions the *bending radius* for the particle trajectory in a magnetic field is

$$\frac{1}{\rho} \; = \; \left| \frac{e}{cp} B \right| \; = \; \left| \frac{e}{\beta E} B \right| \tag{4.23}$$

and the angular frequency of revolution of a particle on a complete orbit normal to the field B is

$$\omega_L \; = \; \left| \frac{e\,c}{E} B \right| \, , \tag{4.24}$$

which is also called the *cyclotron* or *Larmor frequency* [4.2]. To normalize the magnet strength we define the *beam rigidity* by

$$|B\rho| \; = \; \frac{c\,p}{e} \, . \tag{4.25}$$

Using more practical units the expressions for the beam rigidity and curvature become

$$B\rho\,[\mathrm{Tesla\,m}] \; = \; \frac{10}{2.998}\,\beta\,E[\mathrm{GeV}] \tag{4.26}$$

and

$$\frac{1}{\rho}\,[\mathrm{m^{-1}}] \; = \; \frac{|B|}{|B\rho|} \; = \; 0.2998\,\frac{|B[\mathrm{Tesla}]|}{\beta\,E[\mathrm{GeV}]} \, . \tag{4.27}$$

For relativistic particles this expression is further simplified since $\beta \approx 1$.

In this textbook, singly charged particles will be assumed unless otherwise noted. For multiply charged particles like ions, the electrical charge e in all equations must be replaced by $e\,Z$ if, for example, ions of net charge Z are to be considered. Since it is also customary not to quote the total ion energy, but the energy per nucleon, (4.27) becomes for ions

$$\frac{1}{\rho}\,[\mathrm{m^{-1}}] \; = \; \frac{|B|}{|B\rho|} \; = \; 0.2998\,\frac{Z}{A}\,\frac{|B[\mathrm{Tesla}]|}{\beta\,E[\mathrm{GeV/u}]} \, , \tag{4.28}$$

where E is the total energy per nucleon.

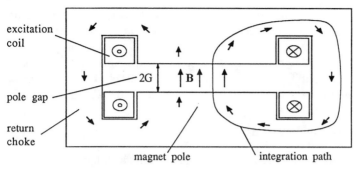

excitation coil

pole gap

return choke

2G

↑B↑

magnet pole

integration path

Fig. 4.3. Cross section of a dipole magnet (schematic)

A transverse magnetic field being constant and homogeneous in space at least in the vicinity of the particle beam, is the lowest order field in beam guidance or beam transport systems. Such a field is called a *dipole field* and can be generated, for example, in an electromagnet as shown in Fig. 4.3.

The magnetic field $\mathbf{B}$ is generated by an electrical current I in current carrying coils surrounding magnet poles. A ferromagnetic return yoke surrounds the excitation coils providing an efficient return path for the magnetic flux. The magnetic field is determined by one of Maxwell's equations

$$\nabla \times \frac{\mathbf{B}}{\mu_\mathrm{r}} = \frac{4\pi}{c}\,\mathbf{j}, \tag{4.29}$$

where μ_r is the relative permeability of the ferromagnetic material and $\mathbf{j}$ is the current density in the coils. Integrating (4.29) along a closed path like the one shown in Fig. 4.3 and using Stoke's theorem gives

$$2\,G\,B_\perp + \int_\mathrm{iron} \frac{\mathbf{B}}{\mu_\mathrm{r}}\,\mathrm{d}\mathbf{s} = \frac{4\pi}{c}\,I_\mathrm{tot}, \tag{4.30}$$

where $B_\perp$ is the magnetic field between and normal to the parallel magnet poles with a gap distance of $2\,G$. The integral term in (4.30) is zero or negligibly small in most cases assuming infinite or a very large permeability within the magnetic iron. I_tot is the total current flowing in the complete cross section of both coils. Solving (4.30) for the total current in each coil we get in more practical units

$$I_\mathrm{tot}\,(\mathrm{Amp}) = \frac{1}{0.4\pi}\,B_\perp[\mathrm{G}]\,G[\mathrm{cm}]. \tag{4.31}$$

The total field *excitation current* in the magnet coils is proportional to the desired magnetic field and the total gap between the magnet poles.

As a practical example, we consider a magnetic field of 10^4 G equivalent to 1 T in a dipole magnet with a gap of 10 cm. From (4.31) we find a total electrical current of about 40,000 A is required in each of two excitation coils to generate this field. Since the coil in general is composed of many turns, the actual electrical current is usually much smaller by a factor equal to the number of turns and the total coil current I_{tot}, therefore, is often measured in units of *Ampere turns*. For example two coils each composed of 40 windings with sufficient cross section to carry an electrical current of 1000 A would provide the total required current of 80,000 A turns.

4.3.2 Particle Beam Focusing

Similar to the properties of light rays, particle beams also have a tendency to spread out due to an inherent beam divergence. To keep the particle beam together and to generate specifically desired beam properties at selected points along the beam transport line, *focusing devices* are required. In photon optics that focusing is provided by glass lenses. The characteristic property of such focusing lenses is that a light ray is deflected by an angle proportional to the distance of the ray from the center of the lens (Fig. 4.4). With such a lens a beam of parallel rays can be focused to a point. The distance of this *focal point* from the lens is called the *focal length*.

Any magnetic field that deflects a particle by an angle proportional to its distance r from the axis of the focusing device will act in the same way as a glass lens does in the approximation of paraxial, geometric optics for visible light. If f is the focal length, the deflection angle α is defined from Fig. 4.4 by

$$\alpha = -\frac{r}{f}. \tag{4.32}$$

A similar focusing property can be provided for charged particle beams by the use of azimuthal magnetic fields $\mathbf{B}_\varphi$ with the property

$$\alpha = -\frac{\ell}{\rho} = -\frac{e}{\beta E}\, B_\varphi \ell = \frac{e}{\beta E}\, g\, r\, \ell, \tag{4.33}$$

where ℓ is the *path length* of the particle trajectory in the magnetic field B_φ and g is the *field gradient* defined by $B_\varphi = gr$ or by $g = \mathrm{d}B_\varphi/\mathrm{d}r$. Here we

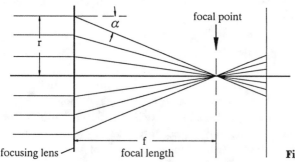

focusing lens   focal length

Fig. 4.4. Principle of focusing

84

have assumed the length ℓ to be short compared to the focal length such that r does not change significantly within the magnetic field. If this is not allowable the product $B_\varphi \ell$ must be replaced by the integral $\int B_\varphi \, ds$.

To get the focusing property (4.32) we require a linear dependence on r of either the magnetic field B_φ or of the magnet length. We choose the magnetic field to increase linearly with the distance r from the axis of the focusing device while the magnet length remains constant.

In beam dynamics it is customary to define an energy independent focusing strength. Similar to the definition of the bending curvature in (4.23) we define a *focusing strength* k by

$$k = \frac{e}{cp} g = \frac{e}{\beta E} g \tag{4.34}$$

and (4.33) becomes

$$\alpha = -k \, r \, \ell. \tag{4.35}$$

The *focal length* of the magnetic focusing device is from (4.32)

$$f^{-1} = k \, \ell. \tag{4.36}$$

In more practical units the focusing strength is given in analogy to (4.27) by

$$k[\text{m}^{-2}] = 0.2998 \, \frac{g[\text{Tesla/m}]}{\beta E[\text{GeV}]}. \tag{4.37}$$

Multiplication with Z/A gives the focusing strength for ions of charge multiplicity Z and atomic weight A.

A magnetic field that provides the required focusing property of (4.33) can be found, for example, in a current carrying conductor. Clearly such a device is not generally useful for particle beam focusing. To improve the "transparency" for particles, Panofsky and Baker [4.3] proposed to use a *plasma lens* "which contains a longitudinal arc of nearly uniform current density" and a similar device has been proposed in [4.4]. Still another variation of this concept is the idea [4.5] to use an evenly distributed array of wires, called the *wire lens*, simulating a uniform longitudinal current distribution. The strength of such lenses, however, is not sufficient for focusing of high energy particles even if we ignore the obviously strong scattering problems. Both issues, however, become irrelevant, where focusing is required in combination with particle conversion targets. Here lithium cylinders, called a *lithium lens*, carrying a large pulsed current are used to focus positrons or antiprotons emerging from conversion targets [4.6, 7].

A different type of focusing device is the *parabolic current sheet lens*. In its simplest form the current sheet lens is shown in Fig. 4.5. The rotationally symmetric lens produces an azimuthal magnetic field which scales inversely proportional to r, $B_\varphi \sim 1/r$. Since the length of the lens scales like $\ell \sim r^2$,

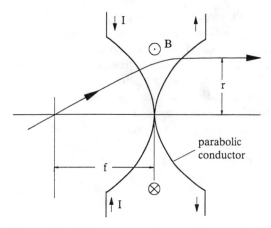

Fig. 4.5. Parabolic current sheet lens (schematic)

the deflection of a particle trajectory increases linear with r as desired for a focusing lens.

The field strength depends on the particular parameter of the paraboloid used for the current sheet and the electrical current. The magnetic field is from Maxwell's equation

$$B_\varphi[\text{G}] = 0.2\,\frac{I[\text{A}]}{r[\text{cm}]} \qquad (4.38)$$

and with $\ell = a\,r^2$ the product of the field gradient $g = \partial B_\varphi/\partial r$ and the length ℓ is

$$g\,\ell[\text{G}] = 0.2\,a[\text{cm}^{-1}]\,I[\text{A}]. \qquad (4.39)$$

A geometric variation of such a system is used in high energy physics to focus a high energy K-meson beam emerging from a target into the forward direction [4.8, 9]. Since the decaying kaon beam produces neutrinos among other particles this device is called a *neutrino horn*. On a much smaller scale compared to the neutrino horn a similar focusing devices can be used to focus positrons from a conversion target into the acceptance of a subsequent accelerator [4.10, 11].

The use of a parabolic shape for the current sheet is not fundamental. Any form with the property $\ell \sim r^2$ will provide the desired focusing properties. This type of lens may be useful for specific applications but cannot be considered as a general focusing device, where an aperture free of absorbing material is required, to let particles pass without being scattered.

The most suitable device that provides a material free *aperture* and the desired focusing field is called a *quadrupole magnet*. The magnetic field can be derived in cartesian coordinates from the *scalar potential*

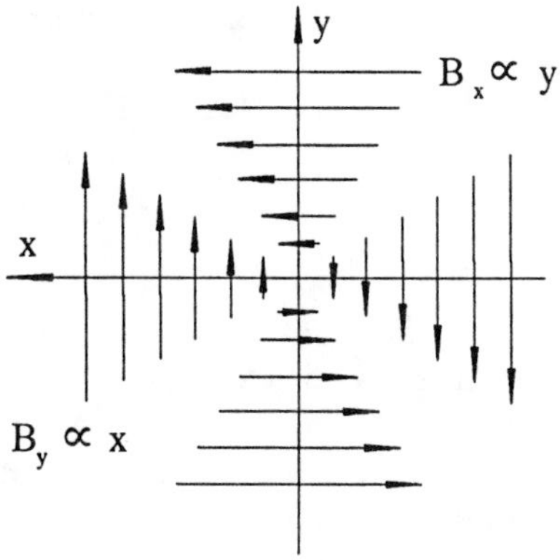

Fig. 4.6. Magnetic field pattern for a quadrupole magnet

$$V = -g\,x\,y \tag{4.40}$$

to be

$$-\frac{\partial V}{\partial x} = B_x = g\,y\,,$$
$$-\frac{\partial V}{\partial y} = B_y = g\,x\,. \tag{4.41}$$

Magnetic *equipotential surfaces* with a profile following the desired scalar potential (4.40) will be suitable to create the desired fields. The field pattern of a quadrupole magnet is shown schematically in Fig. 4.6.

Consistent with the sign convention of Sect. 4.2 the field directions are chosen such that a positron moving along and at a distance $x > 0$ from the s-axis is deflected toward the center (focusing), while the same positron with a vertical offset from the s-axis, $y > 0$, becomes deflected upward (defocusing). Different from previously discussed cylindrically symmetric lenses, a quadrupole magnet is focusing only in one plane and defocusing in the other. This property is a result of Maxwell's equations but does not diminish the usefulness of quadrupole magnets as focusing elements. A combination of quadrupoles can become a system that is focusing in both planes of a cartesian coordinate system. From paraxial light optics it is known that the total focal length of a combination of two lenses with the focal lengths f_1 and f_2 and separated by a distance d is given by

$$\frac{1}{f} = \frac{1}{f_1} + \frac{1}{f_2} - \frac{d}{f_1 f_2}\,. \tag{4.42}$$

A specific solution is $f_1 = -f_2$ and a quadrupole doublet with this property is focusing in both the horizontal and vertical plane with equal focal length $1/f = d/|f_1 f_2|$. Equation (4.42) allows many other solutions different from the simple assumption made above. The fundamental observation here, however, is that there exist indeed combinations of focusing and defocusing quadrupoles which can be made focusing in both planes and are therefore useful for charged particle focusing.

4.4 Multipole Field Expansion

From the discussion in the previous section we learn that specific desired effects on particle trajectories require specific magnetic fields. Dipole fields are the proper fields to bend particle beams and we found a variety of field patterns which could be helpful to focus particle beams. To obtain an explicit formulation of the equations of motion of charged particles in an arbitrary magnetic field we derive the general magnetic fields consistent with Maxwells equations and some desired boundary conditions.

Although we have identified a curvilinear coordinate system moving together with particles to best fit the needs of beam dynamics, we use in this section a fixed, right-handed cartesian coordinate system (x, y, z), where the z-axis coincides with the s-axis of the curvilinear coordinate system. By using the fixed cartesian coordinate system, we neglect the effects of curvature to simplify the derivation of the general magnetic field components in the approximation exhibiting only the main multipole fields. In Vol. II we will derive both the electromagnetic fields and equations of motion in full rigor.

4.4.1 Laplace Equation

In particle beam transport systems a variety of electromagnetic fields are used. Common to such devices is a material free region in the vicinity of the axis of the device to provide a scattering free passage for the particle beam. The electromagnetic fields encountered by the particles, therefore, can be derived from a potential function $V(x, y, z)$ which is determined by the charge free *Laplace equation*

$$\Delta V \equiv 0. \tag{4.43}$$

Generally electromagnetic fields we are interested in will have some degree of symmetry in the plane normal to the particle reference path and a simple field expression may be obtained in a cylindrical coordinate system (r, φ, z) in which the Laplace equation is

$$\Delta V = \frac{\partial^2 V}{\partial r^2} + \frac{1}{r} \frac{\partial V}{\partial r} + \frac{1}{r^2} \frac{\partial^2 V}{\partial \varphi^2} + \frac{\partial^2 V}{\partial z^2} \equiv 0. \tag{4.44}$$

We make an ansatz for the solution in the form of a Taylor expansion with respect to the reference path $(r \equiv 0)$

$$V(r, \varphi, z) = -\frac{cp}{e} \sum_{n>0} \frac{1}{n!} A_n(z) \, r^n \, e^{in\varphi}. \tag{4.45}$$

The field coefficients A_n are derived from the Laplace equation (4.44), excluding negative values for n to avoid nonphysical field singularities for $r \to 0$. The beam rigidity cp/e is factored out to allow later a convenient definition of the coefficients A_n. Inserting (4.45) into the Laplace equation (4.44) we get

$$\sum_n \frac{1}{n!} \left[\frac{n(n-1)+n-n^2}{r^2} A_n(z) \right] r^n e^{in\varphi} = 0. \tag{4.46}$$

For simplicity, we assume the z-dependence of the fields to vanish as is the case in the middle of long magnets. We choose this restriction to two-dimensional transverse fields to simplify the derivation of the basic *multipole fields* but will include the z-dependence of fields later in a more general derivation. Equation (4.46) is true for arbitrary angles φ and nonvanishing strengths of the multipole fields if and only if the expression in the square brackets vanishes for all values of n which is true since the factor $n(n-1)+n-n^2 = 0$. With this condition we find the ansatz (4.45) to be a valid description of a general electromagnetic fields for any field component A_n.

To calculate the Lorentz force on charged particles we are primarily interested in the electromagnetic fields which now can be derived from the potential by differentiation

$$\begin{aligned} \mathbf{E} &= -\frac{cp}{e} \nabla V_e(x,y), \\ \mathbf{B} &= -\frac{cp}{e} \nabla V_m(x,y). \end{aligned} \tag{4.47}$$

Here we distinguish between the electrical potential V_e and the magnetic potential V_m. Since the Laplace equation is valid for both the electric as well as magnetic field in a material free region we need not make a real distinction between both fields. The general field potential in (4.45) may now be decomposed into its independent multipole terms. Since we restrict the discussion in this section to transverse fields only, we find the lowest order term $A_o(z)$ to be an additive value to the potential which vanishes when we derive transverse fields by differentiation of the potential. For $n = 1$ we get

$$V_1(r,\varphi) = -\frac{cp}{e} A_1 r e^{i\varphi} \tag{4.48}$$

or in cartesian coordinates

$$V_1(x,y) = -\frac{cp}{e} A_1 (x + iy). \tag{4.49}$$

Both the real and imaginary parts are two independent solutions of the same Laplace equation and therefore the potential for both components can be written in the form

$$V_1(x,y) = -\frac{cp}{e} A_{10}\, x - \frac{cp}{e} A_{01}\, y\,. \qquad (4.50)$$

Here we recognize that both the real and imaginary solutions are independent solutions with independent coefficients A_{10} and A_{01}. All coefficients A_i are still functions of z although we do not indicate this explicitly. The equipotential lines in the transverse (x, y) - plane for the first order potential are determined by

$$A_{10}\, x + A_{01}\, y = \text{const} \qquad (4.51)$$

and the corresponding electromagnetic field is given in component formulation by the vector

$$\left(\frac{cp}{e} A_{10},\ \frac{cp}{e} A_{01},\ 0 \right)\,. \qquad (4.52)$$

This field is uniform in space and is called a *dipole field*. To simplify the design of beam transport systems it is customary to use dipole fields that are aligned with the coordinate system such as to exert a force on the particles only in the x or y direction. A simple rotation of such dipole fields about the beam axis would create again the general field expressed by (4.50,51). The dipole field of (4.52) together with the sign convention (4.13) defines the coefficients A_{10} and A_{01} as the vertical and horizontal curvatures, respectively

$$
\begin{aligned}
A_{10} &= \frac{e}{cp} B_x = -\kappa_y\,, \\
A_{01} &= \frac{e}{cp} B_y = +\kappa_x\,.
\end{aligned}
\qquad (4.53)
$$

For an arbitrary single higher order multipole $n > 1$ the field components can be derived in a similar way from (4.45)

$$V_n(r,\varphi) = -\frac{cp}{e}\frac{1}{n!} A_n\, r^n\, e^{in\varphi}\,, \qquad (4.54)$$

which, in cartesian coordinates, becomes

$$V_n(x,y) = -\frac{cp}{e}\frac{1}{n!} A_n\, (x + i y)^n\,. \qquad (4.55)$$

Again we note that both the real and imaginary part of the potential are valid independent solutions to the Laplace equation. Expanding the factor $(x + iy)^n$ we get, after some manipulation and after selecting different coefficients for the real and imaginary terms, a generalization of (4.50) for the potential of the nth order multipole,

$$V_n(x,y) = -\frac{cp}{e} \sum_{j=0}^{n} A_{n-j,j}\, \frac{x^{n-j}}{(n-j)!}\, \frac{y^j}{j!}\,. \qquad (4.56)$$

From this equation it is straight forward to extract an expression for the potential of any multipole field satisfying the Laplace equation. Since both electrical and magnetic fields may be derived from the Laplace equation, we need not make any distinction here and may use (4.56) as an expression for the electrical as well as the magnetic potential. Separating the real and imaginary terms of the potential we get

$$\mathrm{Re}\,[V_n(x,y)] = -\frac{cp}{e} \sum_{m=0}^{n/2} A_{n-2m,2m} \frac{x^{n-2m}}{(n-2m)!} \frac{y^{2m}}{(2m)!}. \tag{4.57}$$

The imaginary terms are then

$$\mathrm{Im}[V_n(x,y)] =$$
$$-\frac{cp}{e} \sum_{m=1}^{(n+1)/2} A_{n-2m+1,2m-1} \frac{x^{n-2m+1}}{(n-2m+1)!} \frac{y^{2m-1}}{(2m-1)!}. \tag{4.58}$$

As we will see, it is useful to keep both sets of solutions separate because they describe two distinct orientations of multipole fields. For a particular multipole both orientations can be realized by a mere rotation about the axis.

4.4.2 Magnetic Field Equations

The field equations are obtained in the usual way by differentiation of the potentials with respect to the coordinates. So far no specific assumption has been made as to the nature of the potential and we may assume it to be either the electrical or the magnetic potential. In real beam transport systems we use mostly magnetic fields from magnetic devices which are designed to represent the imaginary solutions (4.58). The magnetic field components for the nth order multipoles derived from the imaginary solution are given by

$$
\begin{aligned}
B_{nx} &= -\frac{\partial}{\partial x} \mathrm{Im} V \\
&= +\frac{cp}{e} \sum_{m=1}^{n/2} A_{n-2m+1,2m-1} \frac{x^{n-2m}}{(n-2m)!} \frac{y^{2m-1}}{(2m-1)!}
\end{aligned} \tag{4.59}
$$

and

$$
\begin{aligned}
B_{ny} &= -\frac{\partial}{\partial y} \mathrm{Im} V \\
&= +\frac{cp}{e} \sum_{m=1}^{(n+1)/2} A_{n-2m+1,2m-1} \frac{x^{n-2m+1}}{(n-2m+1)!} \frac{y^{2m-2}}{(2m-2)!}.
\end{aligned} \tag{4.60}
$$

This asymmetry is not fundamental but only reflects the fact that most beam transport lines are installed in a horizontal plane for which the imaginary solutions provide the appropriate fields for beam dynamics. The real and imaginary solutions differentiate between two classes of magnet orientation. From (4.57,58) we find that only the imaginary solution has what is called *midplane symmetry* with the property that $\mathrm{Im} V_n(x, y) = -\mathrm{Im} V_n(x, -y)$ or $B_{ny}(x, y) = B_{ny}(x, -y)$. In this symmetry we have no horizontal field components in the midplane, $B_{nx}(x, 0) \equiv 0$, and a particle travelling in the horizontal mid plane will remain in this plane. We call the magnets in this class *upright magnets*. The magnets which derive from the real solution of the potential (4.57) we call *rotated magnets* since they differ from the upright magnets only by a rotation about the magnet axis.

The dependence of the electric or magnetic potential on the space coordinates is important for the actual design of electromagnetic bending and focusing devices. It is well known that metallic surfaces are equipotential surfaces for electrical fields. Similarly, surfaces of ferromagnetic unsaturated materials are equipotentials for magnetic fields. To create specific electric or magnetic fields, we will design devices with metallic or ferromagnetic surfaces formed such as to satisfy the desired field potential. We will apply this principle in the next section to design such devices to be used in charged particle beam transport lines. Following the derivation of the field components for the imaginary solution, it is also straightforward to derive the fields for the real solution.

4.5 Multipole Fields for Beam Transport Systems

We are now in a position to determine the field characteristics for any multipole. This will be done in this section for magnetic fields most commonly used in particle transport systems. General multipole fields can be derived from the Laplace equation but in practical cases magnets are designed to produce only a single multipole. Only for very special applications are two or more multipole field components desired in the same magnet.

In the previous section we derived the dipole field potential

$$V_1(x, y) = -\frac{cp}{e}(A_{10}\, x + A_{01}\, y) = -\frac{cp}{e}(-\kappa_y\, x + \kappa_x\, y). \tag{4.61}$$

A purely horizontally deflecting dipole field ($\kappa_y = 0$) has by virtue of the Laplace equation a nonvanishing component only in the nondeflecting plane. Such a horizontally deflecting field is therefore described by the *magnetic field components*

$$\frac{e}{cp}\, B_x = 0 \quad \text{and} \quad \frac{e}{cp}\, B_y = \kappa_x, \tag{4.62}$$

Table 4.1. Magnetic Multipole Potentials

Dipole	$-\frac{e}{cp} V_1 = -\kappa_y\, x + \kappa_x\, y$
Quadrupole	$-\frac{e}{cp} V_2 = -\frac{1}{2}\, \underline{k}\,(x^2 - y^2) + k\,x\,y$
Sextupole	$-\frac{e}{cp} V_3 = -\frac{1}{6}\, \underline{m}\,(x^3 - 3xy^2) + \frac{1}{6}\, m\,(3x^2 y - y^3)$
Octupole	$-\frac{e}{cp} V_4 = -\frac{1}{24}\, \underline{r}\,(x^4 - 6x^2 y^2 + y^4) + \frac{1}{6}\, r\,(x^3 y - xy^3)$
Decapole	$-\frac{e}{cp} V_5 = -\frac{1}{120}\, \underline{d}\,(x^5 - 10x^3 y^2 + 5xy^4)$
	$\qquad\qquad\quad +\frac{1}{120}\, d\,(5x^4 y - 10x^2 y^3 + y^5)$

while a purely vertically deflecting dipole ($\kappa_x = 0$) has the field components

$$\frac{e}{cp}\, B_x = -\kappa_y \qquad \text{and} \qquad \frac{e}{cp}\, B_y = 0. \tag{4.63}$$

Similarly we get for the second order fields from (4.56)

$$V_2(x,y) = -\frac{cp}{e}\,\Big[\,-\tfrac{1}{2}\,\underline{k}\,(x^2 - y^2) + kxy\,\Big], \tag{4.64}$$

where we used $-\underline{k}$ and k for the coefficients of the real and imaginary terms respectively, $A_{20} = -A_{02} = -\underline{k}$ and $A_{11} = +k$.

From this second order potential it is straight forward to derive the field components for an upright oriented quadrupole, used commonly for focusing in particle beam dynamics, by setting $\underline{k} = 0$ and differentiation of (4.64)

$$\frac{e}{cp}\, B_x = ky \qquad \text{and} \qquad \frac{e}{cp}\, B_y = kx. \tag{4.65}$$

Similarly we may get the field components for a *rotated quadrupole* or for all other *multipole potentials*. The results up to fifth order are compiled in Table 4.1.

The coefficients A_{jk} are replaced by more commonly used notations defining particular magnetic multipoles. Each expression for the magnetic potential is composed of both the real and the imaginary contribution. Since both components differ only by a rotational angle, real magnets are generally aligned such that only one or the other component appears. Only due to alignment errors may the other component appear as a field error which can be treated as a perturbation. The correspondence between the coefficients A_{jk} and the commonly used notation for the *multipole strength parameters* are shown in Table 4.2. Here κ_y and the underlined coefficients are the magnet strengths for the rotated multipoles while κ_x and the straight letters are the strength coefficients for the upright multipoles.

Table 4.2. Correspondence between the Potential Coefficients and Multipole Strength Parameters

$$
\begin{array}{ccccccccccc}
 & & & & & A_{00} & & & & & \\
 & & & & A_{10} & & A_{01} & & & & \\
 & & & A_{20} & & A_{11} & & A_{02} & & & \\
 & & A_{30} & & A_{21} & & A_{12} & & A_{03} & & \\
 & A_{40} & & A_{31} & & A_{22} & & A_{13} & & A_{04} & \\
A_{50} & & A_{41} & & A_{32} & & A_{23} & & A_{14} & & A_{05}
\end{array}
$$

$$\Updownarrow$$

$$
\begin{array}{ccccccccccc}
 & & & & & 0 & & & & & \\
 & & & & -\kappa_y & & \kappa_x & & & & \\
 & & & -k & & k & & k & & & \\
 & & -m & & m & & m & & -m & & \\
 & -r & & r & & r & & -r & & -r & \\
-d & & d & & d & & -d & & -d & & d
\end{array}
$$

Table 4.3. Upright Multipole Fields

Dipole
$$\frac{e}{cp}\,B_x = 0 \qquad\qquad \frac{e}{cp}\,B_y = \kappa_x$$

Quadrupole
$$\frac{e}{cp}\,B_x = k\,y \qquad\qquad \frac{e}{cp}\,B_y = k\,x$$

Sextupole
$$\frac{e}{cp}\,B_x = m\,xy \qquad\qquad \frac{e}{cp}\,B_y = \tfrac{1}{2}\,m\left(x^2 - y^2\right)$$

Octupole
$$\frac{e}{cp}\,B_x = \tfrac{1}{6}\,r\left(3x^2y - y^3\right) \qquad \frac{e}{cp}\,B_y = \tfrac{1}{6}\,r\left(x^3 - 3xy^2\right)$$

Decapole
$$\frac{e}{cp}\,B_x = \tfrac{1}{6}\,d\left(x^3y - xy^3\right) \qquad \frac{e}{cp}\,B_y = \tfrac{1}{24}\,d\left(x^4 - 6x^2y^2 + y^4\right)$$

From the expressions for the multipole potentials in Table 4.1 we obtain by differentiation the multipole field components. For the imaginary solutions of the Laplace equation these multipole field components up to decapoles are compiled in Table 4.3.

The *multipole strength parameters* can be related to the derivatives of the magnetic field. Generalizing from (4.23,34) we get for upright multipoles

$$s_n(\text{m}^{-n}) = \frac{e}{cp} \frac{\partial^{n-1} B_y}{\partial x^{n-1}} \bigg|_{\substack{x=0 \\ y=0}} \tag{4.66}$$

or in more practical units

$$s_n(\text{m}^{-n}) = C_m \frac{1}{\beta\, E[\text{GeV}]} \frac{\partial^{n-1} B_y[\text{Tesla}]}{\partial x^{n-1}[\text{m}^{n-1}]}, \tag{4.67}$$

where the coefficient

$$C_m = 0.2997925 \text{ GeV/Tesla/m} \tag{4.68}$$

and s_n is the strength parameter of the nth order multipole, i.e., $s_1 = 1/\rho, s_2 = k, s_3 = m, s_4 = r, s_5 = d$, etc. The lowest order field derivatives are often expressed by special symbols as well. The quadrupole *field gradient*, for example, can be defined by $g = \partial B_y/\partial x$ and the derivative of the field gradient in sextupole magnets by $s = \partial^2 B_y/\partial x^2 = -\partial^2 B_y/\partial y^2$. Derivations with respect to other coordinates in accordance with the field definitions in Table 4.3 can be used as well to define magnet parameters.

The real solutions of the Laplace equation (4.57) describe basically the same field patterns as the imaginary solutions, but the fields are rotated about the z-axis by an angle $\Phi_n = \pi/(2n)$, where n is the order of the multipole. The magnetic field equations in cartesian coordinates as derived from Table 4.1 are compiled in Table 4.4.

Table 4.4. Rotated Multipole Fields

Dipole ($\Phi = 90°$)

$$\frac{e}{cp} B_x = -\kappa y \qquad\qquad B_y = 0$$

Quadrupole ($\Phi = 45°$)

$$\frac{e}{cp} B_x = -\underline{k}\, x \qquad\qquad \frac{e}{cp} B_y = +\underline{k}\, y$$

Sextupole ($\Phi = 30°$)

$$\frac{e}{cp} B_x = -\tfrac{1}{2}\underline{m}\,(x^2 - y^2) \qquad\qquad \frac{e}{cp} B_y = +\underline{m}\, xy$$

Octupole ($\Phi = 22.5°$)

$$\frac{e}{cp} B_x = -\tfrac{1}{6}\underline{r}\,(x^3 - 3xy^2) \qquad\qquad \frac{e}{cp} B_y = +\tfrac{1}{6}\underline{r}\,(3x^2 y - y^3)$$

Decapole ($\Phi = 18.0°$)

$$\frac{e}{cp} B_x = -\tfrac{1}{24}\underline{d}\,(x^4 - 6x^2 y^2 + y^4) \qquad\qquad \frac{e}{cp} B_y = +\tfrac{1}{6}\underline{d}\,(x^3 y - xy^3)$$

The characteristic difference between the two sets of field solutions is that the fields of upright linear magnets in Table 4.3 do not cause coupling for particles traveling in the horizontal or vertical midplane, in contrast to the rotated magnet fields of Table 4.4 which would deflect particles out of the horizontal midplane. In linear beam dynamics, where we use only dipole and upright quadrupole magnets, the particle motion in the horizontal and vertical plane are completely independent. This is a highly desirable "convenience" without which particle beam dynamics would be much more complicated and less predictable. Since there is no particular fundamental reason for a specific orientation of magnets in a beam transport systems, we may as well use that orientation that leads to the simplest and most predictable results. We will therefore use exclusively upright magnet orientation for the main magnets and treat the occasional need for rotated magnets as a perturbation.

The general *magnetic field equation* including only the most commonly used upright multipole elements is given by

$$
\begin{aligned}
B_x &= g\,y + s\,x\,y + \tfrac{1}{6}\,o(3\,x^2\,y - y^3) + \dots \\
B_y &= B_{yo} + g\,x + \tfrac{1}{2}\,s\,(x^2 - y^2) + \tfrac{1}{6}\,o(x^3 - 3\,x\,y^2) + \dots.
\end{aligned}
\tag{4.69}
$$

Sometimes it is interesting to investigate the particle motion only in the horizontal midplane, where $y = 0$. In this case we expect the horizontal field components B_x of all multipoles to vanish and any deflection or coupling is thereby eliminated. In such circumstances, the particle motion is completely contained in the horizontal plane and the general fields to be used are given by

$$
\begin{aligned}
B_x &= 0 \\
B_y &= B_{yo} + g\,x + \tfrac{1}{2}\,s\,x^2 + \tfrac{1}{6}\,o\,x^3 + \dots
\end{aligned}
\tag{4.70}
$$

In terms of multipole strength parameters the second equation in (4.70) becomes

$$
\frac{e}{cp}\,B_y = \kappa_x + k\,x + \tfrac{1}{2}\,m\,x^2 + \tfrac{1}{6}\,r\,x^3 + \tfrac{1}{24}\,d\,x^4 \dots.
\tag{4.71}
$$

In this form the field expansion exhibits the most significant multipole fields in the horizontal midplane as used in accelerator physics and is frequently employed to study and solve beam stability problems or the effects of particular multipole fields on beam parameters.

4.6 Multipole Field Patterns and Pole Profiles

The expressions for the magnetic potentials give us a guide to design devices that generate the desired fields. Multipole fields are generated mostly in one of two ways: by iron dominated magnets , or by proper placement of electrical current carrying conductors. The latter way is mostly used in high field *superconducting magnets*, where fields beyond the general saturation level for iron at about 20 kG are desired.

In iron dominated magnets, fields are determined by the shape of the iron surfaces. As metallic surfaces are equipotential surfaces for electrical fields, so are surfaces of ferromagnetic material, like iron in the limit of infinite *magnetic permeability*, equipotential surfaces for magnetic fields. This approximate property of the iron surfaces can be exploited for the design of unsaturated or only weakly saturated magnets. From Table 4.1 we know the mathematical formulation for the magnetic potentials of the desired multipoles which in turn becomes the equation of the magnetic multipole profile. The fact that iron never reaches infinite permeability does not affect the validity of the assumption that we can produce specific multipoles by forming iron surfaces designed according to the desired magnetic potential. For preliminary design calculations, it is sufficient to assume infinite permeability of the ferromagnetic material. Where effects of finite permeability or magnetic saturation become important, the fields are determined numerically by mathematical relaxation methods. In this text, we will not be able to discuss the details of magnet design and construction but will concentrate only on the main magnet features from a beam dynamics point of view. A wealth of practical experience in the design of iron dominated accelerator magnets, including an extensive list of references, is compiled in a review article by Fischer [4.12].

The potentials of Table 4.1 do not exhibit any s-dependence because we have assumed that the magnets are sufficiently long so that we may ignore end field effects. In this approximation the magnetic potential depends only on the transverse coordinates x and y. It is this dependence which we will use to determine the two-dimensional cross section of the metallic surface for electric field devices or of the magnet iron surface in the transverse x, y-plane. In Chap. 5 we will discuss some of the end field effects. For a horizontally deflecting *dipole magnet* the potential is

$$\frac{e}{cp}V_1 = -\frac{1}{\rho}y. \tag{4.72}$$

An equipotential iron surface is therefore determined by the requirement that the ratio y/ρ be a constant

$$\frac{y}{\rho} = \text{const} \quad \text{or} \quad y = \text{const}. \tag{4.73}$$

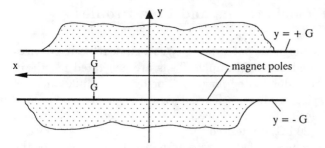

Fig. 4.7. Pole contour for a horizontally deflecting dipole magnet

This equation is true for all values of y especially for $y = G$, where $2G$ is the distance or gap height between the magnet poles as shown in Fig. 4.7. To make the magnet midplane, $y = 0$, a magnetic symmetry plane we have to have a magnetic pole also at $y = -G$. The mathematical equation for the field determining shape of the iron boundary or the dipole magnet *pole profile* is thereby given by

$$y = \pm G. \tag{4.74}$$

Similarly, we would place iron surfaces at

$$x = \pm G. \tag{4.75}$$

to obtain a vertically deflecting magnet with a horizontal gap width $2G$.

The magnet *pole profile* for a *quadrupole* can be derived in a quite similar way. We use the equation for the magnetic potential of a quadrupole from Table 4.1 and define the appropriate iron boundary or equipotential surface by setting

$$k\,x\,y = \text{const}. \tag{4.76}$$

This is the equation of a hyperbola which is the natural shape for the poles in a quadrupole. The inscribed radius of the iron free region is R and the constant in (4.76) therefore is $k\,(R/\sqrt{2})^2 = k\,\tfrac{1}{2}\,R^2$ as shown in Fig. 4.8. The *pole shape* or *pole profile* for a quadrupole with bore radius R is defined by the equation

$$xy = \pm \tfrac{1}{2}\,R^2. \tag{4.77}$$

In a similar way we get for the pole profile of a rotated quadrupole

$$x^2 - y^2 = \pm R^2. \tag{4.78}$$

This is the same hyperbola as (4.77) but rotated by 45° as shown in Fig. 4.9. Both (4.76,77) describe four symmetrically aligned hyperbolas which become the surfaces of the ferromagnetic poles producing an ideal quadrupole

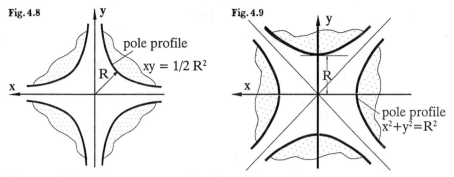

Fig. 4.8. Pole shape of an upright quadrupole magnet

Fig. 4.9. Pole shape of a rotated quadrupole magnet

field. All four poles have the same numerical value for the potential but with alternating signs.

Sometimes a combination of both, the dipole field of a bending magnet and the focusing field of a quadrupole, is desired for compact beam transport lines to form what is called a *synchrotron magnet*. Such a magnet actually is nothing but a transversely displaced quadrupole. The field in a quadrupole displaced by x_0 from the beam axis is $\frac{e}{cp} B_y = k(x - x_0) = k\,x - k\,x_0$ and a particle traversing this quadrupole at $x = 0$ will be deflected onto a curved trajectory with a curvature of $\kappa_x = k\,x_0$. At the same time we observe focusing corresponding to the quadrupole strength k. The pole cross section of such a magnet is shown in Fig. 4.10. The deviation from parallelism of the magnet poles at the reference trajectory is often quantified by the *characteristic length*, defined by

$$\ell_{\text{ch}} = \frac{1}{\rho_0\,k}. \tag{4.79}$$

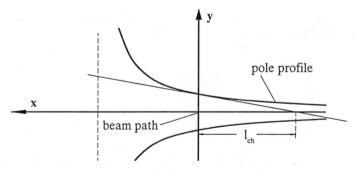

Fig. 4.10. Pole profile for a synchrotron magnet

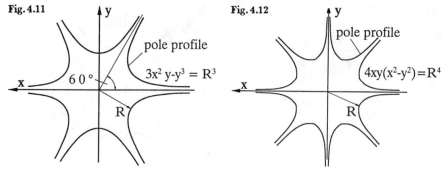

Fig. 4.11. Pole profile for an upright sextupole magnet

Fig. 4.12. Pole profile for an upright octupole magnet

Geometrically this characteristic length is equal to the distance from the reference trajectory to that point at which the tangents from the two magnet poles at the vertical reference plane would touch, Fig. 4.10.

The magnet pole shapes for *sextupole* or *octupole magnets* are derived in a similar way and are shown in Figs. 4.11, 4.12. For odd order multipoles like dipoles, sextupoles, decapoles etc. there exists a central pole along the vertical axis. The profile can be derived directly from the potential equations in Table 4.1. Even order multipoles have no poles along any axis. Only the profile of one pole must be determined since the other poles are generated by simple rotation of the first pole by multiples of the angle $90°/n$, where n is the order of the multipole. Multipoles of higher order than sextupoles are rarely used in accelerator physics but can be derived from the appropriate multipole potentials.

4.7 Equations of Motion
in Charged Particle Beam Dynamics

We use magnetic fields to guide charged particles along a predescribed path or at least keep them close by. This path, or *reference trajectory*, is defined geometrically by straight sections and bending magnets only. In fact it is mostly other considerations, like the need to transport from an arbitrary point A to point B in the presence of building constraints, that determine a particular path geometry. We place dipole magnets wherever this path needs to be deflected and have straight sections in between. Quadrupole and higher order magnets do not influence this path but provide the focusing forces necessary to keep all particles close to the reference path.

The *equations of motion* become very complicated if we would use a fixed cartesian coordinate system. After having defined the reference trajectory

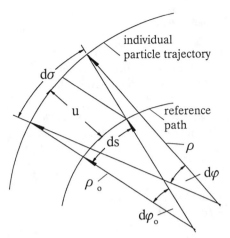

Fig. 4.13. Particle trajectories in deflecting systems. Reference path s and individual particle trajectory σ have in general different bending radii.

we are merely interested in the description of the deviation of an individual particle trajectory from that reference or ideal trajectory.

This is achieved by using a cartesian coordinate system that follows with the particle along the reference path. In other words, we use a curvilinear coordinate system as defined mathematically by (4.17). These curvatures are functions of the coordinate s and are nonzero only where there are bending magnets.

In deriving the equations of motion we limit ourselves to the horizontal plane only. The generalization to both horizontal and vertical plane is straightforward. From (4.9) we calculate the deflection angle of an arbitrary trajectory for an infinitesimal segment of a bending magnet with respect to the ideal trajectory. Using the notation of Fig. 4.13 the deflection angle of the ideal path is $d\varphi_0 = ds/\rho_0$ or utilizing the definition (4.13) for the curvature to preserve the directionality of the deflection

$$d\varphi_0 = \kappa_0 \, ds, \tag{4.80}$$

where κ_0 is the curvature of the ideal path. The deflection angle for an arbitrary trajectory is then given by

$$d\varphi = \kappa \, d\sigma. \tag{4.81}$$

The ideal curvature κ_0 is evaluated along the reference trajectory $u = 0$ for a particle with the ideal momentum $\delta = \Delta c p_0 / c p_0 = 0$ where p_0 is the ideal particle momentum. In linear approximation with respect to the coordinates the *path length* element for an arbitrary trajectory is

$$d\sigma = (1 + \kappa_0 u) \, ds + \mathcal{O}(2), \tag{4.82}$$

where u is the distance of the particle trajectory from the reference trajectory in the deflecting plane.

The magnetic fields depend on s in such a way that the fields are zero in magnet free sections and assume a constant value within the magnets. This assumption results in a step function distribution of the magnetic fields and is referred to as the *hard edge model*, generally used in beam dynamics. The path is therefore composed of a series of segments with constant curvatures. To obtain the equations of motion with respect to the ideal path we subtract from the curvature for an individual particle the curvature of the ideal path at the same location.

Since u is the deviation of a particle from the ideal path, we get for the equation of motion with respect to the ideal path from (4.13,80-82), noting that in the deflection plane $u'' = -(d\varphi/ds - d\varphi_o/ds)$,

$$u'' = -(1 + \kappa_o u)\kappa + \kappa_o, \tag{4.83}$$

where the derivations are taken with respect to s. In particle beam dynamics we generally assume *paraxial beams*, $u'^2 \ll 1$ since the divergence of the trajectories u' is typically of the order of 10^{-3} rad or less and terms in u'^2 can therefore be neglected. Where this assumption leads to intolerable inaccuracies the equation of motion must be modified accordingly.

The equation of motion for charged particles in electromagnetic fields can be derived from (4.83) and the Lorentz force (4.12). In case of horizontal deflection the curvature is $\kappa = \kappa_x$ and expressing the general field by its components from (4.69) we have

$$\kappa_x = \frac{e}{cp} B_y = \frac{e}{cp} \left[B_{yo} + gx + \tfrac{1}{2}s(x^2 - y^2) + \ldots \right]. \tag{4.84}$$

A real particle beam is never monochromatic and therefore effects due to small *momentum errors* must be considered. This can be done by expanding the particle momentum in the vicinity of the ideal momentum p_o as

$$\frac{1}{cp} = \frac{1}{cp_o(1 + \delta)} \approx \frac{1}{cp_o}(1 - \delta + \delta^2 + \ldots). \tag{4.85}$$

We are now ready to apply (4.83) in the horizontal plane, set $u = x$ and $\kappa = \kappa_x$ and get with (4.83,84) and retaining only linear and quadratic terms in δ, x and y, the *equation of motion*

$$\begin{aligned}
x'' + (k_o + \kappa_{xo}^2)x &= \kappa_{xo}(\delta - \delta^2) + (k_o + \kappa_{xo}^2)x\delta \\
&\quad - k_o\kappa_o x^2 - \tfrac{1}{2}m_o(x^2 - y^2) + \mathcal{O}(3).
\end{aligned} \tag{4.86}$$

Here we have used energy independent field strength parameters as defined in (4.66) and Table 4.3 The index $_o$ indicates that the magnet strengths are to be evaluated at the design energy. We note a quadratic focusing term $\kappa^2 x$ which is due to weak focusing in a sector magnet.

The equation of motion in the vertical plane can be derived in a similar way by setting $u = y$ in (4.83) and $\kappa = \kappa_y$. Consistent with the sign convention adopted in Sect. 4.2, (4.84) becomes for the vertical plane

$$\kappa_y = -\frac{e}{cp} B_x = -\frac{e}{cp} (B_{xo} + g\, y + s\, xy + \ldots) \tag{4.87}$$

and the equation of motion in the vertical plane is

$$\begin{aligned}
y'' - (k_o - \kappa_{yo}^2)\, y &= \kappa_{yo}\, (\delta - \delta^2) - (k_o - \kappa_{yo}^2)\, y\delta \\
&\quad + \kappa_o\, k_o\, y^2 + m_o\, xy + \mathcal{O}(3)\,.
\end{aligned} \tag{4.88}$$

In particular we find for cases, where the deflection occurs only in one plane say the horizontal plane, that the equation of motion in the vertical plane becomes simply

$$y'' - k_o\, y = -k_o\, y\delta + m_o\, xy + \mathcal{O}(3)\,, \tag{4.89}$$

which to the order of approximation considered is independent of the strength of the horizontal bending field.

The magnet parameters κ_o, k_o, and m_o are functions of the independent coordinate s. In real beam transport lines, these magnet strength parameters assume constant, non zero values within magnets and become zero in drift spaces between the magnets. The task of beam dynamics is to distribute magnets along the beam transport line in such a way that the solutions to the equations of motion result in the desired beam characteristics.

4.8 General Solution of the Equations of Motion

Equations (4.86,88) are the equations of motion for *strong focusing* beam transport systems [4.13, 14], where the magnitude of the focusing strength is a free parameter. No general analytical solutions are available for arbitrary distributions of magnets. We will, however, develop mathematical tools which make use of partial solutions to the differential equations, of perturbation methods and of particular design concepts for magnets to arrive at an accurate prediction of particle trajectories. One of the most important "tools" in the mathematical formulation of a solution to the equations of motion is the ability of magnet builders and alignment specialists to build magnets with almost ideal field properties and to place them precisely along a predefined ideal path. In addition, the capability to produce particle beams within a very small energy spread is of great importance for the determination of the properties of particle beams. As a consequence, all terms on the right-hand side of (4.86,88) can and will be treated as small perturbations and mathematical perturbation methods can be employed to describe the effects of these perturbations on particle motion.

We further notice that the left-hand side of the equations of motion resembles that of a *harmonic oscillator* although with a time dependent frequency. By a proper transformation of the variables we can , however, express (4.86,88) exactly in the form of the equation for a harmonic oscillator with constant frequency. This transformation is very important because it allows us to describe the particle motion mostly as that of a harmonic oscillator under the influence of weak perturbation terms on the right-hand side. A large number of mathematical tools developed to describe the motion of harmonic oscillators become therefore available for charged particle beam dynamics.

4.8.1 Linear Unperturbed Equation of Motion

In our attempt to solve the equations of motion (4.86,88) we first try to solve the homogeneous differential equation

$$u'' + K u = 0, \tag{4.90}$$

where u stands for x or y and where, for the moment, we assume K to be constant with $K = k + \kappa_x^2$ or $K = -(k - \kappa_y^2)$, respectively. The *principal solutions* of this differential equation are for $K > 0$

$$C(s) = \cos(\sqrt{K}\, s) \qquad \text{and} \qquad S(s) = \frac{1}{\sqrt{K}} \sin(\sqrt{K}\, s), \tag{4.91}$$

and for $K < 0$

$$C(s) = \cosh \sqrt{(|K|}\, s) \qquad \text{and} \qquad S(s) = \frac{1}{\sqrt{|K|}} \sinh(\sqrt{|K|}\, s). \tag{4.92}$$

These linearly independent solutions satisfy the following initial conditions

$$
\begin{aligned}
C(0) &= 1, & C'(0) &= \frac{\mathrm{d}C}{\mathrm{d}s} = 0, \\
S(0) &= 0, & S'(0) &= \frac{\mathrm{d}S}{\mathrm{d}s} = 1.
\end{aligned}
\tag{4.93}
$$

Any arbitrary solution $u(s)$ can be expressed as a linear combination of these two principal solutions

$$
\begin{aligned}
u(s) &= C(s)\, u_\mathrm{o} + S(s)\, u_\mathrm{o}', \\
u'(s) &= C'(s)\, u_\mathrm{o} + S'(s)\, u_\mathrm{o}',
\end{aligned}
\tag{4.94}
$$

where $u_\mathrm{o}, u_\mathrm{o}'$ are arbitrary initial parameters of the particle trajectory and where the derivatives are taken with respect to the independent variable s.

In a general beam transport system, however, we cannot assume that the magnet strength parameter K remains constant and alternative methods of finding a solution for the particle trajectories must be developed. Nonetheless it has become customary to formulate the general solutions for $K = K(s)$ similar to the principal solutions found for the harmonic oscillator with a constant restoring force. Specifically, solutions can be found for any arbitrary beam transport line which satisfy the initial conditions (4.93). These principal solutions are the so-called *sine like solutions* and *cosine like solutions* and we will derive the conditions for such solutions. For the differential equation

$$u'' + K(s)u = 0 \tag{4.95}$$

with a time dependent restoring force $K(s)$, we make an ansatz for the general solutions in the form (4.94). Introducing the ansatz (4.94) into (4.95) we get after some sorting

$$[S''(s) + K(s)S(s)]u_o + [C''(s) + K(s)C(s)]u'_o = 0.$$

This equation must be true for any pair of initial conditions (u_o, u'_o) and therefore the coefficients must vanish separately

$$\begin{aligned} S''(s) + K(s)S(s) &= 0, \\ C''(s) + K(s)C(s) &= 0. \end{aligned} \tag{4.96}$$

The general solution of the equation of motion (4.95) can be expressed by a linear combination of a pair of solutions satisfying the differential equations (4.96) and the boundary conditions (4.93).

It is impossible to solve (4.96) analytically in a general way that would be correct for arbitrary distributions of quadrupoles $K(s)$. Purely numerical methods to solve the differential equations (4.96) maybe practical but are conceptually unsatisfactory since this method reveals little about characteristic properties of beam transport systems. It is therefore not surprising that other more revealing and practical methods have been developed to solve the beam dynamics of charged particle beam transport systems.

The solution (4.94) of the equation of motion (4.95) may be expressed in *matrix formulation*

$$\begin{bmatrix} u(s) \\ u'(s) \end{bmatrix} = \begin{bmatrix} C(s) & S(s) \\ C'(s) & S'(s) \end{bmatrix} \begin{bmatrix} u_o \\ u'_o \end{bmatrix}. \tag{4.97}$$

If we calculate the principal solutions of (4.95) for individual magnets only, we obtain such a *transformation matrix* for each individual element of the beam transport system. Noting that within each of the beam line elements, whether it be a drift space or a magnet, the restoring forces are indeed constant and we may use within each single beam line element the simple solutions (4.91) or (4.92) for the equation of motion (4.95). With these solutions we are immediately ready to form transformation matrices for each beam line element. In matrix formalism we are able to follow a particle trajectory along a complicated beam line by repeated matrix multiplications from element to element. This procedure, discussed in more detail in Chap. 5, is widely used in accelerator physics and lends itself particularly effective for applications in computer programs. With this method we have completely eliminated the need to solve the differential equation (4.95) which we could not have succeeded in doing anyway without applying numerical methods. The simple solutions (4.91,92) will suffice to treat most every beam transport problem.

4.8.2 Wronskian

The transformation matrix just derived has special properties well-known from the theory of linear homogeneous differential equation of second order [4.15]. Only a few properties relevant to beam dynamics shall be repeated here. We consider the linear homogeneous differential equation of second order

$$u'' + v(s)\,u' + w(s)\,u = 0. \tag{4.98}$$

For such an equation the theory of linear differential equations provides us with a set of rules describing the properties of the solutions

there is only one solution that meets the initial conditions
$u(s_o) = u_o$ and $u'(s_o) = u'_o$ at $s = s_o$;
because of the linearity of the differential equation, $c\,u(s)$ is also a solution if $u(s)$ is a solution and if $c = $ const. ;
if $u_1(s)$ and $u_2(s)$ are two solutions, any linear combination thereof is also a solution.

For the two linearly independent solutions $u_1(s)$ and $u_2(s)$ we may form the *Wronskian determinant* or short the *Wronskian*

$$W = \begin{vmatrix} u_1(s) & u_2(s) \\ u'_1(s) & u'_2(s) \end{vmatrix} = u_1\,u'_2 - u_2\,u'_1\,. \tag{4.99}$$

This Wronskian has remarkable properties which become of great fundamental importance in beam dynamics. Both u_1 and u_2 are solutions of (4.98). Multiplying and combining both equations like

$$u''_1 + v(s)\,u'_1 + w(s)\,u_1 = 0 \qquad\qquad |\;-u_2$$
$$u''_2 + v(s)\,u'_2 + w(s)\,u_2 = 0 \qquad\qquad |\;u_1$$

gives

$$(u_1\,u''_2 - u_2\,u''_1) + v(s)\,(u_1\,u'_2 - u_2\,u'_1) = 0\,,$$

which will allow us to derive a single differential equation for the Wronskian. Making use of (4.99) and forming the derivative $\mathrm{d}W/\mathrm{d}s = u_1\,u''_2 - u_2\,u''_1$, we obtain the differential equation

$$\frac{\mathrm{d}W}{\mathrm{d}s} + v(s)\,W(s) = 0\,, \tag{4.100}$$

which can be integrated immediately to give

$$W(s) = W_o\, e^{-\int_{s_o}^{s} v(\tilde{s})\, \mathrm{d}\tilde{s}} .$$
(4.101)

In the case of linear beam dynamics we have $v(s) \equiv 0$ as long as we do not include *dissipating forces* like acceleration or energy losses into synchrotron radiation and therefore $W(s) = W_o = \text{const}$. We use the sine and cosine like solutions as the two independent solutions and get from (4.99) with (4.93)

$$W_o = C_o\, S_o' - C_o'\, S_o = 1 .$$
(4.102)

For the transformation matrix of an arbitrary beam transport line with negligible dissipating forces we finally get the general result

$$W(s) = \begin{vmatrix} C(s) & S(s) \\ C'(s) & S'(s) \end{vmatrix} = 1 .$$
(4.103)

This result will be used repeatedly to prove useful general characteristics of particle beam optics. From the generality of the derivation we conclude that the Wronskian is equal to unity for any arbitrary beam line that is described by (4.98) if $v(s) = 0$ and $w(s) = K(s)$.

4.8.3 Perturbation Terms

The principal solutions of the homogeneous differential equation give us the basic solutions in beam dynamics. We will, however, repeatedly have the need to evaluate the impact of perturbations on basic particle motion. These perturbations are effected by any number of terms on the r.h.s. of the equations of motion (4.86) or (4.88). The principal solutions of the homogeneous equation of motion can be used to find particular solutions $P(s)$ for inhomogeneous differential equations including *perturbations* of the form

$$P''(s) + K(s)\, P(s) = p(s) ,$$
(4.104)

where $p(s)$ stands for any one or more perturbation terms in (4.86,88). For simplicity, only the s-dependence is indicated in the perturbation term although in general they also depend on the transverse particle coordinates. A solution $P(s)$ of this equation can be found from

$$P(s) = \int_o^s p(\tilde{s})\, G(s, \tilde{s})\, \mathrm{d}\tilde{s} ,$$
(4.105)

where $G(s, \tilde{s})$ is a suitable *Green's function* which can be constructed from the principal solutions of the homogeneous equation, i.e.,

$$G(s, \tilde{s}) = S(s)\, C(\tilde{s}) - C(s)\, S(\tilde{s}) .$$
(4.106)

After insertion into (4.105) a particular solution for the perturbation can be found

$$P(s) = S(s) \int_0^s p(\tilde{s}) C(\tilde{s}) \, d\tilde{s} - C(s) \int_0^s p(\tilde{s}) S(\tilde{s}) \, d\tilde{s}. \tag{4.107}$$

The validity of this solution can be verified by evaluating the second derivative of the solution $P''(s)$ and inserting into (4.104). From the first derivative

$$
\begin{aligned}
P'(s) &= S'(s) \int_0^s p(\tilde{s}) C(\tilde{s}) \, d\tilde{s} + S(s) C(s) p(s) \\
&\quad - C'(s) \int_0^s p(\tilde{s}) S(\tilde{s}) \, d\tilde{s} - S(s) C(s) p(s) \\
&= S'(s) \int_0^s p(\tilde{s}) C(\tilde{s}) \, d\tilde{s} - C'(s) \int_0^s p(\tilde{s}) S(\tilde{s}) \, d\tilde{s},
\end{aligned} \tag{4.108}
$$

we obtain the second derivative

$$
\begin{aligned}
P''(s) &= S''(s) \int_0^s p(\tilde{s}) C(\tilde{s}) \, d\tilde{s} + S'(s) C(s) p(s) \\
&\quad - C'(s) \int_0^s p(\tilde{s}) S(\tilde{s}) \, d\tilde{s} - C'(s) S(s) p(s) \\
&= p(s) + S''(s) \int_0^s p(\tilde{s}) C(\tilde{s}) \, d\tilde{s} - C''(s) \int_0^s p(\tilde{s}) S(\tilde{s}) \, d\tilde{s},
\end{aligned} \tag{4.109}
$$

where we have used the property of the Wronskian (4.103) for the principal solutions. Recalling that $S'' = -K\,S$ and $C'' = -K\,C$ we insert (4.107) and (4.109) into (4.104) and verify the validity of the ansatz (4.105) for

$$P''(s) + K(s) P(s) = p(s), \tag{4.110}$$

which is identical to (4.104). The function $P(s)$, therefore, is indeed a particular solution of the inhomogeneous differential equation (4.104).

The general solution of the equations of motion (4.86,88) then is given by the combination of the two principal solutions of the homogenous part of the differential equation and a particular solution for the inhomogeneous differential equation

$$u(s) = a\,C(s) + b\,S(s) + P(s), \tag{4.111}$$

where the coefficients a and b are arbitrary constants to be determined by the initial parameters of the trajectory.

Because of the linearity of the differential equation we find a simple superposition of the general solutions of the homogeneous equation and a particular solution for the inhomogeneous equations for any number of small perturbations. This is an important feature of particle beam dynamics since it allows us to solve the equation of motion up to the precision required by a particular application. While the basic solutions are very simple, corrections can be calculated for each perturbation term separately and applied as necessary. However, these statements, true in general, must be used carefully. In special circumstances even small perturbations may have a great effect on the particle trajectory if there is a resonance or if a particular instability occurs. We will come back to these effects in later chapters and Vol. II. With these caveats in mind, however, one can assume that in a well defined particle beam line with reasonable beam sizes and well designed and constructed magnets the perturbations are generally small and that mathematical perturbations methods are applicable. Specifically, we will in most cases assume that the (x, y) amplitudes appearing in some of the perturbation terms can be replaced by the principal solutions of the homogeneous differential equations.

4.8.4 Dispersion Function

One of the most important perturbations derives from the fact that the particle beams are not monochromatic but have a finite spread of energies about the nominal energy cp. The deflection of a particle with the wrong energy in any magnetic or electric field will deviate from that for a particle with the nominal energy. The variation in the deflection caused by such a *chromatic error* ΔE in bending magnets is the lowest order of perturbation given by the term δ/ρ_o, where $\delta = \Delta p/p_o \ll 1$. We will ignore for now all terms quadratic or of higher order in δ and use the *Green's function method* to solve the perturbed equation

$$u'' + K(s)\, u = \frac{1}{\rho_o}(s)\, \delta. \tag{4.112}$$

In (4.111) we have derived a general solution for the equation of motion for any perturbation and applying this to (4.112) we get

$$\begin{aligned} u(s) &= a\, C(s) + b\, S(s) + \delta\, D(s), \\ u'(s) &= a\, C'(s) + b\, S(s) + \delta\, D'(s), \end{aligned} \tag{4.113}$$

where we have set $P(s) = \delta\, D(s)$ and used (4.107) to obtain

$$D(s) = \int_0^s \frac{1}{\rho_o}(\tilde{s})\, [S(s)\, C(\tilde{s}) - C(s)\, S(\tilde{s})]\, d\tilde{s}. \tag{4.114}$$

We have made use of the fact that like the perturbation the particular solution must be proportional to δ. The function $D(s)$ is called the *dispersion function* and the physical interpretation is simply that the function $\delta\, D(s)$ determines the offset of the reference trajectory from the ideal path for particles with a relative energy deviation δ from the ideal momentum cp_o.

This result shows that the dispersion function generated in a particular bending magnet does not depend on the dispersion at the entrance to the bending magnet which may have been generated by upstream bending magnets. The dispersion generated by a particular bending magnet reaches the value $D(L_m)$ at the exit of the bending magnet of length L_m and propagates from there on through the rest of the beam line just like any other particle trajectory. This can be seen from (4.114), where we have for $s > L_m$

$$
D(s) = S(s) \int_o^{L_m} \frac{1}{\rho}(\tilde{s}) C(\tilde{s}) \, d\tilde{s} - C(s) \int_o^{L_m} \frac{1}{\rho}(\tilde{s}) S(\tilde{s}) \, d\tilde{s}, \tag{4.115}
$$

which has exactly the form of (4.94) describing the trajectory of a particle starting with initial parameters at the end of the bending magnet given by the integrals.

With the solution (4.113) for the dispersion function we can expand the 2×2 matrix in (4.97) into a 3×3 matrix which includes the first order chromatic correction

$$
\begin{bmatrix} u(s) \\ u'(s) \\ \delta \end{bmatrix} = \begin{bmatrix} C(s) & S(s) & D(s) \\ C'(s) & S'(s) & D'(s) \\ 0 & 0 & 1 \end{bmatrix} \begin{bmatrix} u(s_o) \\ u'(s_o) \\ \delta \end{bmatrix}. \tag{4.116}
$$

Here we have assumed that the particle energy and energy deviation remains constant along the beam line. This representation of the first order chromatic aberration will be used extensively in particle beam optics.

4.9 Building Blocks for Beam Transport Lines

With special arrangements of bending and focusing magnets it is possible to construct lattice sections with particular properties. We may desire a lattice section with specific chromatic properties, achromatic or isochronous sections. In the next subsections we will discuss such lattice elements with special properties.

4.9.1 General Focusing Properties

The principal solutions and some elements of transformation matrices through an arbitrary beam transport line can reveal basic beam optical properties of this beam line. A close similarity to paraxial light optics is found in the matrix element $C'(s)$. As shown schematically in Fig. 4.14, parallel trajectories, $u'_o = 0$, are deflected by the focusing system through the matrix element $C'(s)$ and emerge with a slope $u'(s) = C'(s) u_o$.

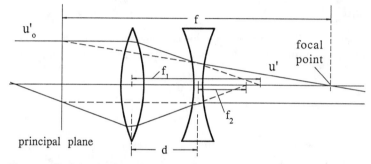

Fig. 4.14. Focusing in a quadrupole doublet

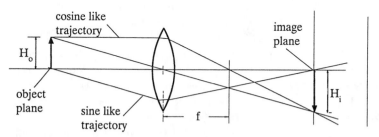

Fig. 4.15. Point to point imaging

From basic principles of light optics we know that the ratio $-u_o/u'(s)$ is defined as the focal length of the system (4.32). In analogy we define therefore also a *focal length* f for a *composite focusing system* by setting

$$f^{-1} = C'(s). \tag{4.117}$$

The focal point is defined by the condition $u(s_f) = 0$ and is, therefore, located where the cosine like solution becomes zero, $C(s_f) = 0$.

More similarities with paraxial light optics can be identified. *Point to point imaging*, for example, is defined in particle beam optics by the sine like function, $S(s)$, starting at the object plane, at $s = s_o$. The image point is located where the sinelike function crosses again the reference axis or where $S(s_i + s_o) = 0$ as shown in Fig. 4.15.

By definition such a section of a beam transport system has a phase advance of $180°$. The beam size or object size H_o at s_o is transformed by the cosine like function to be come at the image point $H(s_i) = |C(s_i + s_o)| H_o$ and the magnification of the beam optical system is given by the absolute value of the cosine like function at the image point

$$M = |C(s_i + s_o)|. \tag{4.118}$$

4.9.2 Chromatic Properties

Very basic features can be derived for the chromatic characteristics of a general beam transport line. We have already derived the dispersion function in (4.114)

$$D(s) = S(s) \int\limits_0^s \frac{1}{\rho_0}(\tilde{s}) C(\tilde{s}) \, d\tilde{s} - C(s) \int\limits_0^s \frac{1}{\rho_0}(\tilde{s}) S(\tilde{s}) \, d\tilde{s}. \qquad (4.119)$$

From this expression we conclude that there is *dispersion* only if at least one of the two integrals in (4.119) is nonzero. That means only dipole fields can cause a dispersion as a consequence of the linear chromatic perturbation term δ/ρ_0. All other perturbation terms in (4.86,88) are of higher order in δ or depend on the transverse particle coordinates and therefore contribute only to higher order corrections of the dispersion function.

Specifically we find from (4.86) the lowest order chromatic quadrupole perturbation to be $k\,x\,\delta$. Since any arbitrary particle trajectory is composed of an energy independent part x_β and an energy dependent part $D\,\delta$, expressed by $x = x_\beta + D\,\delta$ we find the lowest chromatic quadrupole perturbation to the dispersion function to be the second order term $k\,D\,\delta^2$ which does not contribute to the linear dispersion.

While some dispersion cannot be avoided in beam transport systems where dipole magnets are used, it is often desirable to remove this dispersion at least in some parts of the beam line. As a condition for that to happen at say $s = s_d$, we require that $D(s_d) = 0$. From (4.119) this can be achieved if

$$\frac{S(s_d)}{C(s_d)} = \frac{\int_0^{s_d} \frac{1}{\rho_0} S \, d\tilde{s}}{\int_0^{s_d} \frac{1}{\rho_0} C \, d\tilde{s}}, \qquad (4.120)$$

a condition that can be met by proper adjustments of the focusing structure.

4.9.3 Achromatic Lattices

A much more interesting case is the one, where we require both the dispersion and its derivative to vanish, $D(s_d) = 0$ and $D'(s_d) = 0$. In this case we have no dispersion function downstream from the point $s = s_d$ up to the point, where the next dipole magnet creates a new dispersion function. The conditions for this to happen are

$$\begin{aligned}
D(s_d) &= 0 = -S(s_d)\,I_c + C(s_d)\,I_s\,, \\
D'(s_d) &= 0 = -S'(s_d)\,I_c + C'(s_d)\,I_s\,,
\end{aligned} \qquad (4.121)$$

where we have set $I_c = \int_0^{s_d} \frac{1}{\rho_0} C \, d\tilde{s}$ and $I_s = \int_0^{s_d} \frac{1}{\rho_0} S \, d\tilde{s}$. We can solve (4.121) for I_c or I_s and get

$$\begin{aligned}
[C(s_d)\,S'(s_d) - S(s_d)\,C'(s_d)]\,I_c &= 0\,, \\
[C(s_d)\,S'(s_d) - S(s_d)\,C'(s_d)]\,I_s &= 0\,.
\end{aligned} \qquad (4.122)$$

Since $C(s_\mathrm{d})\,S'(s_\mathrm{d}) - S(s_\mathrm{d})\,C'(s_\mathrm{d}) = 1$, the conditions for a vanishing dispersion function are

$$I_\mathrm{c} = \int\limits_0^{s_\mathrm{d}} \frac{1}{\rho_\mathrm{o}}(\tilde{s})\,C(\tilde{s})\,\mathrm{d}\tilde{s} = 0\,,$$

$$I_\mathrm{s} = \int\limits_0^{s_\mathrm{d}} \frac{1}{\rho_\mathrm{o}}(\tilde{s})\,S(\tilde{s})\,\mathrm{d}\tilde{s} = 0\,. \tag{4.123}$$

A beam line is called a *first order achromat* or short an *achromat* if and only if both conditions (4.123)are true. In Vol. II we will derive conditions for a system to be achromatic also in higher order. The physical characteristics of an achromatic beam line is that at the end of the beam line the position and the slope of a particle is independent of the energy.

4.9.4 Isochronous Systems

For the accelerating process we will find that the knowledge of the *path length* is of great importance. The path length L of any arbitrary particle trajectory can be derived by integration of (4.17) to give

$$L = \int\limits_0^{L_\mathrm{o}} \frac{\mathrm{d}\sigma}{\mathrm{d}s}\,\mathrm{d}\tilde{s} = \int\limits_0^{L_\mathrm{o}} \sqrt{x'^2 + y'^2 + (1 + \kappa_x x)^2}\,\mathrm{d}\tilde{s}\,, \tag{4.124}$$

where L_o is the length of the beam line along the ideal reference path. For simplicity we have ignored a vertical deflection of the beam. The path length variation due to a vertical bend would be similar to that for a horizontal bend and can therefore be easily derived form this result. Since $x' \ll 1, y' \ll 1$, and $\kappa_x x \ll 1$ we may expand the square root and get in keeping only terms up to second order

$$L = \int\limits_0^{L_\mathrm{o}} [1 + \kappa_x\,x + \tfrac{1}{2}\,(x'^2 + y'^2 + \kappa_x^2\,x^2)]\,\mathrm{d}\tilde{s} + \mathcal{O}(3)\,. \tag{4.125}$$

Neglecting all quadratic terms we get from (4.125) while utilizing the results (4.116) for the path length difference

$$(L - L_\mathrm{o})_\mathrm{sector} = x_\mathrm{o} \int\limits_0^{L_\mathrm{o}} \kappa_x(\tilde{s})\,C(\tilde{s})\,\mathrm{d}\tilde{s} + x'_\mathrm{o} \int\limits_0^{L_\mathrm{o}} \kappa_x(\tilde{s})\,S(\tilde{s})\,\mathrm{d}\tilde{s}$$

$$+ \delta \int\limits_0^{L_\mathrm{o}} \kappa_x(\tilde{s})\,D(\tilde{s})\,\mathrm{d}\tilde{s}\,. \tag{4.126}$$

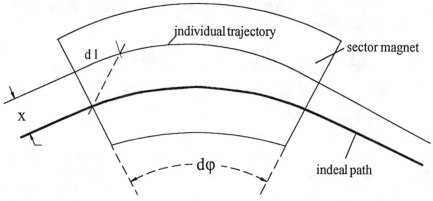

Fig. 4.16. Path length in a sector magnet

The variation of the path length has two contributions. For $\delta = 0$ the path length varies due to the curvilinear coordinate system, where dipole fields exist. This is a direct consequence of the coordinate system which selects a *sector magnet* as its natural bending magnet. The ideal path enters and exits this type of dipole magnet normal to its *pole face* as shown in Fig. 4.16. It becomes obvious from Fig. 4.16 that the path length difference depends on the particle position with respect to the reference path and is in linear approximation

$$d\ell = \ell - \ell_o = (\rho_o + x)\, d\varphi - \rho_o\, d\varphi. \tag{4.127}$$

Figure 4.17 displays the general situation for a *wedge magnet* with arbitrary entrance and exit *pole face angles*. The path length differs from that in a sector magnet on either end of the magnet. The first integral in (4.126) therefore must be modified to take into account the path length elements in the fringe field. For a wedge magnet we have therefore instead of (4.126)

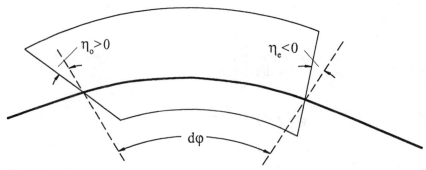

Fig. 4.17. Path length in a wedge magnet

$$(L - L_o)_{\text{wedge}} = x_o \int_o^{L_o} \kappa_x(\tilde{s}) \, C(\tilde{s}) \, d\tilde{s}$$

$$+ [C(s_o) x_o + \rho_o] \eta_o + [C(s_e) x_o + \rho_o] \eta_e$$
$$- x_o \, C(s_o) \tan \eta_o - x_o \, C(s_e) \tan \eta_e \qquad (4.128)$$

$$+ x_o' \int_o^{L_o} \kappa_x(\tilde{s}) \, S(\tilde{s}) \, d\tilde{s} + \delta \int_o^{L_o} \kappa_x(\tilde{s}) \, D(\tilde{s}) \, d\tilde{s}$$

$$\approx (L - L_o)_{\text{sector}} + \mathcal{O}(2).$$

Here $[C(s) x_o + \rho_o] \eta$ is the arc length through the wedge-like deviations from a sector magnet which must be compensated by the decrease or increase $C(s) x_o \tan \eta$ in the adjacent drift space. For small edge angles both terms compensate well and the total path length of a wedge magnet is similar to the equivalent sector magnet. In general we therefore ignore path length variations in wedge magnets with respect to sector magnets as well as those in the adjacent drift spaces. For large edge angles, however, this assumption should be reconsidered.

Equation (4.126) imposes quite severe restrictions on the focusing system if the path length is required to be independent of initial condition and the energy. Since the parameters x_o, x_o' and δ are independent parameters for different particles, all three integrals in (4.126) must vanish separately. A beam transport line which requires the path length to be the same for all particles must be a first order achromat (4.123) with the additional condition that $\int \kappa_x \, D \, d\tilde{s} = 0$.

For highly relativistic particles, where $\gamma = E/mc^2 \gg 1$, this condition is equivalent to being an *isochronous* beam line. In general any beam line becomes isochronous if we require the time of flight rather than the path length to be equal for all particles. In this case we have to take into account the velocity of the particles as well as its variation with energy. The variation of the particle velocity with energy introduces in (4.126) an additional chromatic correction and the variation of the time of flight becomes

$$\beta \, c \, (T - T_o) = x_o \, I_c + x_o' \, I_s + \delta (I_d - \gamma^{-2}). \qquad (4.129)$$

In straight beam lines, where no bending magnets are involved, $\kappa_x = 0$, (4.129) vanishes and higher than linear terms must be considered. From (4.125) we find that the bending independent terms are quadratic in nature and therefore isochronicity cannot be achieved exactly since

$$\beta \, c \, \Delta T = \int_o^{L_o} (x'^2 + y'^2) \, d\tilde{s} > 0. \qquad (4.130)$$

This integral is positive for any particle oscillating with a finite betatron amplitude. A straight beam transport line is therefore an isochronous transport system only in first order.

Problems

Problem 4.1. Derive the geometry of electrodes for a horizontally deflecting electric dipole with an aperture radius of 2 cm which is able to deflect a particle beam with a kinetic energy of 1 GeV by 10 mrad. The dipole be 1 m long and has a minimum distance between electrodes of 10 cm. What is the potential required on the electrodes?

Problem 4.2. Design an electrostatic quadrupole which provides a focal length of 10 m in the horizontal plane for particles with a kinetic energy of 1 GeV. The device shall have an aperture with a diameter of 10 cm and an effective length of 1m. What is the form of the electrodes and the potential?

Problem 4.3. In the text we have derived the fields from a scalar potential. We could also derive the magnetic fields from a vector potential $\mathbf{A}$ through the differentiation $\mathbf{B} = \nabla \times \mathbf{A}$. For purely transverse magnetic fields show that only the longitudinal component $A_z \neq 0$. Derive the vector potential for a dipole and quadrupole field and compare with the scalar potential. What is the difference between the scalar potential and the vector potential?

Problem 4.4. Derive the pole profile (aperture radius $r = 1$cm) for a combined function magnet including a dipole field to produce a bending radius of $\rho = 300$ m, a focusing strength $k = 0.45\,\mathrm{m}^{-2}$, and a sextupole strength of $m = 23.0\,\mathrm{m}^{-3}$.

Problem 4.5. Strong mechanical forces exist between the magnetic poles when a magnet is energized. Are these forces attracting or repelling the poles? Why? Consider a dipole magnet 1 m long with a field of 15 kG. What is the total force between the two magnet poles ?

Problem 4.6. Verify the numerical validity of (4.27).

Problem 4.7. Following the derivation of (4.27) for a bending magnet, derive a similar expression for the electrical excitation current in $A \cdot$ turns of a quadrupole with an aperture radius R and a desired field gradient g. What is the total excitation current necessary in a quadrupole with an effective length of 1 m and $R = 3$ cm to produce a focal length of $f = 10$ m for particles with an energy of $cp = 400$ GeV?

Problem 4.8. Consider a coil in the aperture of a magnet as shown in Fig. 4.18. All n windings are made of very thin wires and are located exactly on

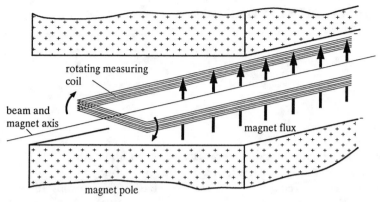

rotating measuring
coil

beam and
magnet axis

magnet flux

magnet pole

Fig. 4.18. Rotating coil in a magnet to determine higher order multipole components

the radius R. We rotate now the coil about its axis at a rotation frequency ν. Such rotating coils are used to measure the multipole field components in a magnet. Show analytically that the recorded signal is composed of harmonics of the rotation frequency ν. What is the origin of the harmonics?

Problem 4.9. Through magnetic measurements the following magnetic multipole field components in a quadrupole are determined. At $x = 1.79$ cm and $y = 0$ cm: $B_2 = 3729$ G, $B_3 = 1.25$ G, $B_4 = 0.23$ G, $B_5 = 0.36$ G, $B_6 = 0.726$ G, $B_7 = 0.020$ G, $B_8 = 0.023$ G, $B_9 = 0.0051$ G, $B_{10} = 0.0071$ G. Calculate the relative multipole strengths at $x = 1$ cm normalized to the quadrupole field at 1 cm. Why are the 12-pole and 20-pole components larger than the other multipole components?

Problem 4.10. Derive the equation for the pole profile of an iron dominated octupole with a bore radius R. To produce a field of 5 kG at the pole tip, ($R = 3$ cm), what total current in the coils is required?

Problem 4.11. Derive the 3×3 transformation matrices for a drift space, a quadrupole, and a bending magnet of constant curvature.

Problem 4.12. Show analytically that the dispersion function for a single bending magnet with a bending angle θ seems to emerge from the middle of the magnet with a slope $D' = \theta$.

5 Linear Beam Dynamics

The general equations of motion, characterized by an abundance of perturbation terms on the right-hand side of, for example, (4.86,88) have been derived in the previous chapter. If these perturbation terms were allowed to become significant in real beam transport systems, we would face almost insurmountable mathematical problems trying to describe the motion of charged particles in a general way. For practical mathematical reasons it is therefore important to design components for particle beam transport systems such that undesired terms appear only as small perturbations of the general motion of particles. With a careful design of beam guidance magnets and accurate alignment of these magnets we can indeed achieve this goal.

Most of the *perturbation terms* are valid solutions of the Laplace equation describing higher order fields components. Virtually all these terms can be minimized to the level of perturbations by proper design of beam transport magnets. Specifically we will see that the basic goals of beam dynamics can be achieved by using only two types of magnets, bending magnets and quadrupoles which sometimes are combined into one magnet. Beam transport systems, based on only these two lowest order magnet types, are called *linear systems* and the resulting theory of particle dynamics in the presence of only such magnets is referred to as *linear beam dynamics*.

In addition to the higher order magnetic field components we also find purely kinematic terms in the equations of motion due to large amplitudes or due to the use of curvilinear coordinates. Some of these terms are generally very small for particle trajectories which stay close to the reference path such that divergences are small, $x' \ll 1$ and $y' \ll 1$. The lowest order kinematic effects resulting from the use of a curvilinear coordinate system, however, cannot generally be considered small perturbations. One important consequence of this choice for the coordinate system is that the natural bending magnet is a sector magnet which has very different beam dynamics properties than a rectangular magnet which would be the natural type of magnet for a cartesian coordinate system. While a specific choice of a coordinate system will not change the physics, we must expect that some features are expressed easier or more complicated in one or the other coordinate system. We have chosen to use the curvilinear system because it follows the ideal path of the particle beam and offers a simple direct way to express particle trajectories deviating from the ideal path. In a fixed cartesian coordinate system we would have to deal with geometric expressions relating the points along the ideal path to an arbitrary reference point. The difference becomes evident for a simple trajectory like a circle of radius r and center at (x_o, y_o) which in a fixed orthogonal coordinate system would be expressed by $(x - x_o)^2 + (y - y_o)^2 = r^2$. In the curvilinear coordinate system this equation reduces to the simple identity $x(s) = 0$.

5.1 Linear Beam Transport Systems

The theory of beam dynamics based on quadrupole magnets for focusing is called *strong focusing* beam dynamics in contrast to the case of weak focusing discussed earlier in Chap. 3. Weak focusing systems utilize the focusing of sector magnets in combination with a small gradient in the bending magnet profile. Such focusing is utilized in circular accelerators like betatrons or some cyclotrons and the first generation of synchrotron. The invention of strong focusing by Christofilos [5.1] and independently by Courant et al. [5.2] changed quickly the way focusing arrangements for large particle accelerators are determined. One of the main attraction for this kind of focusing was the ability to greatly reduce the magnet aperture needed for the particle beam since the stronger focusing confines the particles to a much smaller cross section compared to weak focusing. A wealth of publications and notes have been written during the fifties to determine and understand the intricacies of strong focusing, especially the rising problems of alignment and field tolerances as well as those of resonances. Particle stability conditions from a mathematical point of view have been investigated by Moser [5.3].

Extensive mathematical tools are available to determine the characteristics of linear particle motion. In this chapter we will discuss the theory of linear charged particle beam dynamics and apply it to the development of beam transport systems, the characterization of particle beams, and to the derivation of beam stability criteria.

The bending and focusing function may be performed either in separate magnets or be combined within a *synchrotron magnet*. The arrangement of magnets in a beam transport system, called the *magnet lattice*, is often referred to as a *separated function* or *combined function lattice* depending on whether the lattice makes use of separate dipole and quadrupole magnets or uses combined function magnets respectively.

Linear equations of motion can be extracted from (4.86,88) to treat beam dynamics in first or *linear approximation*. For simplicity and without restricting generality we assume the bending of the beam to occur only in one plane, the x-plane. The *linear magnetic fields* for bending and quadrupole magnets are expressed by

$$B_x = g\,y,$$
$$B_y = B_{yo} + g\,x,$$

(5.1)

where B_{yo} is the dipole field and g the gradient of the quadrupole field. With these field components we obtain from (4.86,88) the equations of motion in the approximation of *linear beam dynamics*

$$x'' + \left(\frac{1}{\rho_o^2} + k_o \right) x = 0,$$

$$y'' - k_o y = 0.$$

(5.2)

The differential equations (5.2) cannot be solved in general because the magnet strength parameters are determined by the actual distribution of the magnets along the beam transport line and therefore are functions of the independent variable s. The equations of motion exhibit, however, directly the focusing power along the beam transport line. Integrating either of (5.2) over a short distance Δs we find the deflection angle $\int y'' \, ds = y' - y_o' = \alpha$. On the other hand $\int k_o y \, ds \approx k_o y \, \Delta s$ and applying the definition (4.36) for the focal length we find the general expressions for the *focal length* of magnetic gradient fields

$$\frac{1}{f_x} = k_o \Delta s = \frac{e}{cp} \frac{\partial B_y}{\partial x} \Delta s,$$

$$\frac{1}{f_y} = -k_o \Delta s = -\frac{e}{cp} \frac{\partial B_x}{\partial y} \Delta s.$$

(5.3)

Knowing the field gradient in any segment of a beam transport line of length Δs we may either immediately determine the focusing power of this segment or formulate the equations of motion. Both, the focusing from the bending magnet and that from a quadrupole may be combined into one parameter

$$K(s) = \frac{1}{\rho_o^2(s)} + k_o(s).$$

(5.4)

So far no distinction has been made between combined or separated function magnets and the formulation of the equations of motion based on the magnet strength parameter K as defined in (5.4), is valid for both types of magnets. For separated function magnets either k_o or ρ_o is set to zero while for combined function magnets both parameters are nonzero.

5.1.1 Nomenclature

Focusing along a beam transport line is performed by discrete quadrupoles placed to meet specific particle beam characteristics required at the end or some intermediate point of the beam line. The dependence of the magnet strength on s is, therefore, a priori indeterminate and is the subject of lattice design in accelerator physics. To describe focusing lattices simple symbols are used to point out location and sometimes relative strength of magnets. In this text we will use symbols from Fig. 5.1 for bending magnets, quadrupoles, and sextupoles or multipoles.

Fig. 5.1. Symbols for magnets in lattice design

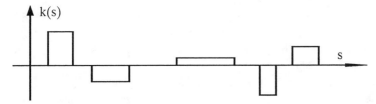

Fig. 5.2. Typical distribution of the quadrupole strength parameter

All magnets are symbolized by squares along the s-axis and the length of the symbol may represent the actual magnetic length. The symbol for pure dipole magnets is a square centered about the s-axis while bending magnets with a gradient are shifted vertically to indicate the sign of the focusing. Positive squares are used to indicate horizontal focusing and negative squares for horizontal defocusing quadrupoles. Similar but vertically higher symbols are used for quadrupoles indicating the sign of the focusing as well.

Using such symbols a typical beam transport line may have general patterns like that shown in Fig 5.2. The sequence of magnets and their strength seems random and is mostly determined by external conditions to be discussed later. More regular magnet arrangements occur for circular accelerators or very long beam transport lines as will be discussed in Chap. 6.

5.2 Matrix Formalism in Linear Beam Dynamics

The seemingly arbitrary distribution of focusing parameters in a beam transport system makes it impossible to formulate a general solution of the differential equations of motion (5.2). To describe particle trajectories analytically through a beam transport line composed of drift spaces, bending magnets, and quadrupoles, we will derive mathematical tools which consist of partial solutions and can be used to describe complete particle trajectories.

In this section we will derive and discuss the *matrix formalism* [5.4] as a method to describe particle trajectories. This method makes use of the fact that the magnet strength parameters are constant at least within each individual magnet. The equations of motion becomes very simple since the restoring force K is constant and the solutions have the form of trigonometric functions. The particle trajectories may now be described by analytical functions at least within each uniform element of a transport line including magnet free drift spaces.

These solutions can be applied to any arbitrary beam transport line, where the focusing parameter K changes in a step like function along the beam transport line as shown Fig. 5.2. By cutting this beam line into its smaller uniform pieces so that $K = $ const. in each of these pieces we will be able to follow the particle trajectories analytically through the whole transport system. This is the model generally used in particle beam optics and is called the *hard edge model.*

In reality, however, since nature does not allow sudden changes of physical quantities (*natura non facit saltus*) the hard edge model is only an approximation, although for practical purposes a rather good one. In a real magnet the field strength does not change suddenly from zero to full value but rather follows a smooth transition from zero to the maximum field. Sometimes the effects due to this smooth field transition or *fringe field* are important and we will derive the appropriate corrections later in this section. We therefore continue using the hard edge model for beam transport magnets and keep in mind that for some cases a correction may be needed to take into account the effects of a smooth field transition at the magnet edges. Such corrections become significant only for short, strong quadrupoles with large apertures.

Using this approximation, where $1/\rho_o$ and k are constants, and ignoring perturbations, the equation of motion is reduced to that of a harmonic oscillator,

$$u'' + K\,u = 0, \quad \text{where} \quad K = \frac{1}{\rho_o^2} + k_o = \text{const}. \tag{5.5}$$

The principal solutions have been derived in Sect. 4.8 and are expressed in matrix formulation by

$$\begin{bmatrix} u(s) \\ u'(s) \end{bmatrix} = \begin{bmatrix} C(s) & S(s) \\ C'(s) & S'(s) \end{bmatrix} \begin{bmatrix} u_o \\ u_o' \end{bmatrix}, \tag{5.6}$$

where u may be used for either x or y. We have deliberately separated the motion in both planes since we do not consider coupling. Formally we could combine the two 2×2 transformation matrices for each plane into one 4×4 matrix describing the transformation of all four coordinates

$$\begin{bmatrix} x(s) \\ x'(s) \\ y(s) \\ y'(s) \end{bmatrix} = \begin{bmatrix} C_x(s) & S_x(s) & 0 & 0 \\ C_x'(s) & S_x'(s) & 0 & 0 \\ 0 & 0 & C_y(s) & S_y(s) \\ 0 & 0 & C_y'(s) & S_y'(s) \end{bmatrix} \begin{bmatrix} x_o \\ x_o' \\ y_o \\ y_o' \end{bmatrix}. \tag{5.7}$$

Obviously the transformations are still completely decoupled but in this form we could include coupling effects, where, for example, the x-motion depends also on the y-motion and vice versa. This can be further generalized to include any particle parameter like the longitudinal position of a particle with respect to a reference particle, or the energy of a particle, the spin vector, or any particle coordinate that may depend on other coordinates. The energy of a particle, for example, has an influence on the actual trajectory which we called the dispersion and in (4.116) we have already made use of an expanded transformation matrix to describe this energy effect on the particle trajectory. For practical reasons, however, it is convenient to limit the order of the transformation matrices to the minimum required for the problem under study.

In the following paragraphs we will derive linear transformation matrices for a variety of beam line elements.

5.2.1 Driftspace

In a *driftspace* and in a weak bending magnet, where $1/\rho_o^2 \ll 1$ and $k_o = 0$, the focusing parameter $K = 0$ and the solution of (5.5) in matrix formulation can be expressed by

$$\begin{bmatrix} u(s) \\ u'(s) \end{bmatrix} = \begin{bmatrix} 1 & s - s_o \\ 0 & 1 \end{bmatrix} \begin{bmatrix} u(s_o) \\ u'_o(s_o) \end{bmatrix}. \tag{5.8}$$

A more precise derivation of the transformation matrices for bending magnets of arbitrary strength will be described later in this chapter. Any drift space of length $\ell = s - s_o$, therefore, is represented by the simple transformation matrix

$$\mathcal{M}_{\mathrm{d}}(\ell|0) = \begin{bmatrix} 1 & \ell \\ 0 & 1 \end{bmatrix}. \tag{5.9}$$

We recognize the expected features of a particle trajectory in a field free drift space. The amplitude u changes only if the trajectory has an original non vanishing slope $u'_o \neq 0$ while the slope itself does not change at all.

5.2.2 Quadrupole Magnet

For a pure quadrupole we set the bending term $1/\rho_o = 0$. The field gradient, however, will not vanish $k_o(s) \neq 0$ and can be positive as well as negative. With these assumptions we solve again (5.5) and determine the integration constants by initial conditions. For $k_o = |k_o| > 0$ we get the transformation for a *focusing quadrupole*

$$\begin{bmatrix} u(s) \\ u'(s) \end{bmatrix} = \begin{bmatrix} \cos\psi & \frac{1}{\sqrt{k_o}}\sin\psi \\ -\sqrt{k_o}\sin\psi & \cos\psi \end{bmatrix} \begin{bmatrix} u(s_o) \\ u'(s_o) \end{bmatrix}, \tag{5.10}$$

where $\psi = \sqrt{k_o}\,(s - s_o)$. This equation is true for any section within the quadrupole as long as both points s_o and s are within the active length of the quadrupole.

For a full quadrupole of length ℓ and strength k_o we set $\varphi = \sqrt{k_o}\,\ell$ and the transformation matrix for a full quadrupole in the focusing plane is

$$
\mathcal{M}_{\mathrm{QF}}(\ell|0) = \begin{bmatrix} \cos\varphi & \dfrac{1}{\sqrt{k_o}}\sin\varphi \\ -\sqrt{k_o}\,\sin\varphi & \cos\varphi \end{bmatrix}. \tag{5.11}
$$

Similarly we get in the other plane with $k_o = -|k_o| < 0$ the solution for a *defocusing quadrupole*

$$
\begin{bmatrix} u(s) \\ u'(s) \end{bmatrix} = \begin{bmatrix} \cosh\psi & \dfrac{1}{\sqrt{|k_o|}}\sinh\psi \\ \sqrt{|k_o|}\,\sinh\psi & \cosh\psi \end{bmatrix} \begin{bmatrix} u(s_o) \\ u'(s_o) \end{bmatrix}, \tag{5.12}
$$

where $\psi = \sqrt{|k_o|}\,(s - s_o)$. The transformation matrix in the defocusing plane through a complete quadrupole of length ℓ with $\varphi = \sqrt{|k_o|}\,\ell$ is therefore

$$
\mathcal{M}_{\mathrm{QD}}(\ell|0) = \begin{bmatrix} \cosh\varphi & \dfrac{1}{\sqrt{|k_o|}}\sinh\varphi \\ \sqrt{|k_o|}\,\sinh\varphi & \cosh\varphi \end{bmatrix} \begin{bmatrix} u_o \\ u'_o \end{bmatrix}. \tag{5.13}
$$

These transformation matrices make it straight forward to follow a particle through a transport line. Any arbitrary sequence of drift spaces, bending magnets and quadrupole magnets can be represented by a series of transformation matrices $\mathcal{M}_i$. The transformation matrix for the whole composite beam line is then just equal to the product of the individual matrices. For example, by multiplying all matrices along the path of a particle in Fig. 5.3 the total transformation matrix $\mathcal{M}$ for the ten magnetic elements of this example is determined by the product

$$
\mathcal{M} = \mathcal{M}_{10} \ldots \mathcal{M}_5\,\mathcal{M}_4\,\mathcal{M}_3\,\mathcal{M}_2\,\mathcal{M}_1 \tag{5.14}
$$

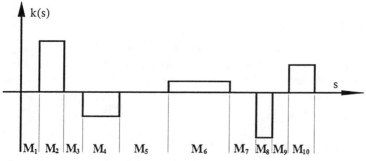

Fig. 5.3. Example of a beam transport line (schematic)

and the particle trajectory transforms through the whole composite transport line like

$$
\begin{bmatrix} u(s) \\ u'(s) \end{bmatrix} = \mathcal{M}(s|s_0) \begin{bmatrix} u(s_0) \\ u'(s_0) \end{bmatrix}, \tag{5.15}
$$

where the starting point s_0 in this case is at the beginning of the drift space $\mathcal{M}_1$ and the end point s is at the end of the magnet $\mathcal{M}_{10}$.

5.2.3 Thin Lens Approximation

As will become more apparent in the following sections, this matrix formalism is widely used to calculate trajectories for individual particle or for a virtual particle representing the central path of a whole beam. The repeated multiplication of matrices, although straightforward, is very tedious and therefore, most beam dynamics calculations are performed on digital computers. In some cases, however, it is desirable to analytically calculate the approximate properties of a small set of beam elements. For these cases it is sufficient to use what is called the *thin lens approximation*. In this approximation it is assumed that the length of a quadrupole magnet is small compared to its focal length ($\ell \ll f$) and we set

$$
\ell \to 0, \tag{5.16}
$$

while keeping the focal strength constant,

$$
f^{-1} = +k_0\, l = \text{const}. \tag{5.17}
$$

As a consequence $\varphi = \sqrt{k_0}\, \ell \to 0$ and the transformation matrices (5.11,13) are the same in both planes except for the sign of the focal length

$$
\begin{bmatrix} u \\ u' \end{bmatrix} = \begin{bmatrix} 1 & 0 \\ -\frac{1}{f} & 1 \end{bmatrix} \begin{bmatrix} u_0 \\ u_0' \end{bmatrix}, \tag{5.18}
$$

where

$$
\begin{aligned}
f^{-1} &= k_0\, \ell > 0 \qquad \text{in the focusing plane} \\
f^{-1} &= k_0\, \ell < 0 \qquad \text{in the defocusing plane}.
\end{aligned} \tag{5.19}
$$

The transformation matrix has obviously become very simple and exhibits only the focusing property in form of the focal length. Quite generally one may regard for single as well as composite systems the matrix element M_{21} as the element that expresses the focal strength of the transformation.

In thin lens approximation it is rather easy to derive focusing properties of simple compositions of quadrupoles. A *quadrupole doublet* composed of two quadrupole magnets separated by a drift space of length d (see Fig 4.14) is described by the total transformation matrix

$$\mathcal{M}_{db}(d|0) = \begin{bmatrix} 1 & 0 \\ -\frac{1}{f_2} & 1 \end{bmatrix} \begin{bmatrix} 1 & d \\ 0 & 1 \end{bmatrix} \begin{bmatrix} 1 & 0 \\ -\frac{1}{f_1} & 1 \end{bmatrix}$$

(5.20)

$$= \begin{bmatrix} 1 - d/f_1 & d \\ -1/f^* & 1 - d/f_2 \end{bmatrix},$$

where we find the well known expression from geometric paraxial light optics

$$\frac{1}{f^*} = \frac{1}{f_1} + \frac{1}{f_2} - \frac{d}{f_1 f_2} .$$

(5.21)

Such a doublet can be made focusing in both planes if, for example, the quadrupole strengths are set such that $f_1 = -f_2 = f$. The total focal length then is $f^* = +d/f^2 > 0$ in both the horizontal and the vertical plane.

This simple result, where the focal length is the same in both planes, is a valid solution only in thin lens approximation. For a doublet of finite length quadrupoles the focal length in the horizontal plane is always different from that in the vertical plane as can be verified by using the transformations (5.10-13) to calculate the matrix $\mathcal{M}_{db}$. Since individual matrices are not symmetric with respect to the sign of the quadrupole field, the transformation matrices for the horizontal plane $\mathcal{M}_{db,x}$ and the vertical plane $\mathcal{M}_{db,y}$ must be calculated separately and turn out to be different. In special composite cases, where the quadrupole distribution is symmetric as shown in Fig. 5.4, the matrices for both of the two symmetric half sections are related in a simple way. If the matrix for one half of the symmetric beam line is

$$\mathcal{M} = \begin{bmatrix} a & b \\ c & d \end{bmatrix}$$

(5.22)

then the *reversed matrix* for the second half of the beam line is

$$\mathcal{M}_r = \begin{bmatrix} d & b \\ c & a \end{bmatrix}$$

(5.23)

and the total symmetric beam line has the transformation matrix

$$\mathcal{M}_{tot} = \mathcal{M}_r \mathcal{M} = \begin{bmatrix} ad + bc & 2bd \\ 2ac & ad + cb \end{bmatrix} .$$

(5.24)

We made no particular assumptions for the lattice shown in Fig. 5.4 except for symmetry and the relations (5.22-24) are true for any arbitrary but symmetric beam line.

The result for the reversed matrix is not to be confused with the *inverse matrix*, where the direction of the particle path is also reversed. In this case we have

$$\mathcal{M}_i = \begin{bmatrix} d & -b \\ -c & a \end{bmatrix} .$$

(5.25)

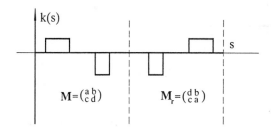

Fig. 5.4. Reversed lattice

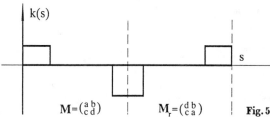

Fig. 5.5. Symmetric quadrupole triplet

Going through an arbitrary section of a beam line and then back to the origin again results in a total transformation matrix equal to the unity matrix

$$\mathcal{M}_{\text{tot}} = \mathcal{M}_{\text{i}}\,\mathcal{M} = \begin{bmatrix} 1 & 0 \\ 0 & 1 \end{bmatrix}. \tag{5.26}$$

These results allow us now to calculate the transformation matrix $\mathcal{M}_{\text{tr}}$ for a *symmetric quadrupole triplet*. With (5.20,22,24) the transformation matrix of a quadrupole triplet as shown in Fig. 5.5 is

$$\mathcal{M}_{\text{tr}} = \mathcal{M}_{\text{r}}\,\mathcal{M} = \begin{bmatrix} 1 - 2L^2/f^2 & 2L\,(1 + L/f) \\ -1/f^* & 1 - 2L^2/f^2 \end{bmatrix}, \tag{5.27}$$

where f^* is defined by (5.21) with $f_1 = -f_2 = f$.

Such a triplet is focusing in both planes as long as $f > L$. Symmetric triplets as shown in Fig. 5.5 have become very important design elements of long beam transport lines or circular accelerators since such a triplet can be made focusing in both planes and can be repeated arbitrarily often to provide a periodic focusing structure called a *FODO-channel*. The acronym is derived from the sequence of focusing (F) and defocusing (D) quadrupoles separated by nonfocusing elements (O) like a drift space or a bending magnet.

5.2.4 Quadrupole End Field Effects

In defining the transformation through a quadrupole we have assumed the strength parameter $k(s)$ to be a step function with a constant nonzero value

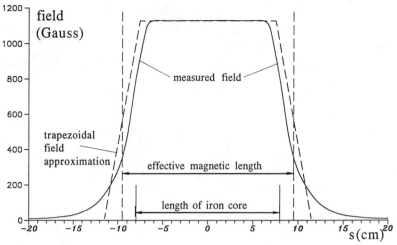

Fig. 5.6. Field profile in a real quadrupole with a bore radius of $R = 3$ cm and an iron length of $\ell_{\text{iron}} = 15.9$ cm

within the quadrupole and zero outside. Such a hard edge field distribution is only approximately true for a real quadrupole. The strength parameter in a real quadrupole magnet varies in a gentle way from zero outside the quadrupole to a maximum value in the middle of the quadrupole. In Fig. 5.6 the measured gradient of a real quadrupole along the axis is shown. The field extends well beyond the length of the iron core and the effective magnetic length, defined by

$$\ell_{\text{eff}} = \frac{\int g \, ds}{g_0}, \tag{5.28}$$

where g_0 is the field gradient in the middle of the quadrupole, is longer than the iron length by about the radius of the bore aperture

$$\ell_{\text{eff}} \approx \ell_{\text{iron}} + R. \tag{5.29}$$

The real field distribution can be approximated by a trapezoid such that $\int g \, ds$ is the same in both cases (see Fig. 5.6). To define the trapezoidal approximation we assume a fringe field extending over a length equal to the bore radius R as shown in Fig. 5.6. End field effects must therefore be expected specifically with quadrupoles of large bore radii and short iron cores.

It is interesting to investigate as to what extend the transformation characteristics for a real quadrupole differ from the hard edge model. The real transformation matrix can be obtained by slicing the whole axial quadrupole field distribution in thin segments of varying strength. Treating these segments as short hard edge quadrupoles the full transformation matrix is the product of the matrices for all segments.

While it is possible to obtain an accurate transformation matrix this way the variations of the matrix elements due to this smooth field distribution turn out to be mostly small and in praxis, therefore, the hard edge model is used to develop beam transport lattices. Nonetheless after a satisfactory solution has been found, these real transformation matrices should be used to check the solution and possibly make small adjustment to the idealized hard edge model design.

In this section, we will discuss an analytical estimate of the correction to be expected for a real field distribution [5.5] by looking for the "effective" hard edge model parameters (k, ℓ) which result in a transformation matrix equal to the transformation matrix for the corresponding real quadrupole. The transformation matrix for the real quadrupole be

$$\mathcal{M}_{\mathbb{Q}} = \begin{bmatrix} C & S \\ C' & S' \end{bmatrix}, \tag{5.30}$$

where the matrix elements are the result of multiplying all "slice" matrices for the quadrupole segments as shown in Fig. 5.7.

We assume now that this real quadrupole can be represented by a hard edge model quadrupole of length ℓ with adjacent drift spaces λ as indicated in Fig. 5.7. The transformation through this system for a focusing quadrupole is given by [5.5]

$$\begin{bmatrix} 1 & \lambda \\ 0 & 1 \end{bmatrix} \begin{bmatrix} \cos\varphi & \frac{1}{\sqrt{k}}\sin\varphi \\ -\sqrt{k}\sin\varphi & \cos\varphi \end{bmatrix} \begin{bmatrix} 1 & \lambda \\ 0 & 1 \end{bmatrix}$$
$$= \begin{bmatrix} \cos\varphi - \sqrt{k}\,\lambda\,\sin\varphi & \lambda(2\cos\varphi + \frac{\sin\varphi}{\sqrt{k}\lambda}) \\ -\sqrt{k}\,\sin\varphi & \cos\varphi - \sqrt{k}\,\lambda\,\sin\varphi \end{bmatrix} \tag{5.31}$$

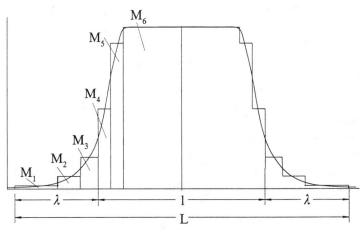

Fig. 5.7. Decomposition of an actual quadrupole field profile into segments of hard edge quadrupoles

with $\varphi = \sqrt{k}\,\ell$. This hard edge transformation matrix must be the same as the actual matrix (5.30) and we will use this equality to determine the effective quadrupole parameters k, ℓ. First we note that the choice of the total length $L = \ell + 2\lambda$ is arbitrary as long as it extends over the whole field profile. Equating (5.31) and (5.30) we can derive two equations which allow us to determine the effective parameters k, ℓ from known quantities

$$
\begin{aligned}
C_{\mathrm{f}} - \tfrac{1}{2} L C_{\mathrm{f}}' &= \cos\varphi_{\mathrm{f}} + \tfrac{1}{2}\varphi_{\mathrm{f}} \sin\varphi_{\mathrm{f}}, \\
C_{\mathrm{f}}' \ell_{\mathrm{f}} &= -\varphi_{\mathrm{f}} \sin\varphi_{\mathrm{f}}.
\end{aligned}
\tag{5.32}
$$

Here we have added the index $_{\mathrm{f}}$ to indicate a focusing quadrupole. The first of these equations can be solved for φ_{f} since the quantities $C_{\mathrm{f}}, C_{\mathrm{f}}'$, and L are known. The second equation then is solved for ℓ_{f} and $k_{\mathrm{f}} = \varphi_{\mathrm{f}}^2/\ell_{\mathrm{f}}$.

Two parameters are sufficient to equate the 2×2 matrices (5.30,31) since two of the four equations are redundant for symmetry reasons, $M_{11} = M_{22} = C = S'$, and the determinant of the matrices on both sides must be unity. Similarly we get for a defocusing quadrupole

$$
\begin{aligned}
C_{\mathrm{d}} - \tfrac{1}{2} L C_{\mathrm{d}}' &= \cosh\varphi_{\mathrm{d}} - \tfrac{1}{2}\varphi_{\mathrm{d}} \sinh\varphi_{\mathrm{d}}, \\
C_{\mathrm{d}}' \ell_{\mathrm{d}} &= -\varphi_{\mathrm{d}} \sinh\varphi_{\mathrm{d}}.
\end{aligned}
\tag{5.33}
$$

Equations (5.32,33) define a hard edge representation of a real quadrupole. From (5.32) and (5.33), however, we note that the effective quadrupole length ℓ and strength k are different from the customary definition, where k_{o} is the actual magnet strength in the middle of the quadrupole and the magnet length is defined by $\ell = \frac{1}{k_{\mathrm{o}}} \int k(s)\,\mathrm{d}s$. We also observe that the effective values ℓ and k are different for the focusing and defocusing plane. Since the endfields are not the same for all quadrupoles but depend on the design parameters of the magnet we cannot determine the corrections in general. In practical cases, however, it turns out that the corrections $\Delta k = k - k_{\mathrm{o}}$ and $\Delta\ell = \ell - \ell_{\mathrm{o}}$ are small for quadrupoles which are long compared to the aperture and are larger for short quadrupoles with a large aperture. In fact the differences Δk and $\Delta\ell$ turn out to have opposite polarity and the thin lens focal length error $\Delta k\,\Delta\ell$ is generally very small.

As an example we use the quadrupole of Fig. 5.6 and calculate the corrections due to end field effects. We calculate the total transformation matrix for the real field profile as discussed above by approximating the actual field distribution by a series of hard edge "slice" matrices in both planes as a function of the focusing strength k_{o} and solve (5,32,33) for the effective parameters $(k_{\mathrm{f}}, \ell_{\mathrm{f}})$ and $(k_{\mathrm{d}}, \ell_{\mathrm{d}})$, respectively. In Fig. 5.8 these relative fringe field corrections to the quadrupole strength $\Delta k/k_{\mathrm{o}}$ and to the quadrupole length $\Delta\ell/\ell_{\mathrm{o}}$ are shown as functions of the strength k_{o}. The effective quadrupole length is longer and the effective quadrupole strength is lower than the pure hard edge values. In addition the corrections are different in both planes. Depending on the sensitivity of the beam transport system these corrections may have to be included in the final optimization.

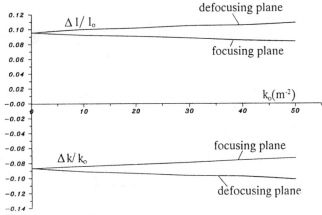

Fig. 5.8. Fringe field correction for the quadrupole of Fig. 5.6 with a bore radius of $R = 3.0$ cm and a steel length of $\ell_{\text{iron}} = 15.9$ cm

5.2.5 Quadrupole Design Concepts

The feasibility of any accelerator or beam transport line design depends fundamentally on the parameters and diligent fabrication of technical components composing the system. Not only need the magnets be designed such as to minimize the undesirable higher order multipole field errors but they also must be designed such that the desired parameters are within technical limits. Most magnets constructed for beam transport lines are electromagnets rather than permanent magnets. Such magnets are excited by electrical current carrying coils around magnet poles or in the case of superconducting magnets by specially shaped and positioned current carrying coils. In this section, we will discuss briefly some fundamental design concepts and limits for most commonly used *iron dominated quadrupole magnets* as a guide for the accelerator designer towards a realistic design. For more detailed discussions on technical quadrupole designs we refer to [5.6].

Iron dominated magnets are the most commonly used magnets for particle beam transport systems. Only where very high particle energies and magnetic fields are required, superconducting magnets are used with maximum magnetic fields of 40 - 60 kG compared to the maximum field in an iron magnet of about 20 kG. Although saturation of ferromagnetic material imposes a definite limit on the strength of iron dominated magnets, most accelerator design needs can be accommodated within this limit.

Quadrupoles together with *bending magnets* are the basic building blocks for charged particle beam transport systems and serve as focusing devices to keep the particle beam close to the desired beam path. In Sect. 4.3.2 the basic principles of particle beam focusing and the characteristics of a *quadrupole magnet* have been discussed. Specifically we found that a quadrupole field can be generated by four symmetrically aligned poles in the form of hyperbolas as shown in Fig. 4.8.

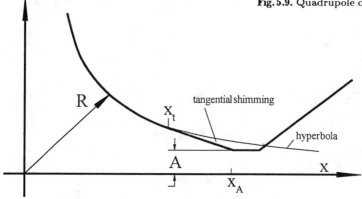

Fig. 5.9. Quadrupole design features

In a real quadrupole we cannot use infinitely wide hyperbolas but must cut off the poles at some width. In Fig. 5.9 some fundamental design features and parameters for a real quadrupole are shown and we note specifically the finite pole width to make space available for the excitation coils. Since only infinitely wide hyperbolic poles create a pure quadrupole field, we expect the appearance of higher multipole field errors characteristic for a finite pole width. While in an ideal quadrupole the field gradient along say the x-axis would be constant we find for a finite pole width a drop off of the field and gradient approaching the corner of the pole. This drop off can be reduced to some extend if the hyperbolic pole profile continues into its tangent close to the pole corner as indicated in Fig. 5.9. The starting point of the tangent determines greatly the final gradient homogeneity in the quadrupole aperture. In Fig. 5.10 the gradient along the x-axis is shown

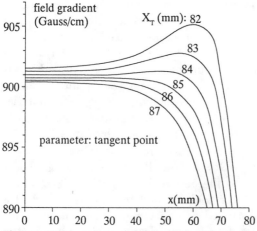

Fig. 5.10. Field gradient and pole profile shimming for a particular quadrupole as determined by numerical simulations with the program MAGNET[5.7]

132

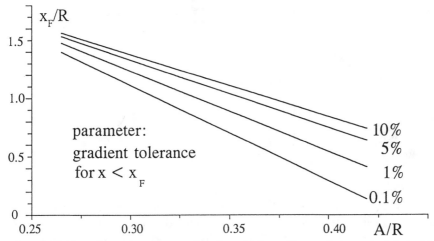

Fig. 5.11. Field gradient tolerances as a function of pole profile parameters calculated with MAGNET [5.7]

for different starting points of the tangent. There is obviously an optimum point for the tangent to minimize the gradient error over a wide aperture.

Application of tangent shimming must be considered as a fine adjustment of the field quality rather than a means to obtain a large good field aperture as becomes apparent from Fig. 5.11. The good field aperture is basically determined by the width of the pole. While optimizing the tangent point we find in Fig. 5.11 an empirical correlation between *gradient tolerance* within an aperture region $x \leq X_F$ and the pole width expressed by the minimum pole gap A. The *good field region* increases as the pole gets wider. For initial design purposes we may use Fig. 5.11 to determine the pole width from A based on the desired good field region X_F and gradient quality.

The final design of a magnet pole profile is made with the help of computer codes which allow the calculation of magnet fields from a given pole profile with saturation characteristics determined from a magnetization curve. Widely used computer codes for magnet design are *MAGNET* [5.7] and *POISSON* [5.8].

Field errors in iron dominated magnets have two distinct sources, the finite pole width and mechanical manufacturing and assembly tolerances. From symmetry arguments we can deduce that field errors due to the finite pole width produce only select multipole components. In a quadrupole, for example, only $(2n + 1)$ 4-pole fields like 12-pole or 20-pole fields are generated. Similarly in a dipole of finite pole width only $(2n + 1)$ 2-pole fields exist. We call these multipole field components often the allowed multipole errors. Manufacturing and assembly tolerances on the other hand do not exhibit any symmetry and can cause the appearance of any multipole field error.

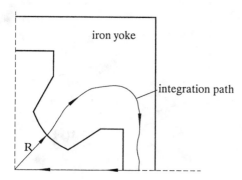

iron yoke

integration path

R

Fig. 5.12. Determination of the field gradient from the excitation current

The particular choice of some geometric design parameters must be checked against technical limitations during the design of a beam transport line. One basic design parameter for a quadrupole is the bore radius R which depends on the aperture requirements of the beam. Addition of some allowance for the vacuum chamber and mechanical tolerance between chamber and magnet finally determines the quadrupole bore radius.

The field gradient is determined by the electrical excitation current in the quadrupole coils. Similar to the derivation for a bending magnet in Sect. 4.3.1 we may derive a relation between field gradient and *excitation current* from Maxwell's curl equation. To minimize unnecessary mathematical complexity we choose an integration path as indicated in Fig. 5.12 which contributes to the integral $\oint \mathbf{B_s}\,\mathrm{d}s$ only in the aperture of the quadrupole. Starting from the quadrupole axis along a path at 45° with respect to the horizontal or vertical plane toward the pole tip, we have

$$\oint \mathbf{B_s}\,\mathrm{d}s \;=\; \int_0^R B_r\,\mathrm{d}r \;=\; \frac{4\pi}{c}\,I_{\mathrm{tot}}\,. \tag{5.34}$$

Since $B_x = g\,y$ and $B_y = g\,x$ the radial field component is $B_r = \sqrt{B_x^2 + B_y^2} = g\,r$ and the excitation current from (5.34) is given by

$$I_{\mathrm{tot}}\,(\mathrm{A\ \ turns}) \;=\; \frac{1}{0.8\pi}\,g\left(\frac{\mathrm{G}}{\mathrm{cm}}\right)R^2(\mathrm{cm})\,. \tag{5.35}$$

The space available for the *excitation coils* or *coil slot* in a real quadrupole design determines the maximum current carrying capability of the coil. Common materials for magnet coils are copper or aluminum. The electrical heating of the coils depends on the current density and a technically feasible balance between heating and cooling capability must be found. As a practical rule the current density in regular beam transport magnets should not exceed about 6 - 8 A/mm². This is more an economical than a technical limit and up to about a factor of two higher current densities could be used for special applications. The total required coil cross section, however, including an allowance for insulation material between coil windings and about 15 - 20 % for water cooling holes in the conductor depends on the electrical losses in the coil. The aperture of the water cooling holes is chosen such that sufficient water cooling can be provided with an allowable water temperature increase which should be kept below about 40°C to avoid boiling of the cooling water at the surface and loss of cooling power. A low temperature rise is achieved if the water is rushed through the coil at high pressure in which case undesirable vibrations of the magnets may occur. The water cooling hole in the conductor must therefore be chosen with all these considerations in mind. Generally the current density averaged over the whole coil cross section is about 60 - 70 % of that in the conductor.

In practical applications we find the required coil cross section to be significant compared to the magnet aperture leading to a long pole length and potential *saturation*. To obtain high field limits due to magnetic saturation, steel with a very low *carbon content* or low carbon steel is used for most magnet applications in particle beam lines. Specifically, we require the carbon content for most high quality magnets to be no more than about 1%. In Fig. 5.13 the *magnetization curve* and the *permeability* as a function of the excitation are shown for a steel with 0.5 % carbon content. We note a steep drop in the permeability above 16 kG reaching full saturation at about 20 kG. A magnet has an acceptable saturation level if the magnetic permeability anywhere over the cross section of the magnet remains large compared to unity, $\mu \gg 1$.

Severe saturation effects at the corners of the magnet pole profile, see Fig. 5.9, can be avoided if the maximum field gradient, as a rule of thumb, is chosen such that the pole tip field does not exceed a value of $B_p = 8$ - 10 kG. This limits the maximum field gradient to $g_{max} = B_p/R$ and the quadrupole length must therefore be long enough to reach the focal length desired in the design of the beam transport line.

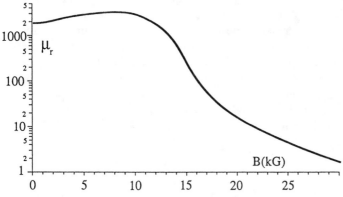

Fig. 5.13. Magnetization and permeability of typical low carbon steel as a function of excitation

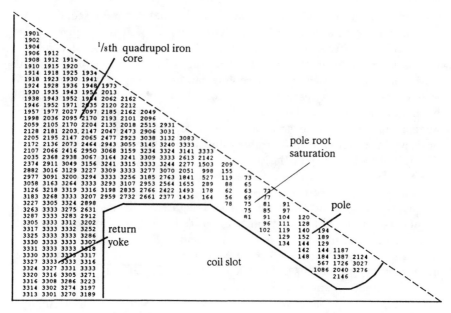

Fig. 5.14. Permeability plot for a highly saturated quadrupole magnet. Permeability values are plotted in a grid over the iron cross section of the quadrupole and we note significantly reduced permeability, ($\mu < 100$), in the pole root while there is negligible saturation elsewhere in the magnet

Saturation of the pole corners introduces higher order multipoles and must therefore be kept to a minimum. Other saturation effects may occur at the *pole root*, where all magnetic flux from one pole including *fringe fields* are concentrated. If the pole root is too narrow the flux density is too high and saturation occurs. This does not immediately affect the field quality but requires much higher excitation currents. A similar effect may occur in the return yokes if the field density is too high because of a small iron cross section. In Fig. 5.14 a permeability plot is shown for a magnet with severe saturation effects. By increasing the width of the pole root and return yoke the saturation is greatly reduced as shown in Fig 5.15. To minimize pole root saturation the pole length should be as short as possible. Unfortunately this also reduces the space available for the excitation coils and this leads to excessively large current densities. To reduce this conflict the pole width is usually increased at the pole root as was done in Fig. 5.15 rather than shortening the pole length.

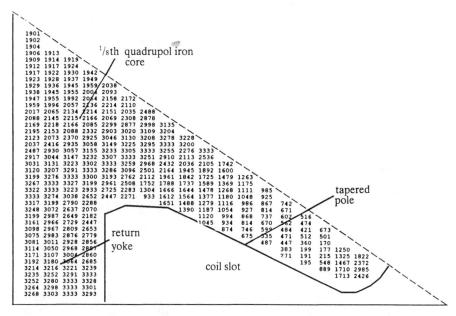

Fig. 5.15. After introducing a tapered width of the pole which widens the pole root the permeability plot shows little saturation at the same excitation current as in Fig. 5.14

5.3 Focusing in Bending Magnets

Bending magnets have been treated so far just like drift spaces as far as focusing properties are concerned. This is a good approximation for weak bending magnets which bend the beam only by a small angle. In cases of larger deflection angles, however, we observe focusing effects which are due to the particular type of magnet and its end fields. In Sect. 4.7 we discussed the geometric focusing term $1/\rho^2$ which appears in sector magnets only. Other focusing terms are associated with bending magnets and we will discuss in this section these effects in a systematic way. Specifically, the focusing of charged particles crossing end fields at oblique angles will be discussed.

The linear theory of particle beam dynamics uses a curvilinear coordinate system following the path of the reference particle and it is assumed that all magnetic fields are symmetric about this path. The "natural" bending magnet in this system is one, where the ideal path of the particles enters and exits normal to the magnet pole faces. Such a magnet is called a *sector magnet* as shown in Fig. 5.16. The total deflection of a particle depends on the distance of the particle path from the ideal path in the deflecting plane which, for simplicity, we assume to be in the horizontal x-plane. Particles following a path at a larger distance from the center of curvature than the ideal path travel a longer distance through this magnet and, therefore, are

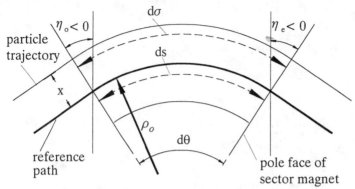

Fig. 5.16. Focusing in a sector magnet, where $\eta_o = \eta_e = d\theta/2$

deflected by a larger angle than a particle on the ideal path. Correspondingly, a particle passing through the magnet closer to the center of curvature is deflected less.

This asymmetry leads to a focusing effect which is purely geometric in nature. On the other hand, we may choose to use a magnet such that the ideal path of the particle beam does not enter the magnet normal to the pole face but rather at an angle. Such a configuration has an asymmetric field distribution about the beam axis and therefore leads to focusing effects. On a more subtle note we find a modification of these focusing effects due to extended fringe field of the magnets. While these fringe fields expand orthogonally to the face of the magnet iron, we obviously have some deflection in these fringe fields which leads to asymmetric field distributions about the particle beam axis and to focusing. In the following subsections we will discuss and formulate more quantitatively these focusing effects.

5.3.1 Sector Magnets

The degree of focusing in a sector magnet can be evaluated in any infinitesimal sector of such a magnet by calculating the deflection angle as a function of the particle position x. With the notation from Fig. 5.16 we get for the deflection angle while keeping only linear terms in x

$$d\theta = \frac{1}{\rho_o} d\sigma = \frac{1}{\rho_o} \left(1 + \frac{x}{\rho_o} \right) ds. \tag{5.36}$$

The first term on the r.h.s. merely defines the ideal path, while the second x-dependent term of the deflection angle in (5.36) describes the particle motion in the vicinity of the ideal path. With respect to the curvilinear coordinate system following the ideal path we get the additional deflection

$$\delta\theta = \frac{x}{\rho_o^2} ds. \tag{5.37}$$

This correction is to be included in the differential equation of motion as an additional focusing term

$$\Delta x'' = -\frac{\delta \theta}{ds} = -\frac{x}{\rho_o^2} \tag{5.38}$$

to the straight quadrupole focusing leading to the equation of motion

$$x'' + \left(k_o + \frac{1}{\rho_o^2}\right) x = 0, \tag{5.39}$$

which is identical to the result obtained in Sect. 4.7.

The differential equation (5.39) has the same form as that for a quadrupole and therefore the solutions must be of the same form. Using this similarity we replace k_o by $(k_o + 1/\rho_o^2)$ and obtain immediately the transformation matrices for a general sector magnet. For $K = k_o + 1/\rho_o^2 > 0$ and

$$\Theta = \sqrt{K}\,\ell \tag{5.40}$$

we get from (5.11) the transformation matrix

$$\boldsymbol{M}_{\mathrm{sy,f}}(\ell|0) = \begin{bmatrix} \cos\Theta & \frac{1}{\sqrt{K}}\sin\Theta \\ -\sqrt{K}\sin\Theta & \cos\Theta \end{bmatrix}, \tag{5.41}$$

where ℓ is the *arc length* of the sector magnet and where both the focusing term k_o and the bending term $1/\rho_o$ may be nonzero. Such a magnet is called a *synchrotron magnet* since this magnet type was first used for lattices of synchrotrons.

For the defocusing case, where $K = k_o + 1/\rho_o^2 < 0$ and $\Theta = \sqrt{|K|}\,\ell$ we get from (5.13)

$$\boldsymbol{M}_{\mathrm{sy,d}}(\ell|0) = \begin{bmatrix} \cosh\Theta & \frac{1}{\sqrt{|K|}}\sinh\Theta \\ \sqrt{|K|}\sinh\Theta & \cosh\Theta \end{bmatrix}. \tag{5.42}$$

Note that the argument Θ is equal to the deflection angle θ only in the limit $k_o \to 0$ because these transformation matrices include bending as well as focusing in the same magnet. Obviously, in the nondeflecting plane $1/\rho_o = 0$ and such a magnet acts just like a quadrupole with strength k_o and length ℓ.

A subset of general sector magnets are pure *dipole sector magnets*, where we eliminate the focusing by setting $k_o = 0$ and get the pure dipole strength $K = 1/\rho_o^2 > 0$. The transformation matrix for a pure sector magnet of length ℓ and bending angle $\theta = \ell/\rho_o$ in the deflecting plane becomes from (5.41)

$$\boldsymbol{M}_{\mathrm{s,\rho}}(\ell|0) = \begin{bmatrix} \cos\theta & \rho_o \sin\theta \\ -\frac{1}{\rho_o}\sin\theta & \cos\theta \end{bmatrix}. \tag{5.43}$$

If we also let $1/\rho_o \to 0$ we arrive at the transformation matrix of a sector magnet in the nondeflecting plane

$$\mathcal{M}_{s,o}(\ell\,|\,0) = \begin{bmatrix} 1 & \ell \\ 0 & 1 \end{bmatrix},$$ (5.44)

which has the form of a drift space. A pure dipole sector magnet therefore behaves in the non-deflecting plane just like a drift space of length ℓ.

Fringe Field Effects: The results obtained above are those for a *hard edge model* and do not reflect modifications caused by the finite extend of the fringe fields. The hard edge model is again an idealization and for a real dipole we consider the gradual transition of the field from the maximum value to zero outside the magnet. The extend of the dipole fringe field is typically about equal to the *gap height* or distance between the magnet poles.

We assume magnet poles which are very wide compared to the gap height and therefore transverse field components in the deflecting plane, here B_x can be neglected. At the entrance into a magnet the vertical field component B_y increases gradually from the field free region to the maximum value in the middle of the magnet (Fig. 5.17). We will now derive the effects on the particle dynamics caused by this fringe field and compare it with the results for a hard edge model magnet.

For the following discussion we consider both a fixed orthogonal cartesian coordinate system (u, v, w) as well as a moving curvilinear system (x, y, s). The origin of the fixed coordinate system is placed at the point P_o where the field starts to rise, Fig. 5.17. At this point both coordinate systems coincide. The horizontal field component vanishes for reasons of symmetry

$$B_u = 0$$ (5.45)

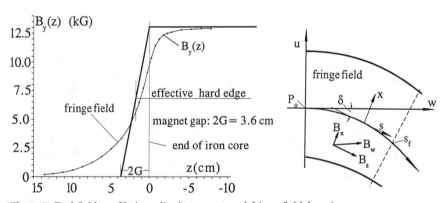

Fig. 5.17. End field profile in a dipole magnet and fringe field focusing

and the vertical field component in the fringe region may be described by

$$B_v = F(w). \tag{5.46}$$

With Maxwell's curl equation $\partial B_w/\partial v - \partial B_v/\partial w = 0$ we get after integration the longitudinal field component $B_w = \int(\partial B_v/\partial w)\,dv$ or

$$B_w = y\,\frac{\partial F(w)}{\partial w}, \tag{5.47}$$

where $y = v$ and where a linear fringe field was assumed with $\partial F(w)/\partial w =$ const.

These field components must be expressed in the curvilinear coordinate system (x, y, s). At the point s within the fringe field the longitudinal field component B_w can be split into B_x and B_s as shown in Fig. 5.17. The horizontal field component is then $B_x = B_w \sin\delta$ where δ is the deflection angle at the point s defined by $\delta = \frac{e}{cp_o} \int\limits_o^s F(\bar{s})\,d\bar{s}$. With

$$B_w = y\,\frac{\partial F(w)}{\partial w} = y\,\frac{\partial F(w)}{\partial s}\,\frac{ds}{dw} \approx y\,\frac{\partial F(s)}{\partial s}\,\frac{1}{\cos\delta} \tag{5.48}$$

we get

$$B_x(s) = y\,F'(s)\,\tan\delta, \tag{5.49}$$

where $F'(s) = dF/ds$. The vertical fringe field component is with $\partial B_x/\partial y - \partial B_y/\partial x = 0$

$$B_y(s) = B_{yo} + x\,F'(s)\,\tan\delta. \tag{5.50}$$

The longitudinal field component is from (5.48) and $B_s = B_w \cos\delta$

$$B_s(s) = y\,F'(s). \tag{5.51}$$

The field components of the fringe field depend linearly on the transverse coordinates and therefore *fringe field focusing* [5.9] must be expected. With the definition of the focal length from (5.3) we get

$$\frac{1}{f} = \int\limits_0^{s_f} K(\bar{s})\,d\bar{s}, \tag{5.52}$$

where $K(s)$ is the focusing strength parameter $K(s) = \kappa^2(s) + k(s)$. In the deflecting plane the fringe field focusing is with $k(s) = (e/cp_o)\,B_y(s)$ and (5.50)

$$\frac{1}{f_x} = \int\limits_0^{s_f} (\kappa'\,\tan\delta + \kappa^2)\,d\bar{s}, \tag{5.53}$$

where we have set $\kappa(s) = (e/cp_o) F(s)$. For small deflection angles δ in the fringe field $\tan \delta \approx \delta = \int_o^{s_f} \kappa \, d\bar{s}$ and after integration of (5.53) by parts through the full fringe field we get the focal length

$$\frac{1}{f_x} = \kappa_o \, \delta_f \,, \tag{5.54}$$

where $\kappa_o = 1/\rho_o$ is the curvature in the central part of the magnet and δ_f is the total deflection angle in the fringe field region.

This result does not deviate from that of the hard edge model, where for a small deflection angle θ we have from (5.43) $1/f_x \approx \kappa_o \theta$ agreeing with the previous result for the fringe field. We obtain therefore the convenient result that in the deflecting plane of a sector magnet there is no need to correct the focusing because of the finite extend of the fringe field.

Finite Pole Gap: In the vertical plane this situation is different since we expect vertical focusing from (5.49) while there is no focusing in the approximation of a hard edge model. Application of (5.52) to the vertical plane gives with $K(s) = -k(s)$ and (5.49)

$$\frac{1}{f_y} = -\int_o^{s_f} \kappa' \tan \delta \, d\bar{s} \approx -\int_o^{s_f} \kappa'(\bar{s}) \, \delta(\bar{s}) \, d\bar{s} \,. \tag{5.55}$$

The fringe field of a sector magnet therefore leads to a defocusing effect which depends on the particular field profile. We may approximate the fringe field by a linear fit over a distance approximately equal to the pole gap $2G$ which is a good approximation for most real dipole magnets. We neglect the nonlinear part of the fringe field and approximate the slope of the field strength by the linear expression $\kappa' = \kappa_o/2G = \text{const}$. The focal length for the full fringe field of length $s_f = 2G$ is therefore with $\kappa(s) = \kappa' s$ and

$$\delta(s) = \int_o^s \kappa' \bar{s} \, d\bar{s} = \frac{\kappa_o}{4G} s^2 \tag{5.56}$$

for $0 \le s \le s_f$ given by

$$\frac{1}{f_y} = -\int_o^{2G} \kappa' \, \delta(\bar{s}) \, d\bar{s} = -\tfrac{1}{3} \kappa_o^2 \, G = -\tfrac{1}{3} \kappa_o \, \delta_f \,, \tag{5.57}$$

where

$$\delta_f = \delta(s_f) = \kappa_o \, G \,. \tag{5.58}$$

This is the focusing due to the fringe field at the entrance of a sector magnet. At the exit we have the same effect since the sign change of κ' is compensated by the need to integrate now from full field to the field free region which is just opposite to the case in the entrance fringe field. Both end field of a sector magnet provide a small vertical defocusing. We note that this defocusing is quadratic in nature, since $\delta_f \propto \kappa$ and therefore independent of the sign of the deflection.

With these results we may now derive a corrected transformation matrix for a sector magnet by multiplying the hard edge matrix (5.44) on either end with thin length fringe field focusing

$$\begin{bmatrix} 1 & 0 \\ -1/f_y & 1 \end{bmatrix} \begin{bmatrix} 1 & \ell \\ 0 & 1 \end{bmatrix} \begin{bmatrix} 1 & 0 \\ -1/f_y & 1 \end{bmatrix} \tag{5.59}$$

and get with (5.57) and $\theta = \ell/\rho_o$ for the transformation matrix in the vertical, non-deflecting plane of a sector magnet instead of (5.44)

$$\mathcal{M}_{s,o}(\ell\,|\,0) = \begin{bmatrix} 1 + \frac{1}{3}\theta\,\delta_f & \ell \\ \frac{2}{3}\frac{\delta_f}{\rho_o} - \frac{1}{9}\frac{\ell}{\rho_o^2}\delta_f^2 & 1 + \frac{1}{3}\theta\,\delta_f \end{bmatrix}. \tag{5.60}$$

The second order term in the M_{21}-matrix element can be ignored for practical purposes but is essential to keep the determinant equal to unity.

5.3.2 Wedge Magnets

In a more general case compared to a sector magnet we will allow the reference path of the particle beam to enter and exit the magnet at an arbitrary angle with the pole face. Magnets with arbitrary *pole face rotation* angles are called *wedge magnets*. Fig. 5.18 shows such a case and we will derive the transformation matrices for wedge magnets. First we note that the fringe field effect is not different from the previous case of a sector magnet except that now the angle $\delta(s)$ must be replaced by a new angle $\eta + \delta(s)$ where the *pole rotation* angle η and the sign convention is defined in Fig. 5.18.

Different from the case of a sector magnet we cannot replace the tangent in (5.49) by its argument since the angle η can be large to prohibit such

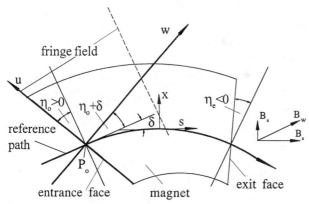

Fig. 5.18. Fringe field focusing in wedge magnets

an approximation. As a further consequence of a large value for η we must take into account the actual path length in the fringe field. To calculate the focal length f_x we have instead of (5.53)

$$\frac{1}{f_x} = \int_0^{s_f} \left[\kappa' \tan(\eta + \delta) + \kappa^2 \right] d\bar{s} . \tag{5.61}$$

Expanding for small angles $\delta \ll 1$ we get $\tan(\eta + \delta) \approx \tan\eta + \delta$. This approximation is true only as long as $\delta \tan\eta \ll 1$ or for entrance angles η not too close to 90° degrees. The argument in the integral (5.61) becomes $\kappa' \tan\eta + \kappa' \delta + \kappa^2$. In addition to the terms for a sector magnet a new term, $\kappa' \tan\eta$ appears and the focal length of the fringe field is

$$\frac{1}{f_x} = \int_0^{s_f} \kappa' \tan\eta \, d\bar{s} + \kappa_o \, \delta_f = \kappa_o \tan\eta + \kappa_o \, \delta_f . \tag{5.62}$$

The integral is to be evaluated through the whole fringe field. Since to first order the path length through the fringe field is

$$s_f = \frac{2G}{\cos\eta} , \tag{5.63}$$

where $2G$ is the pole *gap height*, we have

$$\delta_f = \int_0^{2G/\cos\eta} \kappa \, ds . \tag{5.64}$$

The term $\kappa_o \, \delta_f$ describes again the well-known focusing of a sector magnet in the deflecting plane while the term $\kappa_o \tan\eta$ provides the correction necessary for non-normal entry of the beam path into the magnet. For the case shown in Fig. 5.18, where $\eta > 0$ we obtain beam focusing in the deflecting plane from the fringe field. Similarly, we get a focusing or defocusing effect at the exit fringe field depending on the sign of the pole rotation. The complete transformation matrix of a wedge magnet in the horizontal deflecting plane is obtained by multiplying the matrix of a sector magnet with thin lens matrices to take account of edge focusing. For generality, however, we must assume that the entrance and the exit angle may be different. We will therefore distinguish between the edge focusing for the entrance angle $\eta = \eta_o$ and that for the exit angle $\eta = \eta_e$ and get for the transformation matrix in the deflecting plane

$$\boldsymbol{M}_{w,\rho}(\ell \,|\, 0) = \begin{bmatrix} 1 & 0 \\ -\frac{\tan\eta_e}{\rho_o} & 1 \end{bmatrix} \begin{bmatrix} \cos\theta & \rho_o \sin\theta \\ -\frac{\sin\theta}{\rho_o} & \cos\theta \end{bmatrix} \begin{bmatrix} 1 & 0 \\ -\frac{\tan\eta_o}{\rho_o} & 1 \end{bmatrix} . \tag{5.65}$$

In the vertical plane we have similar to (5.55)

$$\frac{1}{f_y} = - \int_0^{s_f} \kappa' \tan(\eta + \delta) \, d\bar{s} \approx -\kappa_o \tan\eta - \int_0^{s_f} \kappa' \delta \, d\bar{s} . \tag{5.66}$$

144

Again we have the additional focusing term which is now focusing in the vertical plane for $\eta < 0$. For a linear fringe field the focal length is in analogy to (5.57)

$$\frac{1}{f_y} = -\kappa_0 \tan\eta - \tfrac{1}{3} \kappa_0 \delta_f \,, \tag{5.67}$$

where

$$\delta_f = \int\limits_0^{2G/\cos\eta} \kappa(s)\,\mathrm{d}s = \kappa' \frac{2G^2}{\cos^3\eta} = \frac{\kappa_0 G}{\cos^2\eta}\,, \tag{5.68}$$

since $\kappa(s) \approx \kappa'\,s$ and $\kappa' = \kappa_0/(G/\cos\eta)$. The complete transformation matrix in the vertical plane for a horizontally deflecting wedge magnet becomes then

$$\boldsymbol{M}_{w,o}(\ell|0) \begin{bmatrix} 1 & 0 \\ \frac{\tan\eta_e + \delta_{f_e}/3}{\rho_0} & 1 \end{bmatrix} \begin{bmatrix} 1 & \ell \\ 0 & 1 \end{bmatrix} \begin{bmatrix} 1 & 0 \\ \frac{\tan\eta_o + \delta_{f_o}/3}{\rho_0} & 1 \end{bmatrix}. \tag{5.69}$$

Equations (5.65,69) are transformation matrices for bending magnets with arbitrary entrance and exit angles η_0 and η_e respectively. We note specifically that the transformation in the nondeflecting plane becomes different from a simple drift space and find a focusing effect due to the magnet fringe fields which depends on the entrance and exit angles between particle trajectory and pole face.

This general derivation of the focusing properties of a wedge magnet must be taken with caution, where the pole face rotations are very large. In spite of the finite pole rotation angles we have assumed that the particles enter the fringe field at the same location s along the beam line independent of the transverse particle amplitude x. Similarly, the path length of the trajectory in such a wedge magnet depends on the particle amplitude x and slope x'. Obviously these are second order effects but may become significant in special cases.

5.3.3 Rectangular Magnet

A particular case of a symmetric wedge magnet is the *rectangular magnet* which has parallel end faces. If we install this magnet symmetrically about the intended particle trajectory the entrance and exit angles equal to half the bending angle as shown in Fig. 5.19.

For a deflection angle θ we have $\eta_0 = \eta_e = -\theta/2$ and the transformation matrix in the deflecting plane is from (5.65)

$$
\begin{aligned}
\boldsymbol{M}_{r,\rho}(\ell|0) &= \begin{bmatrix} 1 & 0 \\ \frac{\tan\theta/2}{\rho_0} & 1 \end{bmatrix} \begin{bmatrix} \cos\theta & \rho_0 \sin\theta \\ -\frac{\sin\theta}{\rho_0} & \cos\theta \end{bmatrix} \begin{bmatrix} 1 & 0 \\ \frac{\tan\theta/2}{\rho_0} & 1 \end{bmatrix} \\
&= \begin{bmatrix} 1 & \rho_0 \sin\theta \\ 0 & 1 \end{bmatrix}.
\end{aligned}
\tag{5.70}
$$

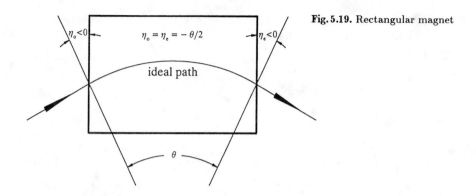

Fig. 5.19. Rectangular magnet

A rectangular dipole magnet transforms in the deflecting plane like a drift space of length $\rho_o \sin \theta$ and does not focus the beam. Note that the "magnet length" ℓ defined by the deflection angle $\theta = \ell / \rho_o$ is the arc length and is related to the straight magnet length L by

$$L = 2\rho_o \sin \frac{\theta}{2} = 2\rho_o \sin \frac{\ell}{2\rho_o}. \tag{5.71}$$

In the vertical plane we observe a focusing with the focal length

$$\frac{1}{f_y} = +\frac{1}{\rho_o} \left(\tan \frac{\theta}{2} - \frac{\delta_{\theta/2}}{3} \right). \tag{5.72}$$

From (5.68) we get $\delta_{\theta/2} = G/[\rho_o \cos(\theta/2)]$ and with (5.71) this becomes $\delta_{\theta/2} = 2G \tan(\theta/2)/L$. Inserting this in (5.72) we obtain for the transformation matrix of a rectangular bending magnet in the nondeflecting plane

$$\mathcal{M}_{\mathrm{r,o}}(\ell|0) = \begin{bmatrix} 1 & 0 \\ -\frac{1}{f_y} & 1 \end{bmatrix} \begin{bmatrix} 1 & \ell \\ 0 & 1 \end{bmatrix} \begin{bmatrix} 1 & 0 \\ -\frac{1}{f_y} & 1 \end{bmatrix}$$

$$= \begin{bmatrix} 1 - \frac{\ell}{f_y} & \ell \\ -\frac{2}{f_y} + \frac{\ell}{f_y^2} & 1 - \frac{\ell}{f_y} \end{bmatrix}, \tag{5.73}$$

where

$$\frac{1}{f_y} = \frac{1}{\rho_o} \tan(\theta/2) \left(1 - \frac{2G}{3L} \right). \tag{5.74}$$

In a rectangular dipole magnet we find just the opposite edge focusing properties compared to a sector magnet. The focusing in the deflecting plane of a sector magnet has shifted to the vertical plane in a rectangular magnet and focusing is completely eliminated in the deflecting plane. Because of the finite extend of the fringe field, however, the focusing strength is reduced by

146

the fraction $2G/(3L)$ where $2G$ is the gap height and L the straight magnet length.

5.4 Particle Beams and Phase Space

The solution of the linear equations of motion allows us to follow a single charged particle through an arbitrary array of magnetic elements. Often, however, it is necessary to consider a beam of many particles and it would be impractical to calculate the trajectory for every individual particle. We, therefore, look for a representation of the particle beam by its boundaries or envelopes.

To learn more about the collective motion of particles we observe their dynamics in phase space. Each particle at any point along a beam transport line is represented by a point in the six-dimensional phase space with coordinates $(x, p_x, y, p_y, \sigma, E)$ where $p_x \approx p_0\, x'$ and $p_y \approx p_0\, y'$ are the transverse momenta with $cp_0 = \beta\, E_0$ the ideal particle energy σ the coordinate along the trajectory, and E the particle energy. Instead of the energy E often also the momentum or the energy deviation from the ideal energy $\Delta E = E - E_0$ or the relative energy deviation $\Delta E/E_0$ may be used. Similarly, in systems, where the beam energy stays constant, we use instead of the transverse momenta rather the slope of the trajectories x', y' which are proportional to the transverse momenta and are generally very small that we may set $\sin x' \approx x'$.

The coupling between the horizontal and vertical plane is being ignored in linear beam dynamics or treated as a perturbation as is the coupling between transverse and longitudinal motion. Only the effect of energy errors on the trajectory will be treated in this approximation. In this paragraph, however, we set $\Delta E = 0$ and represent the beam by its particle distribution in the horizontal x, x' or vertical y, y' phase space separately. Because of the absence of coupling between degrees of freedom in this approximation we may split the six-dimensional phase space into three independent two-dimensional *phase planes*.

5.4.1 Beam Emittance

Particles in a beam occupy a certain region in phase space which is called the *beam emittance* and we define three independent two-dimensional beam emittances. Their numerical values multiplied by π are equal to the area occupied by the beam in the respective phase plane. The beam emittance is a measure of the "transverse or longitudinal temperature" of the beam and depends on the source characteristics of a beam or on other effects like quantized emission of photons into synchrotron radiation and its related excitation and damping effects.

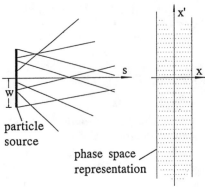

Fig. 5.20. Beam emittance of a diffuse source

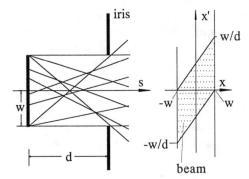

beam
emittance   **Fig. 5.21.** Beam emittance behind the iris

A simple example of a beam emittance and its boundaries is shown in Figs. 5.20,21. There we assume a diffuse source, where the particles can emerge from any point within a distance w from the beam axis and where the direction of the particle trajectories can be anywhere within zero and 90 degrees with respect to the surface of the source. The proper representation in phase space of this beam at the surface of the source is shown in Fig. 5.20. All particles are contained in a narrow strip within the boundaries $x_{\max} = \pm w$ but with a large distribution of transverse momenta ($p_x = p_0 \sin x'$).

Any real beam emerging from its source like in Fig. 5.20 will be clipped by aperture limitations of the vacuum chamber. We assume a simple iris as the aperture limitation located at a distance d from the source and an opening with a radius of $R = w$. The fact that we choose the iris aperture to be the same as the size of the source is made only to simplify the arguments. Obviously many particles emerging from the source will be absorbed at the iris. The part of the beam which passes the iris occupies a phase space area at the exit of the iris like that shown in Fig. 5.21. Among all particles emerging from the source with an amplitude $x = \pm w$ only those will pass the iris for which the slope of the trajectory is between $x' = 0$ and $x' = \mp 2w/\ell$.

Similarly, particles from other points of the source surface will pass the iris only if they emerge from the source within a narrow angle as shown in Fig. 5.21. This beam now has a very specific beam emittance as determined by the iris aperture.

The concept of describing a particle beam in phase space will become very powerful in beam dynamics since we can prove that the density of particles in phase space does not change along a beam transport line, where the forces acting on particles can be derived from macroscopic electric and magnetic fields. In other words particles that are within a closed boundary in phase space at one point of the beam line stay within that boundary. This is Liouville's theorem which we will prove for the fields used in beam dynamics.

5.4.2 Liouville's Theorem

In the previous chapter we have learned to follow individual particles through an arbitrary beam transport line made up of drift spaces, dipole and quadrupole magnets. Since this is true for any particle with known initial parameters in phase space (x, x', y, y') it is in principle possible to calculate trajectories along a beam line for a large number of particles forming a particle beam. Obviously, this is impractical, specifically when the initial coordinates for each particle in phase space are not well known. We are therefore looking for mathematical methods to describe the beam as a whole without concentrating on individual particle trajectories. To this end we make use of methods in statistical mechanics describing the evolution of a large number of particles forming a particle beam.

Liouville's theorem is of specific importance in this respect and we will use it extensively to describe the properties of a particle beam as a whole. This theorem states that under the influence of conservative forces the density of the particles in phase space stays constant. This is specifically true for a system of particles for which the motion in time can be described by contact transformations of canonical coordinates in phase space. Since (5.2) is equivalent to the equation of a free harmonic oscillator we know that the motion of many particles in phase space follow Liousville's theorem. A more direct proof of the validity of Liouville's theorem in particle beam dynamics can be obtained by observing the time evolution of an element in the six-dimensional phase space as we will do in the rest of this paragraph [5.5]. If Ψ is the particle density in phase space, the number of particles within a six-dimensional, infinitesimal element is

$$\Psi(x, y, z, p_x, p_y, p_z)\, \mathrm{d}x\, \mathrm{d}y\, \mathrm{d}z\, \mathrm{d}p_x\, \mathrm{d}p_y\, \mathrm{d}p_z \,. \tag{5.75}$$

The phase space current created by the motion of these particles is

$$\mathbf{j} = (\Psi\, \dot{x}, \Psi\, \dot{y}, \Psi\, \dot{z}, \Psi\, \dot{p}_x, \Psi\, \dot{p}_y, \Psi\, \dot{p}_z), \tag{5.76}$$

where the time derivatives are to be taken with respect to a time τ measured along the trajectory of the phase space element. This time is to be distinguished from the reference time t along the reference orbit in the same way as we distinguish between the coordinates σ and s. We set therefore $\dot{x} = dx/d\tau$, etc. The phase space current must satisfy the continuity equation

$$\nabla \mathbf{j} + \frac{\partial \Psi}{\partial \tau} = 0. \tag{5.77}$$

From this we get with (5.76) and the assumption that the particle location does not depend on its momentum and vice versa

$$\begin{aligned}
-\frac{\partial \Psi}{\partial \tau} &= \nabla_r \left(\Psi \, \dot{\mathbf{r}} \right) + \nabla_p \left(\Psi \, \dot{\mathbf{p}} \right) \\
&= \dot{\mathbf{r}} \nabla_r \Psi + \Psi \left(\nabla_r \, \dot{\mathbf{r}} \right) + \dot{\mathbf{p}} \nabla_p \Psi + \Psi \left(\nabla_p \, \dot{\mathbf{p}} \right),
\end{aligned} \tag{5.78}$$

where ∇_r and ∇_p are the derivatives with respect to (x, y, z) and (p_x, p_y, p_z), respectively. The time derivative of the space vector $\mathbf{r}$

$$\frac{\dot{\mathbf{r}}}{c} = \frac{c\mathbf{p}}{\sqrt{c^2 p^2 + m^2 c^4}},$$

does not depend on the location $\mathbf{r}$ and we have therefore

$$\nabla_r \, \dot{\mathbf{r}} = 0. \tag{5.79}$$

From the Lorentz force equation we get

$$\nabla_p \, \dot{\mathbf{p}} = \frac{e}{c} \nabla_p \left[\dot{\mathbf{r}} \times \mathbf{B} \right] = \frac{e}{c} \mathbf{B} \left(\nabla_p \times \dot{\mathbf{r}} \right) - \frac{e}{c} \dot{\mathbf{r}} \left(\nabla_p \times \mathbf{B} \right). \tag{5.80}$$

The magnetic field $\mathbf{B}$ does not depend on the particle momentum $\mathbf{p}$ and therefore the second term on the right hand side of (5.80) vanishes. For the first term we find $\nabla_p \times \dot{\mathbf{r}} = 0$ because

$$(\nabla_p \times \dot{\mathbf{r}})_x = \frac{\partial \dot{z}}{\partial p_y} - \frac{\partial \dot{y}}{\partial p_z}$$

and

$$\frac{\partial \dot{z}}{\partial p_y} = \frac{\partial}{\partial p_y} \frac{c p_z}{(p^2 + m^2 c^2)^{1/2}} = \frac{c \, p_y \, p_z}{(p^2 + m^2 c^2)^{3/2}} = \frac{\partial \dot{y}}{\partial p_z}, \tag{5.81}$$

where we have used $p^2 = p_x^2 + p_y^2 + p_z^2$. We get a similar result for the other components and have finally for (5.80)

$$\nabla_p \, \dot{\mathbf{p}} = 0. \tag{5.82}$$

With these results we find from (5.78) the total time derivative of the phase space density Ψ to vanish

$$\frac{\partial \Psi}{\partial \tau} + \nabla_r \Psi \, \dot{\mathbf{r}} + \nabla_p \Psi \, \dot{\mathbf{p}} = \frac{d\Psi}{d\tau} = 0, \tag{5.83}$$

which proves the invariance of the phase space density.

Independent from general principles of classical mechanics we have shown the validity of Liouville's theorem for the motion of charged particles under the influence of Lorentz forces. This is obviously also true for that part of the Lorentz force that derives from an electrical field since

$$\nabla_p \dot{\mathbf{p}} = e\nabla_p \mathbf{E} = 0$$

because the electric field $\mathbf{E}$ does not depend on the particle momentum.

The same result can be derived in a different way from the property of the Wronskian in particle beam dynamics. For that we assume that the unit vectors $\mathbf{e}_1, \mathbf{e}_2 \ldots, \mathbf{e}_6$ form a six-dimensional, orthogonal coordinate system. The determinant formed by the components of the six vectors $\mathbf{x}_1, \mathbf{x}_2, \ldots, \mathbf{x}_6$ in this system is equal to the volume of the six-dimensional polygon defined by the vectors $\mathbf{x}_i$. The components of the vectors $\mathbf{x}_i$ with respect to the base vectors $\mathbf{e}_j$ are x_{ij} and the determinant is

$$D = \begin{vmatrix} x_{11} & x_{12} & x_{13} & x_{14} & x_{15} & x_{16} \\ x_{21} & x_{22} & x_{23} & \cdots & \cdots & \cdots \\ x_{31} & x_{32} & \cdots & \cdots & \cdots & \cdots \\ x_{41} & \cdots & \cdots & \cdots & \cdots & \cdots \\ x_{51} & \cdots & \cdots & \cdots & \cdots & \cdots \\ x_{61} & \cdots & \cdots & \cdots & \cdots & x_{66} \end{vmatrix} = |\mathbf{x}_1, \mathbf{x}_2, \mathbf{x}_3, \mathbf{x}_4, \mathbf{x}_5, \mathbf{x}_6|. \tag{5.84}$$

We will derive the transformation characteristics of this determinant considering a transformation

$$\mathbf{y}_i = \boldsymbol{M} \mathbf{x}_j, \tag{5.85}$$

where $\boldsymbol{M} = (a_{ij})$ and the determinant (5.84) then transforms like

$$|\mathbf{y}_1, \mathbf{y}_2 \ldots, \mathbf{y}_6| = \begin{vmatrix} \sum_{j_1=1}^{6} a_{1j_1} \mathbf{x}_{j_1}, & \sum_{j_2=1}^{6} a_{2j_2} \mathbf{x}_{j_2}, \ldots & \sum_{j_6=1}^{6} a_{6j_6} \mathbf{x}_{j_6} \end{vmatrix}$$

$$= \sum_{\substack{j_1=1 \\ j_2=1 \\ \vdots \\ j_6=1}}^{6} a_{1j_1} a_{2j_2} \ldots a_{6j_6} |\mathbf{x}_{j_1}, \mathbf{x}_{j_2}, \ldots \mathbf{x}_{j_6}|. \tag{5.86}$$

The determinant $|\mathbf{x}_{j_1}, \mathbf{x}_{j_2}, \ldots \mathbf{x}_{j_6}|$ is equal to zero if two or more of the indices j_i are equal and further the determinant changes sign if two indices are interchanged. These rules lead to

$$|\mathbf{y}_1, \mathbf{y}_2, \ldots, \mathbf{y}_6| = \sum_{j_i=1}^{6} \epsilon_{j_1 j_2 \ldots j_6}\, a_{1j_1}\, a_{2j_2} \ldots a_{6j_6}\, |\mathbf{x}_1, \mathbf{x}_2, \ldots, \mathbf{x}_6|, \quad (5.87)$$

where

$$\epsilon_{j_1, j_2 \ldots j_6} = \begin{cases} +1 & \text{for even permutations of the indices } j_i; \\ -1 & \text{for odd permutations of the indices } j_i; \\ 0 & \text{if any two indices } j_i \text{ are equal} \end{cases} \quad (5.88)$$

The sum $\sum_{j_i=1}^{6} \epsilon_{j_1 j_2 \ldots j_6}\, a_{1j_1}\, a_{2j_2} \ldots a_{6j_6}$ is just the determinant of the transformation matrix $\mathcal{M}$ and finally we get

$$|\mathbf{y}_1, \mathbf{y}_2, \ldots \mathbf{y}_6| = |\mathcal{M}|\, |\mathbf{x}_1, \mathbf{x}_2 \ldots \mathbf{x}_6|. \quad (5.89)$$

For a particle beam transport line, however, we know that $|\mathcal{M}|$ is the Wronskian with

$$W = |\mathcal{M}| = 1. \quad (5.90)$$

If we now identify this six-dimensional space with the six-dimensional phase space we see from (5.89,90) that the phase space under the class of transformation matrices considered in beam dynamics is constant. Conversely, if $W \neq 1$ we get a change in phase space as we will see when we consider acceleration or synchrotron radiation losses.

5.4.3 Transformation in Phase Space

Liouville's theorem provides a powerful tool to describe a beam in phase space. Knowledge of the area occupied by particles in phase space at the beginning of a beam transport line will allow us to determine the location and distribution of the beam at any other place along the transport line without having to calculate the trajectory of every individual particle. Because it is easy to describe analytically an ellipse in phase space, it has become customary to surround all particles of a beam in phase space by an ellipse called the *phase ellipse* (Fig. 5.22) described by

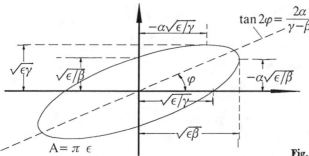

Fig. 5.22. Phase space ellipse

$$\gamma\,x^2 + 2\,\alpha\,x\,x' + \beta\,x'^2 \;=\; \epsilon\,, \tag{5.91}$$

where α, β, γ and ϵ are ellipse parameters. The area enclosed by the ellipse is called the beam emittance ϵ defined by [1]

$$\int_{\text{ellipse}} \mathrm{d}x\,\mathrm{d}x' \;=\; \pi\,\epsilon\,, \tag{5.92}$$

while the parameters α, β and γ determine the shape and orientation of the ellipse.

Since all particles enclosed by the ellipse stay within that ellipse due to Liouville's theorem we only need to know how the ellipse parameters transform along the beam line to be able to describe the whole particle beam. Let the equation

$$\gamma_{\mathrm{o}}\,x_{\mathrm{o}}^2 + 2\alpha_{\mathrm{o}}\,x_{\mathrm{o}}\,x_{\mathrm{o}}' + \beta_{\mathrm{o}}\,x_{\mathrm{o}}'^2 \;=\; \epsilon \tag{5.93}$$

be the equation of the phase ellipse at the starting point $s = 0$ of the beam line. Any particle trajectory transforms from the starting point $s = 0$ to any other point $s \neq 0$ by the transformation

$$\begin{bmatrix} x \\ x' \end{bmatrix} = \begin{bmatrix} C(s) & S(s) \\ C'(s) & S'(s) \end{bmatrix} \begin{bmatrix} x_{\mathrm{o}} \\ x_{\mathrm{o}}' \end{bmatrix} \tag{5.94}$$

and after solving for x_{o} and x_{o}' and inserting into (5.93) we get after sorting coefficients

$$(S'^2\,\gamma_{\mathrm{o}} - 2\,S'\,C'\,\alpha_{\mathrm{o}} + C'^2\,\beta_{\mathrm{o}})\,x^2$$
$$+\,2\,(-S\,S'\,\gamma_{\mathrm{o}} + S'\,C\,\alpha_{\mathrm{o}} + S\,C'\,\alpha_{\mathrm{o}} - C\,C'\,\beta_{\mathrm{o}})\,x\,x' \tag{5.95}$$
$$+\,(S^2\,\gamma_{\mathrm{o}} - 2\,S\,C\,\alpha_{\mathrm{o}} + C^2\,\beta_{\mathrm{o}})\,x'^2 \;=\; \epsilon\,.$$

Equation (5.95) can be brought into the form (5.91) by replacing the coefficients in (5.95) by

$$\gamma = C'^2\,\beta_{\mathrm{o}} - 2S'\,C'\,\alpha_{\mathrm{o}} + S'^2\,\gamma_{\mathrm{o}}\,,$$
$$\alpha = -\,C\,C'\,\beta_{\mathrm{o}} + (S'\,C + S\,C')\,\alpha_{\mathrm{o}} - S\,S'\,\gamma_{\mathrm{o}}\,, \tag{5.96}$$
$$\beta = C^2\,\beta_{\mathrm{o}} - 2S\,C\,\alpha_{\mathrm{o}} + S^2\,\gamma_{\mathrm{o}}\,.$$

The resulting ellipse equation shows still the same area $\pi\,\epsilon$ as we would expect, but due to different parameters, (γ, α, β), the new ellipse has a different orientation and shape. During a transformation along a beam transport line the phase ellipse will therefore continuously change its form and orientation but not its area. In matrix formulation the ellipse parameters which are also called *Twiss parameters* [5.10], transform from (5.96) like

---

[1] The definition of the beam emittance in the literature is not unique. Sometimes the product $\pi\epsilon$ is called the beam emittance.

$$
\begin{bmatrix} \beta \\ \alpha \\ \gamma \end{bmatrix} = \begin{bmatrix} C^2 & -2SC & S^2 \\ -CC' & (S'C + SC') & -SS' \\ C'^2 & -2S'C' & S'^2 \end{bmatrix} \begin{bmatrix} \beta_0 \\ \alpha_0 \\ \gamma_0 \end{bmatrix}. \tag{5.97}
$$

The orientation, eccentricity and area of an ellipse is defined by three parameters, while (5.93) includes four parameters α, β, γ and ϵ. Since the area is defined by ϵ we expect the other three parameters to be correlated. From geometric properties of an ellipse we find that correlation to be

$$
\beta\gamma - \alpha^2 = 1. \tag{5.98}
$$

So far we have used only the (x, x')-phase space, but the results are valid also for the (y, y')-phase space. Equation (5.97) provides the tool to calculate beam parameters anywhere along the beam line from the initial values $\beta_0, \alpha_0, \gamma_0$.

The *phase ellipse* in a drift space, for example, becomes distorted in a clock wise direction without changing the slope of any particle as shown in Fig. 5.23. The angular envelope $A = \sqrt{\epsilon\gamma}$ stays constant. If the drift space is long enough a convergent beam therefore transforms eventually again into a divergent beam. The point s_w at which the beam reaches its minimum size is determined by $\alpha(s_w) = 0$ and we get from (5.97) for the location of a *beam waist* in a drift section.

$$
\ell = s_w - s_0 = \frac{\alpha_0}{\gamma_0}. \tag{5.99}
$$

This point of minimum beam size is up or downstream of $s = s_0$ depending on the sign of α_0 being negative or positive.

Applying these results to a simple drift space of length ℓ we have the transformation

$$
\begin{bmatrix} \beta \\ \alpha \\ \gamma \end{bmatrix} = \begin{bmatrix} 1 & -2\ell & \ell^2 \\ 0 & 1 & -\ell \\ 0 & 0 & 1 \end{bmatrix} \begin{bmatrix} \beta_0 \\ \alpha_0 \\ \gamma_0 \end{bmatrix}, \tag{5.100}
$$

which causes for example a convergent phase ellipse to become a divergent phase ellipse as shown in Fig. 5.23. Particles in the upper half of the phase ellipse move from left to right and particles in the lower half from right to left. During the transition from the convergent to divergent phase ellipse we find an upright ellipse which describes the beam at the location of a waist. The form and orientation of the phase ellipse tells us immediately the characteristics beam behavior. Convergent beams are characterized by a rotated phase ellipse extending from the left upper quadrant to the lower right quadrant while a divergent beam spreads from the left lower to the right upper quadrant. A symmetric phase ellipse signals the location of a waist or symmetry point.

A divergent beam fills, after some distance, the whole vacuum chamber aperture and in order not to lose beam a focusing quadrupole must be

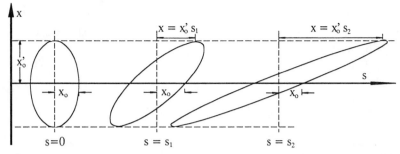

Fig. 5.23. Transformation of a phase space ellipse at different locations along a drift section

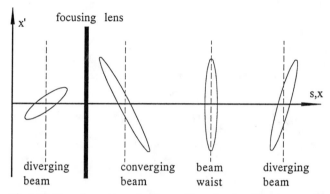

Fig. 5.24. Transformation of a phase ellipse due to a focusing quadrupole. The phase ellipse is shown at different locations along a drift space downstream from the quadrupole.

inserted. During the process of focusing a diverging beam entering a focusing quadrupole reaches a maximum size and then starts to converge again. This transformation, generated by a focusing quadrupole is shown in Fig. 5.24, where we recognize slopes of particle trajectories to reverse signs thus forming a convergent beam again.

After this step the beam may develop as shown for a drift space until the next focusing quadrupole is required. In reality this focusing scenario is complicated by the fact that we need also vertical focusing which requires the insertion of defocusing quadrupoles as well.

Beam Matrix: Particle beams are conveniently described in phase space by enclosing their distribution with ellipses. Transformation rules for such ellipses through a beam transport system have been derived for two-dimensional phase space and we will here expand the discussion of phase space transformations to more dimensions. The equation for an n-dimensional ellipse can be written in the form

$$\mathbf{u}^T \boldsymbol{\sigma}^{-1} \mathbf{u} = 1,$$ (5.101)

where the symmetric matrix $\boldsymbol{\sigma}$ is still to be determined, $\mathbf{u}^T$ is the transpose of the coordinate vector $\mathbf{u}$ defined by

$$\mathbf{u} = \begin{pmatrix} x \\ x' \\ y \\ y' \\ \tau \\ \delta \\ \vdots \end{pmatrix}.$$ (5.102)

The volume of this n-dimensional ellipse is

$$V_n = \frac{\pi^{n/2}}{\Gamma(1 + n/2)} \sqrt{\det \boldsymbol{\sigma}},$$ (5.103)

where Γ is the gamma function. Application of (5.101) to the two-dimensional phase space we get for the ellipse equation

$$\sigma_{11} x^2 + 2 \sigma_{12} x x' + \sigma_{22} x'^2 = 1$$ (5.104)

and comparison with (5.91) defines the *beam matrix* by

$$\boldsymbol{\sigma} = \begin{bmatrix} \sigma_{11} & \sigma_{12} \\ \sigma_{12} & \sigma_{22} \end{bmatrix} = \epsilon \begin{bmatrix} \beta & -\alpha \\ -\alpha & \gamma \end{bmatrix}.$$ (5.105)

This identification of the beam matrix can be expanded to six or arbitrary many dimensions including coupling terms which we have so far neglected. The two-dimensional "volume" or phase space area is

$$V_2 = \pi \sqrt{\det \boldsymbol{\sigma}} = \pi \sqrt{\sigma_{11} \sigma_{22} - \sigma_{12}^2} = \pi \epsilon$$ (5.106)

consistent with the earlier definition of beam emittance, since $\beta\gamma - \alpha^2 = 1$.

Similar to the two-dimensional case, we look for the evolution of the n dimensional phase ellipse along a beam transport line. Let the $n \times n$ transformation matrix from point P_0 to P_1 be $\boldsymbol{M}(P_1|P_2)$ and $\mathbf{u}_1 = \boldsymbol{M}(P_1|P_0) \mathbf{u}_0$. The equation of the phase ellipse at point P_1 is then

$$(\boldsymbol{M}^{-1} \mathbf{u}_1)^T \boldsymbol{\sigma}_0^{-1} (\boldsymbol{M}^{-1} \mathbf{u}_1) = 1$$

or after some manipulations

$$\mathbf{u}_1^T \boldsymbol{\sigma}_1^{-1} \mathbf{u}_1 = 1,$$ (5.107)

where we used the identity

$$(\boldsymbol{M}^T)^{-1} \boldsymbol{\sigma}_0^{-1} \boldsymbol{M}^{-1} = [\boldsymbol{M} \boldsymbol{\sigma}_0 \boldsymbol{M}^T]^{-1}.$$

The beam matrix transforms therefore like

$$\sigma_1 = \mathcal{M}\sigma_0\mathcal{M}^T. \tag{5.108}$$

This formalism will be useful for the experimental determination of beam emittances.

5.4.4 Measurement of the Beam Emittance

The ability to manipulate in a controlled and measurable way the orientation and form of the phase ellipse with quadrupoles gives us the tool to experimentally determine the emittance of a particle beam. Since the beam emittance is a measure of both the beam size and beam divergence, we cannot directly measure its value. While we are able to measure the beam size with the use of a fluorescent screen, for example, the beam divergence cannot be measured directly. If, however, the beam size is measured at different locations or under different focusing conditions such that different parts of the ellipse will be probed by the beam size monitor, the beam emittance can be determined.

Utilizing the definition of the beam matrix in (5.105) we have

$$\sigma_{11}\,\sigma_{22} - \sigma_{12}^2 = \epsilon^2 \tag{5.109}$$

and the beam emittance can be measured, if we find a way to determine the beam matrix. To determine the beam matrix at point, P_0 we consider downstream from P_0 a beam transport line with some quadrupoles and beam size monitors like fluorescent screens at three places P_1 to P_3. From (5.97) and (5.105) we get for the beam sizes at locations P_i three relations of the form

$$\sigma_{i,11} = C_i^2\,\sigma_{11} + 2S_iC_i\,\sigma_{12} + S_i^2\,\sigma_{22}$$

which we may express in matrix formulation by

$$\begin{bmatrix} \sigma_{1,11} \\ \sigma_{2,11} \\ \sigma_{3,11} \end{bmatrix} = \begin{bmatrix} C_1^2 & 2C_1S_1 & S_1^2 \\ C_2^2 & 2C_2S_2 & S_2^2 \\ C_3^2 & 2C_3S_3 & S_3^2 \end{bmatrix} \begin{bmatrix} \sigma_{11} \\ \sigma_{12} \\ \sigma_{22} \end{bmatrix} = \mathcal{M}_\sigma \begin{bmatrix} \sigma_{11} \\ \sigma_{12} \\ \sigma_{22} \end{bmatrix}, \tag{5.110}$$

where C_i and S_i are elements of the transformation matrix from point P_0 to P_i and $\sigma_{i,jk}$ are elements of the beam matrix at P_i. Equation (5.110) can be solved for the beam matrix elements σ_{ij} at P_0

$$\begin{bmatrix} \sigma_{11} \\ \sigma_{12} \\ \sigma_{22} \end{bmatrix} = (\mathcal{M}_\sigma^T\,\mathcal{M}_\sigma)^{-1}\,\mathcal{M}_\sigma^T \begin{bmatrix} \sigma_{1,11} \\ \sigma_{2,11} \\ \sigma_{3,11} \end{bmatrix}, \tag{5.111}$$

where the matrix $\mathcal{M}_\sigma$ is known from the parameters of the beam transport line between P_o and P_i and $\mathbf{M}_\sigma^{\mathrm{T}}$ is the transpose of it. The solution vector can be used in (5.109) to calculate finally the beam emittance.

This procedure to measure the beam emittance is straight forward but requires three beam size monitors at appropriate locations such that the measurements can be conducted with the desired resolution. A much simpler procedure makes use of only one beam size monitor at P_1 and one quadrupole between P_o and P_1. We vary the strength of the quadrupole and measure the beam size at P_1 as a function of the quadrupole strength. These beam size measurements as a function of quadrupole strength are equivalent to the measurements at different locations discussed above and we can express the results of n measurements by the matrix equation

$$
\begin{bmatrix} \sigma_{1,11} \\ \sigma_{2,11} \\ \vdots \\ \sigma_{n,11} \end{bmatrix} = \begin{bmatrix} C_1^2 & 2C_1S_1 & S_1^2 \\ C_2^2 & 2C_2S_2 & S_2^2 \\ \vdots & \vdots & \vdots \\ C_n^2 & 2C_nS_n & S_n^2 \end{bmatrix} \begin{bmatrix} \sigma_{11} \\ \sigma_{12} \\ \sigma_{22} \end{bmatrix} = \mathbf{M}_{\sigma,n} \begin{bmatrix} \sigma_{11} \\ \sigma_{12} \\ \sigma_{22} \end{bmatrix} . \tag{5.112}
$$

We obtain the solution again like in (5.111) by simple matrix multiplications

$$
\begin{bmatrix} \sigma_{11} \\ \sigma_{12} \\ \sigma_{22} \end{bmatrix} = (\mathbf{M}_{\sigma,n}^{T} \mathbf{M}_{\sigma,n})^{-1} \mathbf{M}_{\sigma,n}^{T} \begin{bmatrix} \sigma_{1,11} \\ \sigma_{2,11} \\ \vdots \\ \sigma_{n,11} \end{bmatrix} . \tag{5.113}
$$

An experimental procedure has been derived which allows us to determine the beam emittance through measurements of beam sizes as a function of focusing. Practically the evaluation of (5.113) is performed by measuring the beam size $\sigma_{1,11}(k)$ at P_1 as a function of the quadrupole strength k and comparing the results with the theoretical expectation

$$
\sigma_{1,11}(k) = C^2(k)\sigma_{11} + 2C(k)S(k)\sigma_{12} + S^2(k)\sigma_{22} . \tag{5.114}
$$

By fitting the parameters σ_{11}, σ_{12} and σ_{22} to match the measured curve one can determine the beam emittance from (5.109). However, this procedure does not guarantee automatically a measurement with the desired precision. To accurately fit three parameters we must be able to vary the beam size considerably such that the nonlinear variation of the beam size with quadrupole strength becomes quantitatively significant. An analysis of measurement errors indicates that the beam size at P_o should be large, preferable divergent. In this case variation of the quadrupole strength will dramatically change the beam size at P_1 from a large value when the quadrupole is off to a narrow focal point and again to a large beam size by over focusing.

5.5 Betatron Functions

The trajectory of a particle through an arbitrary beam transport system can be determined by repeated multiplication of transformation matrices through each of the individual elements of the beam line. This method is convenient especially for computations on a computer but it does not reveal characteristic properties of particle trajectories. For deeper insight we attempt to solve the equation of motion analytically. The differential equation of motion is

$$u'' + k(s)u = 0, \qquad (5.115)$$

where u stands for x or y and $k(s)$ is an arbitrary function of s resembling the particular distribution of focusing along a beam line. For a general solution of (5.115) we apply the method of variation of integration constants and use an ansatz with an s-dependent amplitude and phase

$$u(s) = \sqrt{\epsilon}\,\sqrt{\beta(s)}\,\cos[\psi(s) - \psi_o], \qquad (5.116)$$

which is similar to the solution of a harmonic oscillator equation with a constant coefficient k. The quantities ϵ and ψ_o are integration constants. From (5.116) we form the first and second derivatives with the understanding that $\beta = \beta(s), \psi = \psi(s)$ etc.

$$u' = \sqrt{\epsilon}\,\frac{\beta'}{2\sqrt{\beta}}\,\cos(\psi - \psi_o) - \sqrt{\epsilon}\,\sqrt{\beta}\,\sin(\psi - \psi_o)\,\psi',$$

$$u'' = \sqrt{\epsilon}\,\frac{\beta\,\beta'' - 1/2\,\beta'^2}{2\,\beta^{3/2}}\,\cos(\psi - \psi_o) - \sqrt{\epsilon}\,\frac{\beta'}{\sqrt{\beta}}\,\sin(\psi - \psi_o)\,\psi' \qquad (5.117)$$

$$- \sqrt{\epsilon}\,\sqrt{\beta}\,\sin(\psi - \psi_o)\,\psi'' - \sqrt{\epsilon}\,\sqrt{\beta}\,\cos(\psi - \psi_o)\,\psi'^2,$$

and insert into (5.115). The sum of all coefficients of the sine and cosine terms respectively must vanish separately to make the ansatz (5.116) valid for all phases ψ. From this we get the two conditions:

$$\tfrac{1}{2}\left(\beta\,\beta'' - \tfrac{1}{2}\,\beta'^2\right) - \beta^2\,\psi'^2 + \beta^2\,k = 0 \qquad (5.118)$$

and

$$\beta'\,\psi' + \beta\,\psi'' = 0. \qquad (5.119)$$

Equation (5.119) can be integrated immediately since $\beta'\,\psi + \beta\,\psi'' = (\beta\,\psi')'$ giving

$$\beta\,\psi' = \text{const} = 1, \qquad (5.120)$$

where a specific normalization of the phase function has been chosen by selecting the integration constant to be equal to unity. From (5.120) we get for the *phase function*

$$\psi(s) = \int_0^s \frac{d\bar{s}}{\beta(\bar{s})} + \psi_0. \tag{5.121}$$

Knowledge of the function $\beta(s)$ along the beam line obviously allows us to compute the phase function. We insert (5.120) into (5.118) and get the differential equation for the function $\beta(s)$

$$\tfrac{1}{2} \beta \beta'' - \tfrac{1}{4} \beta'^2 + \beta^2 k = 1 \tag{5.122}$$

or with $\alpha = -\tfrac{1}{2}\beta'$ and $\gamma = (1+\alpha^2)/\beta$ (5.122) becomes

$$\beta'' + 2k\beta - 2\gamma = 0. \tag{5.123}$$

With $\alpha' - \tfrac{1}{2}\beta''$ this is equivalent to

$$\alpha' = k\beta - \gamma. \tag{5.124}$$

Before we solve the differential equation (5.123)we try to determine the physical nature of the functions $\beta(s)$, $\alpha(s)$, and $\gamma(s)$. To do that we note first that any solution that satisfies (5.123)together with the phase function $\psi(s)$ can be used to make (5.116) a real solution of the equation of motion (5.115). From that solution

$$u(s) = \sqrt{\epsilon}\,\sqrt{\beta}\,\cos(\psi - \psi_0)$$

and the derivative

$$u'(s) = -\sqrt{\epsilon}\,\frac{\alpha}{\sqrt{\beta}}\,\cos(\psi - \psi_0) - \frac{\sqrt{\epsilon}}{\sqrt{\beta}}\,\sin(\psi - \psi_0)$$

we eliminate the phase $(\psi - \psi_0)$ and obtain the constant of motion which is also called the *Courant-Snyder invariant* [5.2]

$$\gamma u^2 + 2\alpha u u' + \beta u'^2 = \epsilon. \tag{5.125}$$

This invariant expression is equal to the equation of an ellipse with the area $\pi\epsilon$ which we have encountered in the previous section and the particular choice of the letters $\beta, \alpha, \gamma, \epsilon$ for the betatron functions and beam emittance becomes obvious. The physical interpretation of this invariant is that of a single particle trajectory in phase space along the contour of an ellipse with the parameters $\beta, \alpha,$ and γ. Since these parameters are functions of s however, the form of the ellipse is changing constantly but, due to Liouville's theorem, any particle starting on that ellipse will stay on it. The choice of an ellipse to describe the evolution of a beam in phase space is thereby more than a mathematical convenience. We may now select a single particle to define a phase ellipse, knowing that all particles with lesser betatron oscillation amplitudes will stay within that ellipse. The description of the ensemble of all particles within a beam have thereby been reduced to that of a single particle.

The ellipse parameter functions β, α, γ and the *phase function* ψ are called the *betatron functions* or *lattice functions* and the oscillatory motion of a particle along the beam line (5.116) is called the *betatron oscillation*. This oscillation is quasi periodic with varying amplitude and frequency. To demonstrate the close relation to the solution of the harmonic oscillator, we use the betatron and phase function to perform a coordinate transformation

$$(u, s) \qquad \longrightarrow \qquad (w, \psi)$$

by setting

$$w(\psi) = \frac{u(s)}{\sqrt{\beta(s)}} \qquad \text{and} \qquad \psi = \int_0^s \frac{d\bar{s}}{\beta(\bar{s})}, \tag{5.126}$$

where $u(s)$ stands for $x(s)$ and $y(s)$ respectively. These coordinates (w, ψ) are called the *normalized coordinates*. In these coordinates the equation of motion (5.115) transforms into

$$\frac{d^2 w}{d\psi^2} + w = 0, \tag{5.127}$$

which indeed is the equation of a harmonic oscillator with frequency one. This identity will be very important for the treatment of perturbing driving terms that appear on the right hand side of (5.127) and we will discuss this transformation in more detail in Sect. 5.5.3. So far we have tacitly assumed that the betatron function $\beta(s)$ never vanishes or changes sign. This can be shown to be true by setting $q(s) = \sqrt{\beta(s)}$ and inserting into (5.122) and we get with $\beta' = 2 q q'$ and $\beta'' = 2(q'^2 + q q'')$ the differential equation

$$q'' + k q - \frac{1}{q^3} = 0. \tag{5.128}$$

The term $1/q^3$ prevents a change of sign of $q(s)$. Letting $q > 0$ vary toward zero we get $q'' \approx 1/q^3 > 0$. This curvature being positive will become arbitrarily large and eventually turn the function $q(s)$ around before it reaches zero. Similarly, the function $q(s)$ stays negative along the whole beam line if it is negative at one point. Since the sign of the betatron function is not determined and does not change it has became customary to use only the positive solution.

The beam emittance parameter ϵ appears as an amplitude factor in the equation for the trajectory of an individual particle. This amplitude factor is equal to the beam emittance only for particles traveling on an ellipse that just encloses all particles in the beam. In other words, a particle traveling along a phase ellipse with amplitude $\sqrt{\epsilon}$ defines the emittance of that part of the total beam which is enclosed by this ellipse or for all those particles whose trajectories satisfy

$$\beta u'^2 + 2\alpha u u' + \gamma u^2 \leq \epsilon. \tag{5.129}$$

Since it only leads to confusion to use the letter ϵ as an amplitude factor we will from now on use it only when we want to define the whole beam and set $\sqrt{\epsilon} = a$ for all cases of individual particle trajectories.

5.5.1 Beam Envelope

To describe the beam as a whole a *beam envelope* can be defined. All particles on the beam emittance defining ellipse follow trajectories described by

$$x_i(s) = \sqrt{\epsilon}\sqrt{\beta(s)}\,\cos[\psi(s) + \delta_i]\,,$$

where δ_i is an arbitrary phase constant for the particle i. By selecting at every point along the beam line that particle i for which $\cos[\psi(s) + \delta_i] = \pm 1$ we can construct an *envelope* of the beam containing all particles

$$E(s) = \pm\sqrt{\epsilon}\sqrt{\beta(s)}\,. \tag{5.130}$$

Here the two signs indicate only that there is an envelope on either side of the beam center. We note that the beam envelope is determined by the beam emittance ϵ and the betatron function $\beta(s)$. The beam emittance is a constant of motion and resembles the transverse "temperature" of the beam. The betatron function reflects exterior forces from focusing magnets and is highly dependent on the particular arrangement of quadrupole magnets. It is this dependence of the beam envelope on the focusing structure that lets us design beam transport systems with specific properties like small or large beam sizes at particular points.

5.5.2 Beam Dynamics in Terms of Betatron Functions

Properties of betatron functions can now be used to calculate the parameters of individual particle trajectories anywhere along a beam line. Any particle trajectory can be described by

$$u(s) = a\sqrt{\beta}\,\cos\psi + b\sqrt{\beta}\,\sin\psi \tag{5.131}$$

and the amplitude factors a and b can be determined by setting at $s = 0$

$$\begin{aligned} \psi &= 0\,, & \beta &= \beta_0\,, & u(0) &= u_0\,, \\ & & \alpha &= \alpha_0\,, & u'(0) &= u_0'\,. \end{aligned} \tag{5.132}$$

With these boundary conditions we get

$$\begin{aligned} a &= \frac{u_0}{\sqrt{\beta_0}}\,, \\ b &= \sqrt{\beta_0}\,u_0' + \frac{\alpha_0}{\sqrt{\beta_0}}\,u_0 \end{aligned} \tag{5.133}$$

and after insertion into (5.131) the particle trajectory and its derivative is

$$u(s) = \sqrt{\frac{\beta}{\beta_\mathrm{o}}}\,(\cos\psi + \alpha_\mathrm{o}\,\sin\psi)\,u_\mathrm{o} + \sqrt{\beta\,\beta_\mathrm{o}}\,\sin\psi\,u_\mathrm{o}',$$

$$u'(s) = \frac{1}{\sqrt{\beta_\mathrm{o}\beta}}\,[(\alpha_\mathrm{o} - \alpha)\,\cos\psi - (1 + \alpha\,\alpha_\mathrm{o})\,\sin\psi]\,u_\mathrm{o} \qquad (5.134)$$

$$+ \sqrt{\frac{\beta_\mathrm{o}}{\beta}}\,(\cos\psi - \alpha\,\sin\psi)\,u_\mathrm{o}'.$$

These equations can immediately be expressed in matrix formulation and with (5.94) we get the general transformation matrix

$$\begin{bmatrix} C(s) & S(s) \\ C'(s) & S'(s) \end{bmatrix} =$$

$$\begin{bmatrix} \sqrt{\frac{\beta}{\beta_\mathrm{o}}}\,(\cos\psi + \alpha_\mathrm{o}\,\sin\psi) & \sqrt{\beta\beta_\mathrm{o}}\,\sin\psi \\ \frac{\alpha_\mathrm{o} - \alpha}{\sqrt{\beta\beta_\mathrm{o}}}\,\cos\psi - \frac{1 + \alpha\alpha_\mathrm{o}}{\sqrt{\beta\beta_\mathrm{o}}}\,\sin\psi & \sqrt{\frac{\beta_\mathrm{o}}{\beta}}\,(\cos\psi - \alpha\,\sin\psi) \end{bmatrix}. \qquad (5.135)$$

Knowledge of the betatron functions along a beam line allows us to calculate individual particle trajectories. The betatron functions can be obtained by either solving numerically the differential equation (5.122) or by using the matrix formalism (5.97) to transform phase ellipse parameters. Since the ellipse parameters in (5.97) and the betatron functions are equivalent, we have found a straightforward way to calculate their values anywhere once we have the initial values at the start of the beam line. This method is particularly convenient when using computers to perform the matrix multiplication.

Transformation of the betatron functions becomes particularly simple in a drift space of length s where the transformation matrix is

$$\begin{bmatrix} C & S \\ C' & S' \end{bmatrix} = \begin{bmatrix} 1 & s \\ 0 & 1 \end{bmatrix}. \qquad (5.136)$$

The betatron functions at the point s is from (5.100)

$$\begin{aligned} \beta(s) &= \beta_\mathrm{o} - 2\alpha_\mathrm{o}\,s + \gamma_\mathrm{o}\,s^2, \\ \alpha(s) &= \alpha_\mathrm{o} - \gamma_\mathrm{o}\,s, \qquad\qquad\qquad (5.137) \\ \gamma(s) &= \gamma_\mathrm{o}, \end{aligned}$$

with initial values $\beta_\mathrm{o}, \alpha_\mathrm{o}, \gamma_\mathrm{o}$ taken at the beginning of the drift space.

We note that $\gamma(s) = $ const in a drift space. This result can be derived also from the differential equation (5.123) which for $k = 0$ becomes $\beta'' = 2\gamma$ and the derivative with respect to s is $\beta''' = 2\gamma'$. On the other hand, we calculate from the first equation (5.137) the third derivative of the betatron function with respect to s to be $\beta''' = 0$. Obviously both results are correct only if the γ-function is a constant in a drift space where $k = 0$.

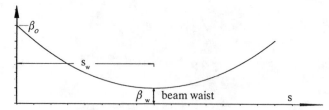

Fig. 5.25. Betatron function in a drift space

The location of a beam waist is defined by $\alpha = 0$ and occurs from (5.137) at $s_\mathrm{w} = \alpha_0/\gamma_0$. The betatron function increases quadratically with the distance from the beam waist, see Fig. 5.25, and can be expressed by

$$\beta(s - s_\mathrm{w}) = \beta_\mathrm{w} + \frac{(s - s_\mathrm{w})^2}{\beta_\mathrm{w}}, \tag{5.138}$$

where β_w is the value of the betatron function at the waist and $s - s_\mathrm{w}$ is the distance from the waist. From (5.138) we note that the magnitude of the betatron function away from the waist reaches large values for both large and small betatron functions at the waist. We may therefore look for conditions to obtain the minimum value for the betatron function anywhere in a drift space of length $2L$. For this we take the derivative of β with respect to β_w and get from $(\mathrm{d}\beta/\mathrm{d}\beta_\mathrm{w} = 0)$

$$\beta_\mathrm{w,opt} = L. \tag{5.139}$$

At either end of the drift space we have then

$$\beta(L) = 2\,\beta_\mathrm{w,opt}. \tag{5.140}$$

This is the optimum solution of the betatron function in a drift space resulting in a minimum aperture requirement. The phase advance in a drift space we calculate from (5.138) starting the integration at the waist $\bar{s} = s - s_\mathrm{w} = 0$ and get

$$\psi(L) = \int_0^L \frac{\mathrm{d}\ell/\beta_\mathrm{w}}{1 + (\ell/\beta_\mathrm{w})^2} = \arctan\frac{L}{\beta_\mathrm{w}} \rightarrow \frac{\pi}{2} \quad \text{for} \quad \frac{L}{\beta_\mathrm{w}} \rightarrow \infty. \tag{5.141}$$

The phase advance through a drift space is therefore never larger than π and actually never quite reaches that value

$$\Delta\psi_\mathrm{drift} \le \pi. \tag{5.142}$$

5.5.3 Beam Dynamics in Normalized Coordinates

The form and nomenclature of the differential equation (5.115) resembles very much that of a harmonic oscillator and indeed this is not accidental since in both cases the restoring force increases linearly with the oscillation amplitude. In particle beam dynamics we find an oscillatory solution with varying amplitude and frequency and by a proper coordinate transformation we are able to make the motion of a particle look mathematically exactly like that of a harmonic oscillator. This kind of formulation of beam dynamics will be very useful in the evaluation of perturbations on particle trajectories since all mathematical tools that have been developed for harmonic oscillators will be available for particle beam dynamics.

We introduce *Floquet's coordinates*, or *normalized coordinates* through the transformation

$$w = \frac{u}{\sqrt{\beta}} \tag{5.143}$$

and

$$\varphi = \int_{0}^{s} \frac{d\bar{s}}{\nu \, \beta(\bar{s})} . \tag{5.144}$$

Note that we used in (5.144) a different normalization than that selected in (5.120) to adapt more appropriately to the issues to be discussed here. With this transformation we get for the first derivative

$$u' = \dot{w} \frac{\sqrt{\beta}}{\nu \, \beta} + u \frac{\beta'}{2\sqrt{\beta}} = \frac{\dot{w}}{\nu \sqrt{\beta}} - w \frac{\alpha}{\sqrt{\beta}} \tag{5.145}$$

and for the second derivative

$$u'' = \frac{\ddot{w}}{\nu^2 \, \beta^{3/2}} - w \frac{\alpha'}{\sqrt{\beta}} - w \frac{\alpha^2}{\beta^{3/2}} , \tag{5.146}$$

where dots indicate derivatives with respect to the phase $\dot{w} = dw/d\varphi$ etc. We insert these expressions into (5.115) and get the general equation of motion expressed in normalized coordinates

$$u'' + k u = \frac{1}{\nu^2 \, \beta^{3/2}} \left(\ddot{w} + \left[\tfrac{1}{2} \beta \, \beta'' - \alpha^2 + k \, \beta^2 \right] \nu^2 \, w \right)$$
$$= p(x, y, z), \tag{5.147}$$

where the right-hand side represents a general perturbation term $p(x, y, z)$ which was neglected so far. The square bracket is equal to unity according to (5.122) and the equation of motion takes the simple form of a harmonic oscillator with some perturbation

$$\ddot{w} + \nu^2 \, w = \nu^2 \, \beta^{3/2} \, p(x, y, s). \tag{5.148}$$

Since the parameter ν is constant we have in the case of vanishing perturbations $p \equiv 0$ the exact equation of a harmonic oscillator and particles perform in this representation periodic sinelike oscillations with the frequency ν

$$w = w_o \cos(\nu\varphi + \delta). \tag{5.149}$$

The transformation matrix in these variables is given by

$$\mathcal{M}(s|0) = \begin{bmatrix} C(\varphi) & S(\varphi) \\ C'(\varphi) & S'(\varphi) \end{bmatrix} = \begin{bmatrix} \cos\nu\varphi & \frac{1}{\nu}\sin\nu\varphi \\ -\nu\sin\varphi & \cos\nu\varphi \end{bmatrix} \tag{5.150}$$

as can easily be derived from (5.149). The equations of motion (5.148) can be derived from the *Hamiltonian*

$$H_o = \tfrac{1}{2}\dot{w}^2 + \tfrac{1}{2}\nu^2 w^2 - \int \nu^2 \beta^{3/2} p(x,y,z)\,dw \tag{5.151}$$

with the conjugate coordinates $(w, \dot{w})$ by differentiation

$$\frac{\partial H_o}{\partial w} = -\ddot{w} = \nu^2 w - \nu^2 \beta^{3/2} p(x,y,z). \tag{5.152}$$

The use of normalized coordinates not only allows us to treat particle beam dynamics equivalent to a harmonic oscillator but is also convenient in the discussions of perturbations or aberrations. In phase space each particle performs closed trajectories in the form of an ellipse which we called the phase ellipse. In cartesian coordinates this ellipse, however, continuously changes its shape and orientation and correlations between two locations are not always obvious. If we use normalized coordinates the unperturbed phase ellipse becomes an invariant circle as shown in Fig. 5.26.

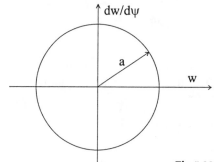

Fig. 5.26. Ideal phase ellipse in normalized coordinates

From (5.143) we get with $u(s) = a\sqrt{\beta(s)}\cos\psi(s)$ where $\psi(s) = \nu\varphi(s)$

$$w(\varphi) = \frac{u}{\sqrt{\beta}} = a\cos\psi,$$

$$\frac{dw}{d\psi} = \sqrt{\beta}\, u' + \frac{\alpha}{\sqrt{\beta}}\, u = -a\sin\psi, \tag{5.153}$$

and after elimination of the phase the Courant-Snyder invariant becomes

$$w^2 + \left(\frac{dw}{d\psi}\right)^2 = a^2, \qquad (5.154)$$

where a is the betatron oscillation amplitude.

In the absence of perturbations, particles follow phase trajectories in the form of a circles. Conversely, it is now easy to identify the existence and degree of perturbations or aberrations as deviations from the circular trajectory. We have chosen the betatron phase, ψ as the independent variable and the particles therefore cover one full turn along the phase "ellipse" for each betatron oscillation. This is a convenient way of representation in beam transport systems. For circular accelerators we may find it more appropriate to use $\varphi = \psi/\nu$ as the independent variable in which case the particle rotation frequency in phase space is the same as that in the ring.

The representation of particle trajectories in phase space may also be extended to the path or orbit of a particle beam in which case the ideal path or orbit is described by the origin of the coordinate system, $w = 0$ and $dw/d\psi = 0$. Any path distortion would show up as a deviation from this origin and betatron oscillations about the ideal or distorted path are represented by circles of radius a about the location of the path. Consider, for example, an off momentum particle oscillating about the ideal path of a straight beam line. At some location we insert a bending magnet and the off momentum particle will be deflected by a different angle with respect to particles with correct momentum. The difference in the deflection angle appears as a displacement in phase space from the center to a finite value $\Delta \dot{w} = D(s)/\sqrt{\beta}\,\delta$. From here on the off momentum reference path follows the dispersion function $D(s)\,\delta$ and the particle performs betatron oscillations in the form of circles with radius a until another bending magnet further modifies or compensates this motion (Fig. 5.27).

In case a second bending magnet is placed half a betatron oscillation downstream from the first causing the same difference in the deflection angle the effect of the first magnet can be compensated completely and

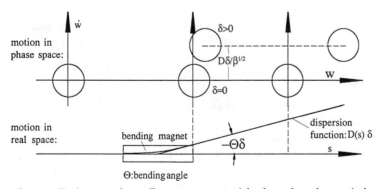

Fig. 5.27. Trajectory of an off momentum particle through a chromatic beam transport section

the particle continues to move along the ideal path. A section of a beam transport line with this property is called an *achromat*. We may apply this representation not only to off momentum particles but also to the whole beam if we need to produce a beam bump by displacing the beam locally by a small amount. In this case, a bending magnet or a *steering magnet* is placed on either side of the location, where the beam bump is to occur and half a betatron wave length apart from each other. Choosing the deflection angles $\sqrt{\beta_1}\,\theta_1 = \sqrt{\beta_2}\,\theta_2$ we obtain a perfectly corrected beam bump between the steering magnets.

5.6 Dispersive Systems

In particle dynamics the beam guidance and focusing is performed by applying Lorentz forces. The effects of these fields on particle trajectories depend on the momentum of the particles. So far, we have derived beam dynamics for particles with ideal momentum for which the beam transport system is designed. To properly describe the dynamics of a real particle beam we must include chromatic effects caused by an error in the beam energy or by a spread of energies within the particle beam. In Sect. 4.8.4 the perturbation due to a momentum error has been derived and expressed in terms of a dispersion. Continuing the formulation of beam dynamics in terms of transformation matrices we derive in this section transformation matrices for particles with a momentum error.

5.6.1 Analytical Solution

Equation (4.114),

$$D(s) \;=\; \int\limits_{0}^{s} \frac{1}{\rho}(\bar{s})\,[S(s)\,C(\bar{s}) - C(s)\,S(\bar{s})]\,\mathrm{d}\bar{s}\,, \tag{4.114}$$

describes the dispersion function in a beam transport line in a very general form. There is no contribution to the dispersion function unless there is at least one bending magnet in the beam line. Knowledge of the location and strength of bending magnets, together with the principal solutions of the equations of motion, we may calculate the dispersion anywhere along the beam transport line by integration of (4.114).

Similar to the matrix formalism for betatron oscillations we would also like to apply the same formalism for the dispersion function. For this we note that the particle deviation u from the reference path is composed of the betatron motion and a displacement due to an energy error $u = u_\beta + u_\delta$. The transformation matrix is therefore a composite of both contributions and can be expressed by

$$
\begin{bmatrix} u(s) \\ u'(s) \\ \delta \end{bmatrix} = M \begin{bmatrix} u_\beta(s_0) \\ u'_\beta(s_0) \\ \delta \end{bmatrix} + M \begin{bmatrix} u_\delta(s_0) \\ u'_\delta(s_0) \\ \delta \end{bmatrix} , \tag{5.155}
$$

where M is the 3×3 transformation matrix, δ the relative momentum error and $u_\delta(s) = D(s)\delta$ and $u'_\delta(s) = D'(s)\delta$ the displacement and slope respectively of the reference path for particles with a momentum error δ. Equation (5.155) can also be applied to the dispersion function alone by setting the betatron oscillation amplitudes to zero and the momentum error $\delta = 1$ for

$$
\begin{bmatrix} D(s) \\ D'(s) \\ 1 \end{bmatrix} = M \begin{bmatrix} D(s_0) \\ D'(s_0) \\ 1 \end{bmatrix} . \tag{5.156}
$$

By determining the transformation matrices for individual bending magnets we are in a position to calculate in matrix formulation the dispersion function anywhere along a beam transport line.

In the deflecting plane of a *pure sector magnet* the principal solutions are with $K = 1/\rho_0^2$

$$
\begin{bmatrix} C(s) & S(s) \\ C'(s) & S'(s) \end{bmatrix} = \begin{bmatrix} \cos\frac{s}{\rho_0} & \rho_0 \sin\frac{s}{\rho_0} \\ -\frac{1}{\rho_0}\sin\frac{s}{\rho_0} & \cos\frac{s}{\rho_0} \end{bmatrix} . \tag{5.157}
$$

With $\rho(s) = \rho_0 = \text{const}$ we get from 4.114 and (5.157) for the dispersion function within the magnet

$$
\begin{aligned}
D(s) &= \frac{1}{\rho_0} \int_0^s \left[\rho_0 \sin\frac{s}{\rho_0} \cos\frac{\bar{s}}{\rho_0} - \rho_0 \cos\frac{s}{\rho_0} \sin\frac{\bar{s}}{\rho_0} \right] \mathrm{d}\bar{s} \\
&= \rho_0 \left(1 - \cos\frac{s}{\rho_0} \right)
\end{aligned} \tag{5.158}
$$

$$
D'(s) = \sin\frac{s}{\rho_0} .
$$

Particles with momentum error δ follow an equilibrium path given by $D(s)\delta$ which can be determined experimentally by observing the beam path for two different values of the beam momentum δ_1 and δ_2. The difference of the two paths divided by the momentum difference is the dispersion function $D(s) = \Delta u/(\delta_2 - \delta_1)$. In practical applications this is done either by changing the beam energy or by changing the strength of the bending magnets. In circular electron accelerators, however, only the first method will work since the electrons always adjust the energy through damping to the energy determined by the magnetic fields. In circular electron accelerators we determine the dispersion function by changing the rf frequency which enforces a change in the particle energy as we will discuss later in Chap. 8.

5.6.2 3×3 - Transformation Matrices

From (5.157) and (5.158) we may form now 3×3 transformation matrices. In the deflecting plane of a pure sector magnet of arc length ℓ such a transformation matrix is

$$
\mathcal{M}_{s,\rho}(\ell|0) = \begin{bmatrix} \cos\theta & \rho_o \sin\theta & \rho_o(1-\cos\theta) \\ -\frac{1}{\rho_o}\sin\theta & \cos\theta & \sin\theta \\ 0 & 0 & 1 \end{bmatrix} , \tag{5.159}
$$

where $\theta = \ell/\rho_o$ is the deflection angle of the magnet. In the non deflecting plane the magnet behaves like a drift space with $1/\rho_o = 0$ and $k = 0$

$$
\mathcal{M}_{s,o}(\ell|0) = \begin{bmatrix} C(s) & S(s) & 0 \\ C'(s) & S'(s) & 0 \\ 0 & 0 & 1 \end{bmatrix} . \tag{5.160}
$$

For a *synchrotron magnet* of the sector type we get from (5.27) in analogy to (5.158) replacing $1/\rho_o$ by $\sqrt{K}$ and with $\psi = \sqrt{k_o + 1/\rho_o^2}\,\ell$ for the case of a *focusing synchrotron magnet*

$$
\mathcal{M}_{sy,f}(\ell|0) = \begin{bmatrix} \cos\psi & \frac{\sin\psi}{\sqrt{K}} & \frac{1-\cos\psi}{\rho K} \\ -\sqrt{K}\sin\psi & \cos\psi & -\frac{\sin\psi}{\rho\sqrt{K}} \\ 0 & 0 & 1 \end{bmatrix} \tag{5.161}
$$

and for a *defocusing synchrotron magnet*

$$
\mathcal{M}_{sy,d}(\ell|0) = \begin{bmatrix} \cosh\psi & \frac{\sinh\psi}{\sqrt{|K|}} & -\frac{1-\cosh\psi}{\rho|K|} \\ \sqrt{|K|}\sinh\psi & \cosh\psi & +\frac{\sinh\psi}{\rho\sqrt{|k|}} \\ 0 & 0 & 1 \end{bmatrix} , \tag{5.162}
$$

where $\psi = \sqrt{|k_o + 1/\rho_o^2|}\,\ell$.

In case of a *rectangular magnets* without field gradient we proceed by multiplying the matrix for a sector magnet by the transformation matrices for endfield focusing. Since these end effects act like quadrupoles we have no new contribution to the dispersion and the transformation matrices for each endfield are

$$
\mathcal{M}_e = \begin{bmatrix} 1 & 0 & 0 \\ \frac{1}{\rho_o}\tan(\theta/2) & 1 & 0 \\ 0 & 0 & 1 \end{bmatrix} . \tag{5.163}
$$

With these endfield matrices the chromatic transformation matrix for a rectangular bending magnet in the deflecting plane is obtained from (5.158) with $\mathcal{M}_{r,\rho} = \mathcal{M}_e \cdot \mathcal{M}_{s,\rho} \cdot \mathcal{M}_e$

$$\mathcal{M}_{r,\rho}(\ell|0) = \begin{bmatrix} 1 & \rho_0 \sin\theta & \rho_0 (1 - \cos\theta) \\ 0 & 1 & 2\tan(\theta/2) \\ 0 & 0 & 1 \end{bmatrix}. \tag{5.164}$$

Similarly we can derive the transformation matrices for rectangular synchrotron magnets.

5.6.3 Linear Achromat

Frequently it is necessary in beam transport systems to deflect the particle beam. If this is done in an arbitrary way an undesirable finite dispersion function will remain at the end of the deflection section. Special magnet arrangements exist which allow to bend a beam without generating a residual dispersion. Such magnet systems composed of only bending magnets and quadrupoles are called *linear achromats*.

Fig. 5.28 displays an achromatic section proposed by Panofsky [5.11] which may be used as a building block for curved transport lines or circular accelerators. This section is composed of a symmetric arrangement of two bending magnets with a quadrupole in in the center and is also know as a *double bend achromat* or a *Chasman Green lattice* [5.12].

General conditions for linear achromats have been discussed in Sect. 4.9 and we found that the integrals

$$I_s = \int_0^s \frac{S(\tilde{s})}{\rho(\tilde{s})} \, d\tilde{s} = 0,$$

$$\tag{5.165}$$

$$I_c = \int_0^s \frac{C(\tilde{s})}{\rho(\tilde{s})} \, d\tilde{s} = 0,$$

must vanish for a lattice section to become achromatic. For a double bend achromat this can be accomplished by a single parameter or quadrupole if adjusted such that the betatron phase advance between the vertex points of the bending magnet is 180°. Applying the conditions for achromaticity, Steffen [5.5] derived the relationship

$$\frac{1}{\sqrt{k}} \cot(\varphi/2) = \rho \tan(\theta/2) + d \tag{5.166}$$

between the magnet deflection angle θ, the bending radius ρ, the drift space d and the quadrupole strength $\varphi = \sqrt{k}\,\ell$ while the dispersion function reaches a maximum in the quadrupole center of

$$D_{\max} = -\frac{1}{\sqrt{k}} \frac{\sin\theta}{2\sin(\varphi/2)}. \tag{5.167}$$

A variation of this lattice, the *triple bend achromat* [5.13,14] is shown in Fig. 5.29, where a third bending magnet is inserted for practical reasons to provide more locations to install sextupoles for chromatic corrections.

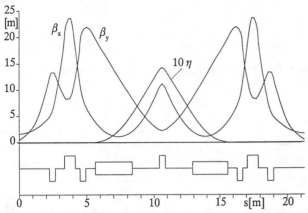

Fig. 5.28. Double bend achromat [5.12]

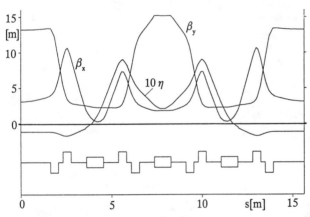

Fig. 5.29. Triple bend achromat [5.13]

Magnet arrangements as shown in Figs. 5.28,29 are dispersion free deflection units or linear achromats. This achromat is focusing only in the deflecting plane but defocusing in the nondeflecting plane which must be compensated by external quadrupole focusing or, since there are no special focusing requirements for the nondeflecting plane, by either including a field gradient in the pole profile of the bending magnets [5.15] or additional quadrupoles between the bending magnets. In a beam transport line this achromat can be used for diagnostic purposes to measure the energy and energy spread of a particle beam as will be discussed in more detail in Sect. 5.6.4.

A further variation of the lattice in Fig. 5.28 has been proposed by Steffen [5.5] which allows to translate a beam as shown in Fig. 5.30.

In this case the total phase advance must be 360° because the integral I_c would not vanish anymore for reason of symmetry. We use therefore stronger

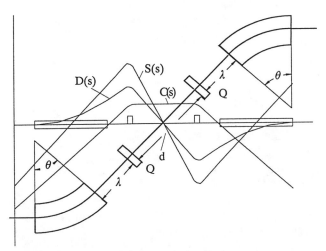

Fig. 5.30. Achromatic beam translation [5.5]

focusing to make I_c vanish because both the bending angle and the cosine like function change sign. Achromatic properties are obtained again for parameters meeting the condition [5.5]

$$\rho \tan(\theta/2) + \lambda = \frac{1}{\sqrt{k}} \frac{d\sqrt{k}\,\cos\varphi + 2\,\sin\varphi}{d\sqrt{k}\,\sin\varphi - 2\,\cos\varphi}, \tag{5.168}$$

where $\varphi = \sqrt{k}\,\ell$ and k, ℓ the quadrupole strength and length, respectively. The need for beam translation occurs frequently during the design of beam transport lines. Solutions exist to perform such an achromatic translation but the required focusing is much more elaborate and may cause significantly stronger aberrations compared to a simple one directional beam deflection of the double bend achromat type.

Utilizing symmetric arrangements of magnets, deflecting achromats can be composed from bending magnets only [5.5]. One version has become particularly important for synchrotron radiation sources, where wiggler magnets are used to produce a high intensity of radiation. Such *wiggler magnets* are composed of a row of alternately deflecting bending magnets which do not introduce a net deflection on the beam. A unit or period of such a wiggler magnet is shown in Fig. 5.31.

The transformation of the dispersion through half a wiggler unit is the superposition of the dispersion function from the first magnet at the end of the second magnet plus the contribution of the dispersion from the second magnet. In matrix formulation and for hard edge rectangular magnets the dispersion at the end of half a wiggler period is

$$\begin{bmatrix} D_w \\ D'_w \end{bmatrix} = \begin{bmatrix} -\rho\,(1-\cos\theta) \\ -2\tan(\theta/2) \end{bmatrix} + \begin{bmatrix} 1 & \ell_w \\ 0 & 1 \end{bmatrix} \begin{bmatrix} \rho\,(1-\cos\theta) \\ 2\tan(\theta/2) \end{bmatrix}, \tag{5.169}$$

173

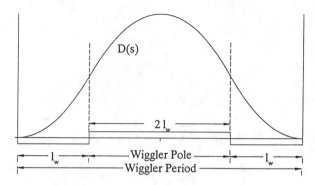

Fig. 5.31. Wiggler achromat

where $\rho > 0$, $\theta = \ell_w \rho$ and ℓ_w the length of one half wiggler pole (see Fig. 5.31). Evaluation of (5.169) gives the simple result

$$
\begin{aligned}
D_w &= 2\,\ell_w\,\tan(\theta/2)\,, \\
D'_w &= 0\,.
\end{aligned}
\tag{5.170}
$$

The dispersion reaches a maximum in the middle of the wiggler period and vanishes again for reasons of symmetry at the end of the period. For sector magnets we would have obtained the same results. A full wiggler period is therefore from a beam optics point of view a linear achromat. Such an arrangement can also be used as a spectrometer by placing a monitor in the center, where the dispersion is large. For good momentum resolution, however, beam focusing must be provided in the deflecting plane upstream of the bending magnets to produce a small focus at the beam monitors for a monochromatic beam as will be discussed in the next section.

The examples of basic lattice designs discussed in this section are particularly suited for analytical treatment. In praxis modifications of these basic lattices are required to meet specific boundary conditions making, however, analytical treatment much more complicated. With the availability of computers and numerical lattice design codes, it is prudent to start with basic lattice building blocks and then use a fitting program for modifications to meet particular design goals.

5.6.4 Spectrometer

Although the dispersion has been treated as a perturbation it is a highly desired feature of a beam line to determine the energy or energy distribution of a particle beam. Such a beam line is called a *spectrometer* for which many different designs exist. A specially simple and effective spectrometer can be made with a single 180° sector magnet [5.16– 18] . For such a spectrometer the transformation matrix is from (5.159)

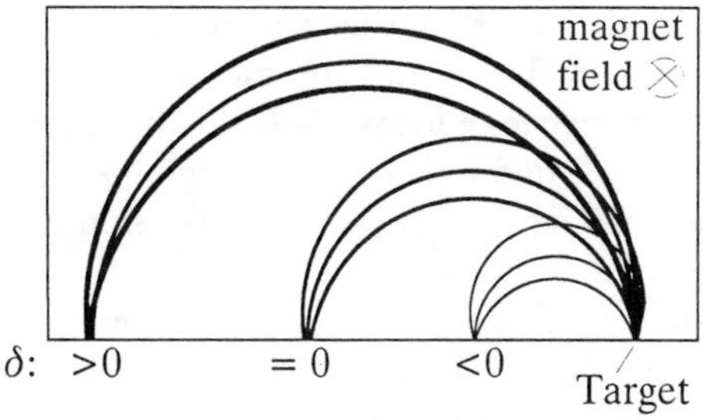

Fig. 5.32. 180° Spectrometer

$$\begin{bmatrix} -1 & 0 & 2\rho_{\mathrm{o}} \\ 0 & -1 & 0 \\ 0 & 0 & 1 \end{bmatrix}. \tag{5.171}$$

In this spectrometer all particles emerging from a small target (Fig. 5.32) are focused to a point again at the exit of the magnet. The focal points for different energies, however, are separated spatially due to dispersion. Mathematically, this is evident since the particle trajectories at the end of the magnet are given by

$$x = -x_{\mathrm{o}} + 2\rho_{\mathrm{o}}\,\delta\,. \tag{5.172}$$

and show different positions x for different energies δ.

The image point is independent of x_{o}' and only proportional to δ with a large proportionality factor which allows a large energy resolution. While this spectrometer seems to have almost ideal features it is also an example of the limitations of perturbation methods. For larger values of δ of the order of several percent higher order terms cannot be neglected anymore. Inclusion of such terms, for example, will first tilt and then bend the focal plane at the end of the magnet.

More sophisticated spectrometers including focusing to accept large emittance beams have been devised with special efforts to reduce the effects of aberrations. It is not the intend of this text to discuss in detail such designs. More comprehensive overviews for spectrometers with further references can be found for example in [5.5,19].

In the treatment of this spectrometer we have ignored the nondeflecting plane. Since there is no focusing, particles are widely spread out in this plane at the end of the magnet. Practical versions of this spectrometer, therefore, include a focusing term in the nondeflecting plane in such a way that the resulting focusing is the same in both planes [5.17, 20].

Measurement of Beam Energy Spectrum: Frequently it is desirable to determine experimentally the particle energy and energy spread. Basically only one bending magnet is needed to perform this experiment. We will see, however, that the finite beam size of the monochromatic part of the beam will greatly influence the resolution of the energy measurement. Optimum resolution is achieved if some focusing is included and the measurement is performed at a location, where the beam reaches a focus while the dispersion is large. In Fig. 5.33 particle beams with different energies are shown in phase space, where both beam centers are separated by the dispersion and its slope.

In reality no such separation exists since we have a spread of energies rather than two distinct energies. This energy spread is mixed with the spread in phase space of the beam emittance and beams of different energies can only be separated completely if the relative energy difference is at least

$$\delta_{min} \geq \frac{2\,E_b}{D} = 2\,\frac{\sqrt{\epsilon\,\beta}}{D}\,, \tag{5.173}$$

where $E_b = \sqrt{\epsilon\,\beta}$ is the beam envelope. To maximize the energy resolution the beam size E_b should be small and the dispersion $D(s)$ large. From Fig. 5.33 we note, therefore, that for a given beam emittance and dispersion the energy resolution can be improved significantly if the measurement is performed at or close to a beam waist, where β reaches a minimum.

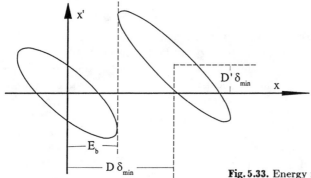

Fig. 5.33. Energy resolution in phase space

To derive mathematical expressions for the energy resolution and conditions for the maximum energy resolution $1/\delta_{min}$ we assume a beam line as shown in Fig. 5.34 with the origin of the coordinate system $s = 0$ in the center of the bending magnet. The salient features of this beam line are the quadrupole followed by a bending magnet. With this sequence of magnets we are able to focus the particle beam in the deflection plane while leaving the dispersion unaffected. In case of a reversed magnet sequence the dispersion function would be focused as well compromising the energy resolution.

Transforming the dispersion (5.158) back from the end of the sector bending magnet to the middle of the magnet we get the simple result

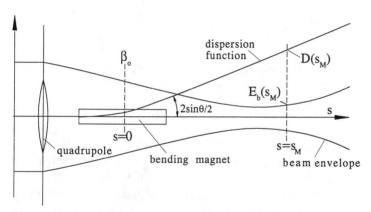

Fig. 5.34. Measurement of the energy spectrum

$$
\begin{bmatrix} D_o \\ D'_o \end{bmatrix} = \begin{bmatrix} \cos(\theta/2) & -\rho_o \sin(\theta/2) \\ \frac{\sin(\theta/2)}{\rho_o} & \cos(\theta/2) \end{bmatrix} \begin{bmatrix} \rho_o(1 - \cos\theta) \\ \sin\theta \end{bmatrix}
$$
$$
= \begin{bmatrix} 0 \\ 2\sin(\theta/2) \end{bmatrix} .
\tag{5.174}
$$

The dispersion seems to originate in the middle of the magnet with a slope $D'_o = 2\sin\theta/2$. At a distance s from the middle of the bending magnet the betatron function is given by $\beta(s) = \beta_o - 2\alpha_o s + \gamma_o s^2$ where $(\beta_o, \alpha_o, \gamma_o)$ are the values of the betatron functions in the middle of the bending magnet, and the dispersion is $D(s) = 2\sin(\theta/2) s$. Inserting these expressions into (5.173) we can find the location s_M for maximum momentum resolution by differentiating δ_{min} with respect to s. Solving $d\delta_{min}/ds = 0$ for s we get

$$
s_M = \frac{\beta_o}{\alpha_o}
\tag{5.175}
$$

and the maximum momentum resolution is

$$
\delta_{min}^{-1} = \frac{\sqrt{\beta_o}\,\sin(\theta/2)}{\sqrt{\epsilon}} .
\tag{5.176}
$$

The best momentum resolution for a beam with emittance ϵ is achieved if both the bending angle θ and the betatron function β_o in the middle of the bending magnet are large. From condition (5.175) we also find $\alpha_o > 0$ which means that the beam must be converging to make a small spot size at the observation point downstream of the bending magnet. With (5.137) we find that $s_M = \beta_o/\alpha_o = -\beta_M/\alpha_M$ and from the beam envelope $E_b^2 = \epsilon\beta_M$ at $s = s_M$ we get the derivative $2E_b E'_b = \epsilon\beta'_M = -2\epsilon\alpha_M$. With this and $D/D' = s$ we get for the optimum place to measure the energy spread of a particle beam

177

$$s_M = \frac{D(s_M)}{D'(s_M)} = \frac{E_b(s_M)}{E_b'(s_M)}. \tag{5.177}$$

It is interesting to note that the optimum location s_M is not at the beam waist, where $\beta(s)$ reaches a minimum, but rather somewhat beyond the beam waist, where $D/\sqrt{\beta}$ is maximum.

At this point we may ask if it is possible through some clever beam focusing scheme to improve this resolution. Analogous to the previous derivation we look for the maximum resolution $\delta_{min}^{-1} = D(s)/[2\sqrt{\epsilon\beta(s)}]$. The dispersion is expressed in terms of the principal solution $D(s) = S(s) D'(0)$ and $D'(s) = S'(s) D'(0)$ since $D(0) = 0$. The betatron function is given by $\beta(s) = C^2(s)\beta_o - 2S(s)\alpha_o + S^2(s)\gamma_o$ and the condition for maximum resolution turns out to be $\alpha/\beta = -D'/D$. With this we get the resolution

$$\delta_{min}^{-1} = \frac{D(s)}{2\sqrt{\epsilon\beta}} = \frac{S(s) D_o'}{2\sqrt{\epsilon\beta}} = \frac{\sin(\theta/2)}{\sqrt{\epsilon\beta}} S(s).$$

and finally with $S(s) = \sqrt{\beta_o \beta(s)} \sin\psi(s)$

$$\delta_{min}^{-1} = \frac{\sqrt{\beta_o} \sin(\theta/2)}{\sqrt{\epsilon}} \sin\psi(s) \le \frac{\sqrt{\beta_o} \sin(\theta/2)}{\sqrt{\epsilon}}, \tag{5.178}$$

which is at best equal to result (5.176) for $\psi(s) = 90°$. The momentum resolution is never larger than in the simple setup of Fig. 5.34 no matter how elaborate a focusing lattice is employed. However, if more than one bending magnet is used the resolution may be increased if the betatron phases between the magnets $\psi(s_i)$ and the place of the measurement $\psi(s_M)$ are chosen correctly. The resolution then is

$$\delta_{min}^{-1} = \frac{1}{\sqrt{\epsilon}} \sum_i \sqrt{\beta_{oi}} \sin(\theta_i/2) \sin[\psi(s_M) - \psi(s_i)], \tag{5.179}$$

where the sum is taken over all magnets i. Such an energy resolving system is often used in beam transport lines to filter out a small energy band of a particle beam with a larger energy spread. In this case a small slit is placed at the place for optimum energy resolution ($s = s_M$).

5.7 Path Length and Momentum Compaction

The existence of different reference paths implies that the path length between two points of the beam transport line may be different as well for different particle momenta. We will investigate this since the path length is of great importance as will be discussed in detail in Chap. 8. In preparation for this discussion, we derive here the functional dependencies of the path length on momentum and focusing lattice.

The path length along a straight section of the beam line depends on the angle of the particle trajectory with the reference path. Since, however, in this chapter we are interested only in linear beam dynamics we may neglect such second order corrections to the path length. The only linear contribution to the path length comes from the curved sections of the beam transport line. The total path length is therefore given by

$$L = \int (1 + \kappa x)\,\mathrm{d}s\,, \tag{5.180}$$

where $\kappa = 1/\rho$.

We evaluate (5.180) along the reference path, where $x = D(s)\,\delta$. First we find the expected result $L_o = \int \mathrm{d}s$ for $\delta = 0$ which is the ideal design length of the beam line or the design circumference of a circular accelerator. The deviation from this ideal length is then

$$\Delta L = \delta \int \frac{D(s)}{\rho(s)}\,\mathrm{d}s\,. \tag{5.181}$$

The variation of the path length with momentum is determined by the momentum compaction factor, defined by

$$\alpha_c = \frac{\Delta L/L_o}{\delta} \qquad \text{with} \qquad \delta = \frac{\Delta p}{p}\,. \tag{5.182}$$

Its numerical value can be calculated with (5.181)

$$\alpha_c = \frac{1}{L_o} \int_o^{L_o} \frac{D(s)}{\rho(s)}\,\mathrm{d}s = \left\langle \frac{D(s)}{\rho} \right\rangle\,. \tag{5.183}$$

In this approximation the path length variation is determined only by the dispersion function in bending magnets and the path length depends only on the energy of the particles. To prepare for the needs of phase focusing in Chap. 8 we will not only consider the path length but also the time it takes a particle to travel along that path. If L is the path length, the travel time is given by

$$\tau = \frac{L}{c\beta}\,. \tag{5.184}$$

Here $\beta = v/c$ is the velocity of the particle in units of the velocity of light and is not to be confused with the betatron function. The variation of τ gives by logarithmic differentiation

$$\frac{\Delta\tau}{\tau} = \frac{\Delta L}{L} - \frac{\Delta\beta}{\beta}\,. \tag{5.185}$$

With $\Delta L/L = \alpha_c\,\delta$ and $cp = \beta E$ we get $\mathrm{d}p/p = \mathrm{d}\beta/\beta + \mathrm{d}E/E$ and with $\mathrm{d}E/E = \beta^2\,\mathrm{d}p/p$ we can solve for $\mathrm{d}\beta/\beta = (1/\gamma^2)\,\mathrm{d}p/p$, where $\gamma = E/mc^2$

179

is the energy of the particles in units of the rest energy mc^2. From (5.185) we have then

$$\frac{\Delta\tau}{\tau} = -\left(\frac{1}{\gamma^2} - \alpha_c\right)\frac{dp}{p} = \eta_c\frac{dp}{p} \qquad (5.186)$$

and call the combination

$$\eta_c = (\gamma^{-2} - \alpha_c) \qquad (5.187)$$

the *momentum compaction*. The energy

$$\gamma_t = \frac{1}{\sqrt{\alpha_c}} \qquad (5.188)$$

for which the momentum compaction vanishes is called the *transition energy* which will play an important role in phase focusing. Below the transition energy the arrival time is determined by the actual velocity of the particles while above the transition energy the particle speed is so close to the speed of light that the arrival time of a particle with respect to other particles depends more on the path length than on its speed. For a circular accelerator we may relate the time τ_r a particle requires to follow a complete orbit to the *revolution frequency,* ω_r and get from (5.186)

$$\frac{d\omega_r}{\omega_r} = -\frac{d\tau_r}{\tau_r} = \eta_c\frac{dp}{p}. \qquad (5.189)$$

For particles above the transition energy this quantity is negative which means a particle with a higher energy needs a longer time for one revolution than a particle with a lower energy. This is because the dispersion function causes particles with a higher energy to follow an equilibrium orbit with a larger average radius compared to the radius of the ideal orbit.

By special design of the lattice one could generate an oscillating dispersion function in such a way as to make the momentum compaction η_c to vanish. Such a ring would be isochronous to the approximation used here. Due to higher order aberrations, however, there are nonlinear terms in the dispersion function which together with an energy spread in the beam cause a spread of the revolution frequency compromising the degree of isochronicity. We will discuss these effects in more detail in Vol. II.

Problems

Problem 5.1.   a) If $\begin{bmatrix} a & b \\ c & d \end{bmatrix}$ is the transformation matrix for an arbitrary beam line then show that $\begin{bmatrix} d & b \\ c & a \end{bmatrix}$ is the transformation matrix for the same beam line in reversed order.

b) Show also that $\begin{bmatrix} d & -b \\ -c & a \end{bmatrix}$ is the transformation matrix for the same transport line going backwards in the inverse direction. Show that going forward and then back again to the starting point indeed gives the unity transformation.

Problem 5.2. a) Calculate the focal length of a quadrupole doublet with $|f_1| = |f_2| = 5$ m and a distance between the magnets of $d = 1$m. Plot for this doublet the focal length as a function of particle momentum $-5\% < \Delta p/p_o < 5\%$.

b) Use a parallel beam of radius r_o and calculate the beam radius r at the focal point of this doublet.

c) Plot the magnification r/r_o as a function of momentum $-5\% < \Delta p/p_o < 5\%$. What is the chromatic aberration $(r - r_o)/r_o$ of the spot size?

Problem 5.3. Assume two thin quadrupoles to form a quadrupole doublet which is used for a 50 GeV electron beam to produce a combined focal length of 8 m in both the horizontal and vertical plane. The distance of the quadrupoles is $d = 2$ m. Both quadrupoles are to be of the same type.

a) The aperture of the quadrupoles has a radius of $R = 2$ cm and the magnetic length is $L_q = 1.0$ m. What is the magnetic field gradient and what is the pole tip field which is the field at the pole profile, where the pole profile and the 45 degree line cross) ?

b) Calculate the required Ampere turns per coil.

Problem 5.4. Use the quadrupole of Fig. 5.6 but with a reduced iron length of $\ell_{iron} = 5.0$ cm and calculate for $k_o = 50\text{m}^{-2}$ and $k_o = 30\text{m}^{-2}$ the corrections for the quadrupole length and strength as discussed in Sect. 5.2.5 and compare with the results shown in Fig. 5.8, where $\ell_{iron} = 15.9$ cm.

Problem 5.5. Explain why a quadrupole with finite pole width does not produce a pure quadrupole field. What are the other allowed multipole field components (ignore mechanical tolerances) and why?

Problem 5.6. Derive the relation (5.166) between deflection angle, strength of the central quadrupole and distance between the two bending magnets in Fig. 5.28with the boundary condition that the section between and including the bending magnets becomes an achromat.

Problem 5.7.

a) Design a symmetric thin lens triplet with a focal point for both planes at the same point $s = s_f$.

b) Calculate and plot the betatron function for the quadrupole triplet and drift space just beyond the focal point. The value for the betatron function be $\beta_o = 8$ m at the entrance to the triplet $s = 0$ where we also assume $\alpha_o = 0$.

c) Derive the phase advance in one plane between $s = 0$ and $s = s_f$ both from the elements of the transformation matrix and by integrating the betatron function. Both method should give the same results. How does this phase advance change if $\beta_o = 20$ m and $\alpha_o = -1.0$? Prove your statement two ways.

Problem 5.8. Consider a combined function sector magnet with nonparallel pole faces to produce a field gradient.

a) Determine the field gradient to produce equal focusing in both the horizontal and vertical plane.

b) What is the relationship between the field index n the bending radius ρ and the focusing strength k for this combined function magnet. What is the field index for a sector magnet with equal focusing in both planes?

c) Derive the equations of motion for both the deflecting and nondeflecting plane in terms of field index and bending radius. State the conditions for the field index n to obtain stable particle oscillations in both planes. Assume a circular accelerator constructed of a uniform sector magnet with a stable field index n. What is the number of betatron oscillations per turn in both planes? Derive the equations of motion in both the deflecting and nondeflecting plane?

Problem 5.9. Sector and rectangular magnets have opposite focusing properties. Determine the geometry of a wedge magnet with equal focusing in both planes. What is the focal length? Compare this focal length with that obtained in problem 5.8.

Problem 5.10. A wiggler magnet is composed of a series of equal rectangular dipoles with alternating polarity. Derive the linear transformation matrices in both planes for a single wiggler magnet pole. For the field distribution assume a sinusoidal field $B_y(s) = B_{yo} \sin ks$ where $k = 2\pi/\lambda_p$ and λ_p is the wiggler magnet period. Define a hard edge model for a wiggler pole with the same deflection angle and a bending radius $1/\rho_o$. What is the equivalent length of this hard edge pole in units of the wiggler period and what is the focal length of the edge field focusing. Compare with the result of the sinusoidal field distribution. By adjusting both the hard edge effective magnetic length and strength it is possible to match both the deflection angle and the focal length of the sinusoidal wiggler field.

Problem 5.11. In an arbitrary beam transport line we assume that at the point s_o the particle beam is kicked in the horizontal or vertical plane by the deflection angle ϑ. What is the betatron amplitude for the beam at any point s downstream from s_o? To maximize the betatron amplitude at s how should the lattice functions, betatron function and/or phase, be chosen at s_o and s?

Problem 5.12. Particle trajectories in phase space assume the shape of an ellipse. Derive a transformation of the phase space coordinates (u, u') to coordinates (x, y) such that the particle trajectories are circles with the radius $\beta\epsilon$.

Problem 5.13. Use (5.91) for the phase ellipse and prove by integration that the area enclosed by the ellipse is indeed equal to $\pi\epsilon$.

Problem 5.14. For most particle beams the particle distribution can be approximated by a gaussian distribution. In the horizontal plane we have therefore

$$n(u)\,du = \frac{N}{\sqrt{2\pi}\,\sigma_u}\,e^{-u^2/(2\sigma_u^2)}\,du\,, \tag{5.190}$$

where $u = x$ or x' and N is the total number of particles in the beam. Derive a similar expression for the particle distribution function $n(x, x')\,dx\,dx'$ in phase space. (Hint. Transform first to a circular distribution)

Problem 5.15. Show that the transformation of the beam matrix (5.105) is consistent with the transformation of the lattice functions.

6 Periodic Focusing Systems

The fundamental principles of charged particle beam dynamics as discussed in previous chapters can be applied to almost every beam transport need. Focusing and bending devices for charged particles are based on magnetic or electric fields which are specified and designed in such a way as to allow the application of fundamental principles of beam optics leading to predictable results.

Beam transport systems can be categorized into two classes: The first group is that of *beam transport lines* which are designed to guide charged particle beams from point A to point B. In the second class we find beam transport systems or magnet lattices used in circular accelerators. The physics of beam optics is the same in both cases but in the design of actual solutions different boundary conditions may become necessary. Basic linear building blocks in a beam transport line are the beam deflecting bending magnets, quadrupoles to focus the particle beam, and field free drift spaces between magnets. The transformation matrices for all three types of elements have been derived in Chap. 5 and we will apply these results to compose more complicated beam transport systems. The arrangement of magnets along the desired beam path is called the *magnet lattice* or short the *lattice*.

Beam transport lines can be made up of an irregular array of magnets or a repetitive sequence of a special magnet arrangement. Such a repetitive magnet sequence is called a *periodic magnet lattice* and if the magnet arrangement within one period is symmetric this lattice is called a *symmetric magnet lattice*. By definition a circular accelerator lattice is a periodic lattice with the circumference being the period length. To simplify the design and theoretical understanding of beam dynamics it is customary, however, to segment the full circumference of a circular accelerator into sectors which are repeated a number of times to form the complete ring. Such sectors are called *superperiods* and include usually all salient features of the accelerator in contrast to much smaller periodic segments called *cells* which include only a few magnets.

In this chapter we will concentrate on the study of periodic focusing structures. For long beam transport lines and specifically for circular accelerators it is prudent to consider focusing structures that repeat periodically. In this case one can apply beam dynamics properties of one periodic lattice structure as many times as necessary with known characteristics. In circular particle accelerators such periodic focusing structures not only simplify the determination of beam optics properties in a single turn but we will also be able to predict the stability criteria for particles orbiting an indefinite number of revolutions around the ring.

To achieve focusing in both planes we will have to use both focusing and defocusing quadrupoles in a periodic sequence such that we can repeat a lattice period any number of times to form an arbitrary long beam line which provides the desired focusing in both planes.

6.1 FODO Lattice

The most simple periodic lattice would be a sequence of equidistant focusing quadrupoles of equal strength. This arrangement is unrealistic with magnetic quadrupole fields which do not focus in both the horizontal and vertical plane in the same magnet. The most simple and realistic compromise is therefore a periodic lattice like the symmetric quadrupole triplet which was discussed in Sect. 5.2. This focusing structure is composed of alternating focusing and defocusing quadrupoles as shown in Fig. 6.1.

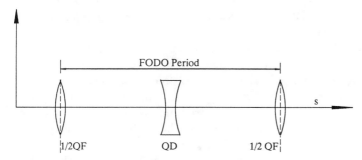

Fig. 6.1. FODO - lattice (QF:focusing quadrupole; QD: defocusing quadrupole)

Each half of such a lattice period is composed of a focusing (F) and a defocusing (D) quadrupole with a drift space (O) in between. Combining such a sequence with its mirror image as shown in Fig. 6.1 results in a periodic lattice which is called a *FODO lattice* or a *FODO channel*. By starting the period in the middle of a quadrupole and continuing to the middle of the next quadrupole of the same sign not only a periodic lattice but also a symmetric lattice is defined. Such an elementary unit of focusing is called a *lattice unit* or in this case a *FODO cell*. The FODO lattice is the most widely used lattice in accelerator systems because of its simplicity, flexibility, and its beam dynamical stability.

6.1.1 Scaling of FODO Parameters

To determine the properties and stability criteria for a FODO period we restrict ourselves to *thin lens approximation*, where we neglect the finite length of the quadrupoles. This approximation greatly simplifies the mathematical expressions while retaining the physical properties to a very high degree. A simple periodic and symmetric transformation matrix for a FODO cell can be derived by starting in the middle of one quadrupole and going to the middle of the next quadrupole of the same type. Any other point in the FODO lattice could be used as the starting point with the analogous point one period downstream to be the ending point, but the mathematical expressions would be unnecessarily complicated without adding more physical insight. The FODO period therefore can be expressed symbolically by the sequence $1/2$ QF, D, QD, D, $1/2$ QF, where the symbol D represents a drift space and the symbols QF and QD focusing or defocusing quadrupoles. In either case we have a triplet structure for which the transformation matrix (5.24) has been derived in Sect. 5.2.4

$$
\mathcal{M}_{\text{FODO}} = \begin{bmatrix} 1 - 2\frac{L^2}{f^2} & 2L \cdot (1 + \frac{L}{f}) \\ -\frac{1}{f^*} & 1 - 2\frac{L^2}{f^2} \end{bmatrix} .
\tag{6.1}
$$

Here $f_{\text{f}} = -f_{\text{d}} = f$, $1/f^* = 2 \cdot (1 - L/f) \cdot (L/f^2)$ and L the distance between quadrupoles. Such a FODO lattice is called a *symmetric FODO lattice* since it uses quadrupoles of equal strength.

From the transformation matrix (6.1) we can deduce an important property for the betatron function. The diagonal elements are equal as they always are in any symmetric lattice. Comparison of this property with elements of the transformation matrix expressed in terms of betatron functions (5.135) shows that the solution of the betatron function is periodic and symmetric since $\alpha = 0$ both at the beginning and the end of the lattice period. We therefore have symmetry planes in the middle of the quadrupoles for the betatron functions in the horizontal as well as in the vertical plane. The betatron functions then have the general periodic and symmetric form as shown in Fig. 6.2.

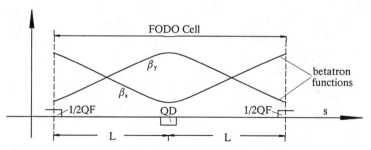

Fig. 6.2. Periodic betatron functions in a FODO channel

From (5.97) and (6.1), we can derive the analytical expression for the periodic and symmetric betatron function by setting $\beta_o = \beta$, $\alpha_o = 0$ and $\gamma_o = 1/\beta$ and get

$$\beta = \left(1 - 2 \cdot \frac{L^2}{f^2}\right)^2 \cdot \beta + 4L^2 \cdot \left(1 + \frac{L}{f}\right)^2 \cdot \frac{1}{\beta}, \tag{6.2}$$

where $f > 0$ and β is the value of the betatron function in the middle of the focusing quadrupole, QF. Solving for β we get after some manipulations.

$$\beta^+ = L \cdot \frac{\frac{f}{L} \frac{f}{L+1}}{\sqrt{\frac{f^2}{L^2 - 1}}} = L \cdot \frac{\kappa \cdot (\kappa + 1)}{\sqrt{\kappa^2 - 1}}, \tag{6.3}$$

where we define the *FODO parameter* κ by

$$\kappa = \frac{f}{L} > 1 \tag{6.4}$$

and set $\beta = \beta^+$ to indicate the solution in the center of the focusing quadrupole. Had we started at the defocusing quadrupole we would have to replace f by $-f$ and get analogous to (6.3) for the value of the betatron function in the middle of the defocusing quadrupole

$$\beta^- = L \frac{\kappa \cdot (\kappa - 1)}{\sqrt{\kappa^2 - 1}}. \tag{6.5}$$

These are the solutions for both the horizontal and the vertical plane. In the middle of the horizontally focusing quadrupole QF ($f > 0$) we have $\beta_x = \beta^+$ and $\beta_y = \beta^-$ and in the middle of the horizontally defocusing quadrupole QD ($f < 0$) we have $\beta_x = \beta^-$ and $\beta_y = \beta^+$. From the knowledge of the betatron functions at one point in the lattice, it is straightforward to calculate the value at any other point by proper matrix multiplications as discussed earlier. In open arbitrary beam transport lines the initial values of the betatron functions are not always known and there is no process other than measurements of the actual particle beam in phase space to determine the values of the betatron functions. The betatron functions in a periodic lattice in contrast are completely determined by the requirement that the solution be periodic with the periodicity of the lattice. It is not necessary that the focusing lattice be symmetric to obtain a unique, periodic solution. Equation (5.97) can be used for any periodic lattice requiring only the equality of the betatron functions at the beginning and at the end of the periodic structure. Of course, not any arbitrary although periodic arrangement of quadrupoles will lead to a viable solution and we will therefore derive conditions for periodic lattices to produce stable solutions.

The betatron phase for a FODO cell can be derived by applying (5.135) to a symmetric lattice. With $\alpha_o = \alpha = 0$ and $\beta_o = \beta$ this matrix is

$$\begin{bmatrix} \cos\phi & \beta\sin\phi \\ \frac{1}{\beta}\sin\phi & \cos\phi \end{bmatrix},$$ (6.6)

where ϕ is the betatron phase advance through a full symmetric period. Since the matrix (6.6) must be equal to the matrix (6.1) the phase must be

$$\cos\phi = 1 - 2\cdot\frac{L^2}{f^2} = \frac{\kappa^2 - 2}{\kappa^2}$$

or

$$\sin\frac{\phi}{2} = \frac{1}{\kappa}.$$ (6.7)

For the solution (6.7) to be real the parameter κ must be larger than unity, a result which also becomes obvious from (6.3,5). This condition is equivalent to stating that the focal length of half a quadrupole in a FODO lattice must be longer than the distances to the next quadrupole.

The solutions for the betatron functions depend strongly on the choice of the strength of the quadrupole. Specifically we observe that (6.3) for β^+ has minimum characteristics. For $\kappa \to 1$ as well as for $\kappa \to \infty$ we get $\beta \to \infty$, and therefore we expect a minimum between these extremes. Taking the derivative $d\beta^+/d\kappa = 0$ we get from (6.3)

$$\kappa_o^2 - \kappa_o - 1 = 0,$$ (6.8)

which can be solved for

$$\kappa_o = \frac{1}{2} \pm \sqrt{\frac{1}{4}+1} = 1.6180...$$ (6.9)

The optimum phase advance per FODO cell is therefore

$$\phi_o = 76.345\ldots^\circ.$$ (6.10)

The maximum value of the betatron function reaches a minimum for a FODO lattice with a phase advance of about 76.3° per cell. Since beam sizes scale with the square root of the betatron functions, a lattice with this phase advance per cell requires the minimum beam aperture.

This criteria, however, is true only for a flat beam when $\epsilon_x \gg \epsilon_y$ or $\epsilon_y \gg \epsilon_x$. For a round beam $\epsilon_x \approx \epsilon_y$ and maximum beam acceptance is obtained by minimizing the beam diameter or $E_x^2 + E_y^2 \sim \beta_x + \beta_y$, where E_x and E_y are the beam envelopes in the horizontal and vertical plane, respectively, see Fig. 6.3. This minimum is determined by $d(\beta_x+\beta_y)/d\varphi = 0$, or by

$$\kappa_{opt} = \sqrt{2}$$ (6.11)

and the optimum betatron phase per cell is

$$\phi_{opt} = 90°.$$ (6.12)

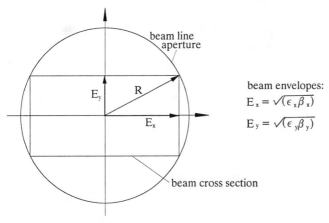

Fig. 6.3. Maximum acceptance of a FODO lattice with a circular aperture of radius R

This solution requires the minimum aperture in quadrupoles with a circular aperture of radius R for a beam with equal beam emittances in both planes $\epsilon_x = \epsilon_y = \epsilon$. The betatron functions are then simply

$$
\begin{aligned}
\beta_{\text{opt}}^+ &= L \cdot (2 + \sqrt{2}), \\
\beta_{\text{opt}}^- &= L \cdot (2 - \sqrt{2}).
\end{aligned}
\tag{6.13}
$$

The beam envelopes are $E_x^2 = \beta_{\text{opt}}^+ \cdot \epsilon$ and $E_y^2 = \beta_{\text{opt}}^- \cdot \epsilon$ and the maximum beam emittance to fit an aperture of radius R or the *acceptance* of the aperture can be determined from

$$
E_x^2 + E_y^2 = R^2 = \epsilon \cdot (\beta^+ + \beta^-).
\tag{6.14}
$$

From (6.13) we find $\beta^+ + \beta^- = 4L$ and the acceptance of a FODO channel with an aperture radius of R becomes

$$
\epsilon_{\max} = \frac{R^2}{4L}.
\tag{6.15}
$$

With this optimum solution we can develop general scaling laws for the betatron functions in a FODO lattice. We need not know the values of the betatron functions at all points of a periodic lattice to characterize the optical properties. It is sufficient to know these values at characteristic points like the symmetry points in a FODO channel, where the betatron functions reach maximum or minimum values. From (6.3,13) the betatron functions at these symmetry points are given by

$$
\begin{aligned}
\frac{\beta^+}{\beta_{\text{opt}}} &= \frac{\kappa \cdot (\kappa + 1)}{(2 + \sqrt{2}) \cdot \sqrt{\kappa^2 - 1}}, \\
\frac{\beta^-}{\beta_{\text{opt}}} &= \frac{\kappa \cdot (\kappa - 1)}{(2 - \sqrt{2}) \cdot \sqrt{\kappa^2 - 1}}.
\end{aligned}
\tag{6.16}
$$

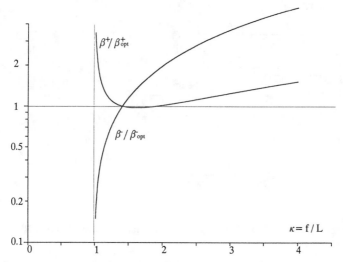

Fig. 6.4. Scaling of horizontal and vertical betatron functions in a FODO lattice

The scaling of the betatron function is independent of L and depends only on the ratio of the focal length to the distance between quadrupoles $\kappa = f/L$. In Fig. 6.4 the betatron functions β^+ and β^- are plotted as a function of the FODO parameter κ.

The distance L between quadrupoles is still a free parameter and can be adjusted to the needs of the particular application. We observe, however, that the maximum value of the betatron function varies linear with L and, therefore, the maximum beam size in a FODO lattice scales like $\sqrt{L}$.

6.2 Betatron Motion in Periodic Structures

For the design of circular accelerators it is of fundamental importance to understand the long term stability of the beam over many revolutions. Specifically we need to know if the knowledge of beam dynamics in one periodic unit can be extrapolated to many periodic units. In the following sections we discuss stability criteria as well as characteristic beam dynamics in periodic lattices.

6.2.1 Stability Criterion

The periodic solution for one period of a FODO lattice has been derived in the last section and we expect that such periodic focusing cells can be repeated indefinitely. Following the classic paper by Courant and Snyder [6.1] we will derive the stability conditions for an indefinite number of periodic but not necessarily symmetric focusing cells. The structure of the focusing cells can be arbitrary but must be periodic. If $\mathcal{M}(s + 2L|s)$ is the transformation matrix for one periodic cell we have for N cells

$$\mathcal{M}(s + N \cdot 2L|s) = [\mathcal{M}(s + 2L|s)]^N . \tag{6.17}$$

Stable solutions are obtained if all elements of the total transformation matrix stay finite as N increases indefinitely. To find the conditions for this we calculate the *eigenvalues* λ of the characteristic matrix equation. The eigenvalues λ are a measure for the magnitude of the matrix elements and therefore finite values for the eigenvalues will be the indication that the transformation matrix stays finite as well. We derive the eigenvalues from the characteristic matrix equation

$$(\mathcal{M} - \lambda \cdot \mathcal{I}) \cdot \mathbf{x} = 0 , \tag{6.18}$$

where $\mathcal{I}$ is the unity matrix. For nontrivial values of the eigenvectors $(\mathbf{x} \neq 0)$ the determinant

$$|\mathcal{M} - \lambda \cdot \mathcal{I}| = \begin{vmatrix} C - \lambda & S \\ C' & S' - \lambda \end{vmatrix} = 0$$

must vanish and with $CS' - SC' = 1$ we get the eigenvalue equation

$$\lambda^2 - (C + S') \cdot \lambda + 1 = 0 . \tag{6.19}$$

The solutions for the eigenvalues are

$$\lambda_{1,2} = \tfrac{1}{2}(C + S') \pm \left[\tfrac{1}{4}(C + S')^2 - 1\right]^{1/2}$$

or with the substitution $\tfrac{1}{2} \cdot (C + S') = \cos \phi$

$$\lambda_{1,2} = \cos \phi \pm \mathrm{i} \cdot \sin \phi = e^{\mathrm{i}\phi} . \tag{6.20}$$

At this point we require that the betatron phase ϕ be real or that the trace of the matrix $\mathcal{M}$ be

$$\mathrm{Tr}\{\mathcal{M}\} = C + S' \leq 2 . \tag{6.21}$$

On the other hand, we have for a full lattice period the transformation matrix

$$\mathcal{M} = \begin{bmatrix} \cos \phi + \alpha \cdot \sin \phi & \beta \cdot \sin \phi \\ -\gamma \cdot \sin \phi & \cos \phi - \alpha \cdot \sin \phi \end{bmatrix} , \tag{6.22}$$

which can be expressed with $\mathcal{J} = \begin{bmatrix} \alpha & \beta \\ -\gamma & -\alpha \end{bmatrix}$ by

$$\mathcal{M} = \mathcal{I} \cdot \cos \phi + \mathcal{J} \cdot \sin \phi . \tag{6.23}$$

This matrix has the form of *Euler's formula* for a complex exponential.

Since the determinant of $\boldsymbol{M}$ is unity we get $\gamma\beta - \alpha^2 = 1$ and $\boldsymbol{J}^2 = -\boldsymbol{I}$. Similar to *Moivre's formula* we get for N equal periods

$$\boldsymbol{M}^N = (\boldsymbol{I} \cdot \cos\phi + \boldsymbol{J} \cdot \sin\phi)^N = \boldsymbol{I} \cdot \cos N\phi + \boldsymbol{J} \cdot \sin N\phi \qquad (6.24)$$

and the trace for N periods is bounded if $\cos\phi < 1$ or if (6.21) holds

$$\mathrm{Tr}(\boldsymbol{M}^N) = 2\cos(N\phi) \le 2. \qquad (6.25)$$

This result is called the *stability criterion* for periodic beam transport lattices. At this point we note that the trace of the transformation matrix $\boldsymbol{M}$ does not depend on the reference point s. To show this we consider two different reference points s_1 and s_2, where $s_1 < s_2$, for which the following identities hold

$$\boldsymbol{M}(s_2 + 2L|s_1) = \boldsymbol{M}(s_2|s_1)\,\boldsymbol{M}(s_1 + 2L|s_1) = \boldsymbol{M}(s_2 + 2L|s_2)\,\boldsymbol{M}(s_2|s_1)$$

and solving for $\boldsymbol{M}(s_2 + 2L|s_2)$ we get

$$\boldsymbol{M}(s_2 + 2L|s_2) = \boldsymbol{M}(s_2|s_1)\,\boldsymbol{M}(s_1 + 2L|s_1)\,\boldsymbol{M}^{-1}(s_2|s_1). \qquad (6.26)$$

This is a similarity transformation and, therefore, both transformation matrices $\boldsymbol{M}(s_2 + 2L|s_2)$ and $\boldsymbol{M}(s_1 + 2L|s_1)$ have the same trace and eigenvalues independent of the choice of the location s.

6.2.2 General FODO Lattice

So far we have considered FODO lattices, where both quadrupoles have equal strength, $f_1 = -f_2 = f$. Since we made no use of this in the derivation of the stability criterion for betatron functions we expect that stability can also be obtained for unequal quadrupoles strengths. In this case the transformation matrix of half a FODO cell is

$$\boldsymbol{M}_{1/2} = \begin{bmatrix} 1 & 0 \\ -\frac{1}{f_2} & 1 \end{bmatrix} \cdot \begin{bmatrix} 1 & L \\ 0 & 1 \end{bmatrix} \cdot \begin{bmatrix} 1 & 0 \\ -\frac{1}{f_1} & 1 \end{bmatrix} = \begin{bmatrix} 1 - \frac{L}{f_1} & L \\ -\frac{1}{f^*} & 1 - \frac{L}{f_2} \end{bmatrix}, (6.27)$$

where $1/f^* = +1/f_1 + 1/f_2 - L/(f_1 \cdot f_2)$. Multiplication with the reverse matrix gives for the full transformation matrix of the FODO cell

$$\boldsymbol{M} = \begin{bmatrix} 1 - 2\frac{L}{f^*} & 2L \cdot (1 - \frac{L}{f_2}) \\ -\frac{2}{f^*} \cdot (1 - \frac{L}{f_1}) & 1 - 2\frac{L}{f^*} \end{bmatrix}. \qquad (6.28)$$

The stability criterion

$$\mathrm{Tr}\{\boldsymbol{M}\} = \left| 2 - \frac{4L}{f^*} \right| < 2 \qquad (6.29)$$

is equivalent to

$$0 < \frac{L}{f^*} < 1. \qquad (6.30)$$

To determine the region of stability in the (u, v)-plane, where $u = L/f_1$ and $v = L/f_2$, we use $L/f^* = +L/f_1 + L/f_2 - L^2/(f_1 \cdot f_2)$ and get with (6.30) the condition

$$0 < u + v - uv < 1, \tag{6.31}$$

where u and v can be positive or negative. Solving the second inequality for either u or v we find the conditions $|u| < 1$ and $|v| < 1$. With this the first inequality can be satisfied only if u and v have different signs. The boundaries of the stability region are therefore given by the four equations

$$|u| = 1, \qquad\qquad |v| = \frac{|u|}{1 - |u|},$$

$$\tag{6.32}$$

$$|v| = 1, \qquad\qquad |u| = \frac{|v|}{1 + |v|},$$

defining the stability region shown in Fig. 6.5 which is also called the *necktie diagram* because of its shape. Due to the full symmetry in $|u|$ and $|v|$ the shaded area in Fig. 6.5 is the stability region for both the horizontal and vertical plane.

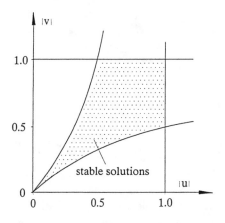

Fig. 6.5. Necktie diagram in thin lens approximation

For convenience, we used the thin lens approximation to calculate the necktie diagram. Nothing fundamentally will, however, change when we use the transformation matrices for real quadrupoles of finite length except for a small variation of the stability boundaries depending on the degree of deviation from the thin lens approximation. With the general transformation matrix for a full FODO period

$$\mathcal{M} = \begin{bmatrix} C & S \\ C' & S' \end{bmatrix}$$

the periodic solution for the betatron function is

$$\beta^2 = \frac{S^2}{1 - C^2} \tag{6.33}$$

and the stability condition:

$$\text{Tr}(\boldsymbol{M}) = |C + S'| < 2. \tag{6.34}$$

The stability diagram has still the shape of a necktie although the boundaries are slightly curved.

The general transformation matrix for half a FODO cell can be obtained in matrix formalism with $\psi = \sqrt{k} \cdot \ell$ by multiplying the matrices

$$
\boldsymbol{M}_{1/2} = \begin{bmatrix} \cosh \psi_2 & \frac{\ell_2}{\psi_2} \cdot \sinh \psi_2 \\ \frac{\psi_2}{\ell_2} \cdot \sinh \psi_2 & \cosh \psi_2 \end{bmatrix} \cdot \begin{bmatrix} 1 & L \\ 0 & 1 \end{bmatrix}
$$
$$
\cdot \begin{bmatrix} \cos \psi_1 & \frac{\ell_1}{\psi_1} \cdot \sin \psi_1 \\ -\frac{\psi_1}{\ell_1} \cdot \sin \psi_1 & \cos \psi_1 \end{bmatrix}, \tag{6.35}
$$

where now L is not the half cell length but just the drift space between two adjacent quadrupoles of finite length and the indices refer to the first and the second quadrupole, respectively. From this we get the full period transformation matrix by multiplication with the reverse matrix

$$\boldsymbol{M} = \begin{bmatrix} C & S \\ C' & S' \end{bmatrix} = \boldsymbol{M}_{1/2,r} \cdot \boldsymbol{M}_{1/2}. \tag{6.36}$$

Obviously the mathematics becomes elaborate although straight forward and it is prudent to use computers to find the desired results.

As reference examples to study and discuss a variety of accelerator physics issues in this text we consider a range of different FODO lattices, see Table 6.1 which are of some but definitely not exhaustive practical interest. Other periodic lattices are of great interest as well specifically for synchrotron radiation sources but are less accessible to analytical discussions than a FODO lattice. All examples except #2 are separated function lattices.

Example #1 is that for a 10 GeV electron synchrotron at DESY [6.2,3] representing a moderately strong focusing lattice with a large stability range as is commonly used if no extreme beam parameters are required as is the case for synchrotrons used to inject into storage rings. Figure 6.6 shows the betatron functions for this lattice. We note small deviations from a regular FODO lattice which is often required to make space for other components. Such deviations from a regular lattice cause only small perturbations in the otherwise periodic betatron functions. Strong focusing is required for beam transport lines in linear electron positron collider facilities to minimize increase of the beam emittance due to synchrotron radiation. As example

Table 6.1. FODO cell parameters

example	# 1	# 2	# 3	# 4
energy, $E(\mathrm{GeV})$	10	50	4	20,000
half cell length, $L(\mathrm{m})$	6.0	2.6	3.6	114.25
quadrupole length, $\ell_q(\mathrm{m})$	0.705	1.243	0.15	3.64
bending radius, $\rho(\mathrm{m})$	27.12	279.38	152.8	10,087
bending magnet length, $\ell_b(\mathrm{m})$	3.550	2.486	2.50	99.24
phase advance per cell, ψ	101.4	108.0	135.0	90.0
quadrupole strength[†], $k(\mathrm{m}^{-2})$	....	....	....	....
lattice type* (FODO)	sf	cf	sf	sf

[†] these parameters will be determined in problem 6.1
* sf: separated function; cf: combined function lattice.

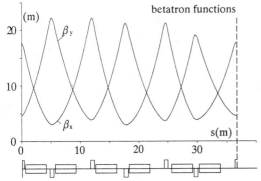

Fig. 6.6. FODO lattice for one octant of a synchrotron [6.2,3] (example #1 in Table 6.1)

#2 we use the lattice for the long curved beam transport lines leading the 50 GeV beam from the linac to the collision area at the Stanford Linear Collider [6.4]. This lattice exhibits the greatest deviation from a thin lens FODO channel as shown in Fig. 6.7. Example #3 resembles a theoretical lattice for an extremely small beam emittance used to study fundamental limits of beam stability and control of aberrations [6.6]. Lattices for future very high energy hadron colliders in the TeV range use rather long FODO cells leading to large values of the betatron and dispersion functions and related high demands on magnet field and alignment tolerances. Arc lattice parameters for the 20 TeV Superconducting Super Collider, SSC [6.7] are compiled as example #4.

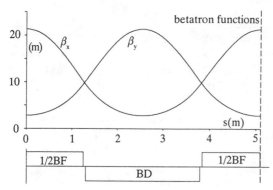

Fig. 6.7. FODO cell for a linear collider transport line [6.4, 5]   (example #2 in Table 6.1)

6.3 Beam Dynamics in Periodic Closed Lattices

In the previous section we discussed the beam dynamics in a general periodic lattice and FODO lattice in particular and we will use such periodic lattices to construct a closed path for circular accelerators like synchrotrons and storage rings. The term "circular" is used in this context rather loosely since such accelerators are generally composed of both circular and straight sections giving the ring the appearance of a circle, a polygon or racetrack. Common to all these rings is the fact that the reference path must be a closed path or *orbit* so that at least the total circumference of the ring constitutes a periodic lattice that repeats turn for turn.

6.3.1 Hill's Equation

The motion of particles or more massive bodies in periodic external fields has been studied extensively by astronomers in the last century specially in connection with the three body problem. In particle beam dynamics we find the equation of motion in periodic lattices to be similar to those studied in the last century by the astronomer Hill. We will discuss in this chapter the equation of motion, called *Hill's equation* its solutions and properties.

Particle beam dynamics in periodic systems is determined by the equation of motion

$$u'' + K(s) \cdot u = 0, \tag{6.37}$$

where $K(s)$ is periodic with the period L_p

$$K(s) = K(s + L_p). \tag{6.38}$$

The length of a period L_p may be the circumference of the circular accelerator lattice or the length of a superperiod which is repeated several times around the circumference. The differential equation (6.37) with the periodic coefficient (6.38) has all the characteristics of a Hill's differential equation [6.8]. The solutions of Hill's equation and their properties have been formulated by *Floquet's theorem*:

- two independent solutions exist of the form

$$u_1(s) = w(s) \cdot e^{i\mu s/L_p},$$
$$u_2(s) = w^*(s) \cdot e^{-i\mu s/L_p};$$

(6.39)

- $w^*(s)$ is the complex conjugate solution to $w(s)$. For all practical cases of beam dynamics we have only real solutions and $w^*(s) = w(s)$;
- the function $w(s)$ is unique and periodic in s with period L_p

$$w(s + L_p) = w(s);$$

(6.40)

- μ is a characteristic coefficient defined by

$$\cos \mu = \tfrac{1}{2} \operatorname{Tr} \big[\mathcal{M}(s + L_p | s) \big];$$

(6.41)

- the trace of the transformation matrix $\mathcal{M}$ is independent of s

$$\operatorname{Tr} \big[\mathcal{M}(s + L_p | s) \big] \neq f(s);$$

(6.42)

- the determinant of the transformation matrix is equal to unity

$$\det \mathcal{M} = 1;$$

(6.43)

- the solutions remain finite for

$$\tfrac{1}{2} \cdot \operatorname{Tr} \big[\mathcal{M}(s + L_p | s) \big] < 1.$$

(6.44)

The amplitude function $w(s)$ and the characteristic coefficient μ can be correlated to quantities we have derived earlier for general beam transport systems using different methods. The transformation of a trajectory u through one lattice period of length L_p must be equivalent to the multiplication by the transformation matrix (6.22) for that period which gives

$$u(s + L_p) = (\cos \psi + \alpha \cdot \sin \psi) \cdot u(s) + \beta \cdot \sin \psi \cdot u'(s),$$

where u stands for any of the two solutions (6.39) and ψ is the betatron phase advance for the period. From (6.39,40) we get on the other hand

$$u(s + L_p) = u(s) \cdot e^{\pm i\mu} = u(s) \cdot (\cos \mu \pm i \cdot \sin \mu).$$

Comparing the coefficients for the sine and cosine terms we get

$$\cos \psi = \cos \mu \quad \text{or} \quad \psi = \mu$$

(6.45)

and

$$\alpha \cdot u(s) + \beta \cdot u'(s) = \pm i \cdot u(s).$$

(6.46)

The first equality, (6.45), can also be derived from (6,22,41). Equation (6.46) can be further simplified if we perform a logarithmic differentiation

$$\frac{u''}{u'} - \frac{u'}{u} = -\frac{\beta'}{\beta} - \frac{\alpha'}{\pm i - \alpha} \cdot$$

On the other hand we can construct from (6.37,46) the expression

$$\frac{u''}{u'} - \frac{u'}{u} = \frac{-K\beta}{\pm i - \alpha} - \frac{\pm i - \alpha}{\beta} \cdot$$

and equating the r.h.s. of both expressions we find

$$(1 - \alpha^2 - K \cdot \beta^2 + \alpha' \cdot \beta - \alpha \cdot \beta') \pm i \cdot (2\alpha + \beta') = 0 \,,$$

where all functions in the brackets are real as long as we have stability and, therefore, both brackets must be equal to zero separately

$$\beta' = -2\alpha \,, \tag{6.47}$$

and

$$\alpha' = K \cdot \beta - \gamma \,. \tag{6.48}$$

Equation (6.47) can be used in (6.46) for

$$\frac{u'}{u} = \frac{\pm i - \alpha}{\beta} = \pm \frac{i}{\beta} + \frac{1}{2}\frac{\beta'}{\beta} \,, \tag{6.49}$$

which can be integrated immediately for

$$\log\frac{u}{u_o} = \pm i \int_o^s \frac{ds}{\beta} + \frac{1}{2} \cdot \log\frac{\beta}{\beta_o} \,, \tag{6.50}$$

where $u_o = u(s_o)$ and $\beta_o = \beta(s_o)$ for $s = s_o$. Solving for $u(s)$ we get the well known solution

$$u(s) = a \sqrt{\beta(s)}\, e^{\pm i\,\psi} \,, \tag{6.51}$$

where $a = u_o/\sqrt{\beta_o}$ and

$$\psi(s - s_o) = \int_{s_o}^s \frac{d\tau}{\beta(\tau)} \,. \tag{6.52}$$

With $\psi(L_p) = \mu$ and

$$\sqrt{\beta(s)} = \frac{w(s)}{a} \tag{6.53}$$

we find the previous definitions of the betatron functions to be consistent with the coefficients of Floquet's solutions in a periodic lattice. In the next section we will apply the matrix formalism to determine the solutions of the betatron functions in periodic lattices.

6.3.2 Periodic Betatron Functions

Having determined the existence of stable solutions for particle trajectories in periodic lattices we will now derive periodic and unique betatron functions. For this we take the transformation matrix of a full lattice period

$$\mathcal{M}(s + L_\mathrm{p} \,|\, s) = \begin{bmatrix} C & S \\ C' & S' \end{bmatrix} \tag{6.54}$$

and construct the well known transformation matrix for betatron functions.

$$\begin{bmatrix} \beta \\ \alpha \\ \gamma \end{bmatrix} = \begin{bmatrix} C^2 & -2SC & S' \\ -CC' & SC' + S'C & -SS' \\ C'^2 & -2S'C' & S'^2 \end{bmatrix} \cdot \begin{bmatrix} \beta_\mathrm{o} \\ \alpha_\mathrm{o} \\ \gamma_\mathrm{o} \end{bmatrix} = \mathcal{M}_\beta \cdot \begin{bmatrix} \beta_\mathrm{o} \\ \alpha_\mathrm{o} \\ \gamma_\mathrm{o} \end{bmatrix} . \tag{6.55}$$

Lattice functions are not changed for a lattice segment with a *unity transformation matrix*. In this case any set of lattice functions $\boldsymbol{\beta} = (\beta, \alpha, \gamma)$ is also a periodic solution. Because of the quadratic nature of the matrix elements we find the same result in case of a 180° phase advance for the lattice segment. Any such lattice segment with a phase advance of an integer multiple of 180° is neutral to the transformation of lattice functions. This feature can be used to create irregular insertions in a lattice that do not disturb the lattice functions outside the *insertions*.

To obtain from (6.55) a general periodic solution for the betatron functions we simply solve the *eigenvector equation*

$$(\mathcal{M}_\beta - \mathcal{I}) \cdot \boldsymbol{\beta} = 0 . \tag{6.56}$$

The solution can be obtained from the component equations of (6.56)

$$\begin{aligned} (C^2 - 1) \cdot \beta - 2\,SC \cdot \alpha + S^2 \cdot \gamma &= 0 , \\ CC' \cdot \beta - (S'C + CS' - 1) \cdot \alpha + SS'\gamma &= 0 , \\ C'^2 \cdot \beta - 2S'C' \cdot \alpha + (S'^2 - 1) \cdot \gamma &= 0 . \end{aligned} \tag{6.57}$$

A particular simple solution is obtained if the periodic lattice includes a symmetry point. In this case we define this symmetry point as the start of the periodic lattice, set $\alpha = 0$, and get the simple solutions

$$\beta^2 = \frac{S^2}{1 - C^2} \qquad \alpha = 0 \qquad \gamma = \frac{1}{\beta} . \tag{6.58}$$

The transformation matrix for a superperiod or the full circumference of a ring becomes then simply from (5.135)

$$M = \begin{bmatrix} \cos\mu & \beta \cdot \sin\mu \\ -\frac{1}{\beta} \cdot \sin\mu & \cos\mu \end{bmatrix}, \tag{6.59}$$

where μ is the phase advance for the full lattice period. The solutions are stable as long as the trace of the transformation matrix meets the *stability criterion* (6.34) or if $\mu \neq n \cdot \pi$, where n is an integer.

Different from an open transport line, well determined and unique starting values for the periodic betatron functions exist in a closed lattice due to the periodicity requirement allowing us to determine the betatron function anywhere else in the lattice. Although (6.58) allows both a positive and a negative solution for the betatron function, we choose only the positive solution for the definition of the betatron function.

Stable general periodic solutions for asymmetric but periodic lattices, where $\alpha \neq 0$, can be obtained in a straightforward way from (6.57) as long as the determinant $|M_p - I| \neq 0$.

The betatron phase for a full turn around a circular accelerator of circumference C is from (6.52)

$$\mu(C) = \int_{s}^{s+C} \frac{d\tilde{s}}{\beta(\tilde{s})}. \tag{6.60}$$

If we divide this equation by 2π we get a quantity ν which is equal to the number of betatron oscillations executed by particles traveling once around the ring. This number is called the *tune* or *operating point* of the circular accelerator. Since there are different betatron functions in the horizontal plane and in the vertical plane, we also get separate tunes in a circular accelerator for both planes

$$\nu_{x,y} = \frac{1}{2\pi} \oint \frac{d\tilde{s}}{\beta_{x,y}(\tilde{s})}. \tag{6.61}$$

This definition is equivalent to having chosen the integration constant equal to $1/2\pi$ in (5.120) instead of unity. Yet another normalization can be obtained by choosing $1/\nu$ for the integration constant in (5.120), in which case the phase defined as

$$\varphi(s) = \frac{\psi(s)}{\nu} = \int_{0}^{s} \frac{d\tilde{s}}{\nu\,\beta(\tilde{s})} \tag{6.62}$$

varies between 0 and 2π along the circumference of a ring lattice. This normalization will become convenient when we try to decompose periodic field errors in the lattice into Fourier components to study their effects on beam stability.

Equation (6.61) can be used to get an approximate expression for the relationship between the betatron function and the tune. If $\overline{\beta}$ is the average value for the betatron function around the ring we have $\mu(C) = 2\pi\nu \approx C/\bar{\beta} \approx 2\pi R/\bar{\beta}$ or

$$\bar{\beta} = \frac{R}{\nu}. \tag{6.63}$$

Eq. (6.63) is amazingly accurate for most rings and is, therefore, a useful tool for a quick estimate of the average betatron function or for the tunes and is often referred to as the *smooth approximation*. This expression and approximation is specially useful to derive scaling laws.

In a circular accelerator three tunes are defined for the three degrees of freedom, the horizontal, the vertical and the longitudinal motion. In Fig. 6.8 the measured frequency spectrum is shown for a particle beam in a circular accelerator. The electric signal from an isolated electrode in the vacuum chamber is recorded and connected to a frequency analyzer. The signal amplitude depends on the distance of the passing beam to the electrode and therefore includes the information of beam oscillations as a modulation of the revolution frequency.

Coupled oscillations of particles about a longitudinal reference point and about the ideal particle momentum are called longitudinal oscillations. Such oscillations cannot be measured with transversely selective electrodes unless the electrode is installed at a location, where the dispersion function is finite and the variation of the particle energy transforms into transverse positions. Analogous to the transverse motion a longitudinal tune, ν_s, is defined as the number of oscillations per revolution.

We note a number of frequencies in the observed spectrum of the storage ring SPEAR as shown in Fig. 6.8. At the low frequency end two frequencies indicate the *longitudinal tune* ν_s and its first harmonic at $2\nu_s$. The other two large signals are the horizontal and vertical tunes of the accelerator. Since the energy oscillation affects the focusing of the particles, we also observe two weak *satellites frequencies* on one of the transverse tunes at a distance of $\pm\nu_s$. The actual frequencies observed are not directly equal to $\nu \cdot \omega_0$, where ω_0 is the *revolution frequency*, but are only the nonintegral part of the tune $\Delta\nu \cdot \omega_0$, where $\Delta\nu = \pm\nu$ mod 1.

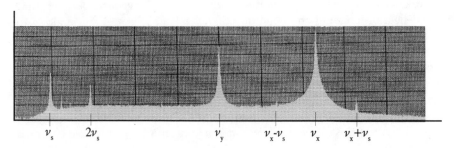

Fig. 6.8. Frequency spectrum from a circulating particle beam, ν_s synchrotron tune, ν_x, ν_y betatron tunes, $\nu_x \pm \nu_s$ satellites

6.4 Periodic Dispersion Function

The dispersion function can be periodic if the lattice is periodic. In this section we will determine the periodic solution of the dispersion function first for the simple lattice building block of a FODO channel and then for general but periodic lattice segments.

6.4.1 Scaling of the Dispersion in a FODO Lattice

Properties of a FODO lattice have been discussed in detail for a monochromatic particle beam only and no chromatic effects have been taken into account. To complete this discussion we now include chromatic effects which cause, in linear approximation, a dispersion proportional to the energy spread in the beam and are caused by bending magnets. We have used the transformation matrix for a symmetric quadrupole triplet as the basic FODO cell. The bending magnet edge focusing was ignored and so were chromatic effects. In the following we still ignore the quadratic edge focusing effects of the bending magnets, but we cannot ignore any longer linear effects of energy errors. For simplicity we assume again thin lenses for the quadrupoles and get for the chromatic transformation matrix through half a FODO cell, $1/2$ QF - B - $1/2$ QD:

$$\mathcal{M}_{1/2\,FODO} = \begin{bmatrix} 1 & 0 & 0 \\ 1/f & 1 & 0 \\ 0 & 0 & 1 \end{bmatrix} \cdot \begin{bmatrix} 1 & L & L^2/(2\rho) \\ 0 & 1 & +L/\rho \\ 0 & 0 & 1 \end{bmatrix} \cdot \begin{bmatrix} 1 & 0 & 0 \\ -1/f & 1 & 0 \\ 0 & 0 & 1 \end{bmatrix}$$

or after multiplication

$$\mathcal{M}_{1/2\,FODO} = \begin{bmatrix} 1-L/f & L & L^2/(2\rho) \\ -L/f^2 & 1+L/f & \frac{L}{\rho(1+L/2f)} \\ 0 & 0 & 1 \end{bmatrix} . \tag{6.64}$$

The absolute value of the *focal length* f is the same for both quadrupoles but since we start at the symmetry point in the middle of a quadrupole this focal length is based only on half a quadrupole. We have also assumed that the deflection angle of the bending magnet is small, $\theta \ll 1$, in analogy to thin lens approximation for quadrupoles.

In Sect. 5.6 dispersive elements of transformation matrices have been derived. In periodic lattices, however, we find also a particular solution which is periodic with the periodicity of the focusing lattice. We look for such a *periodic solution* of the *dispersion unction* and label it by $\eta(s)$ in distinction from the ordinary, generally nonperiodic dispersion function $D(s)$. The typical form of the *periodic dispersion function* in a FODO lattice is shown in Fig. 6.9.

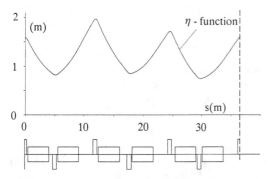

Fig. 6.9. Dispersion function in FODO cells (example #1 in Tab. 6.1)

In addition to the periodicity we also find this η-*function* to be symmetric with respect to the symmetry points in the middle of the FODO quadrupoles, where the derivative of the η-function vanishes. The transformation through one half FODO cell is

$$
\begin{bmatrix} \eta^- \\ 0 \\ 1 \end{bmatrix} = \mathcal{M}_{1/2\,\text{FODO}} \cdot \begin{bmatrix} \eta^+ \\ 0 \\ 1 \end{bmatrix} , \tag{6.65}
$$

where we have set $\delta = 1$ in accordance with the definition of dispersion functions.

In the particular arrangement of quadrupoles chosen in (6.64) the focusing quadrupole is the first element and, therefore, the dispersion function reaches a maximum value η^+ there. In the center of the defocusing quadrupole the dispersion function is reduced to a minimum value η^-. The opposite sequence of the quadrupoles would lead to similar results. From (6.65) we get the two equations

$$
\begin{aligned}
\eta^- &= \left(1 - \frac{L}{f}\right)\eta^+ + \frac{L^2}{2\rho} , \\
0 &= -\frac{L}{f^2}\eta^+ + \frac{L}{\rho}\cdot\left(1 + \frac{L}{2f}\right) .
\end{aligned} \tag{6.66}
$$

Solving the equations of (6.66) for the periodic dispersion function in the middle of the FODO quadrupoles, where $\eta' = 0$, we get in the focusing or defocusing quadrupole respectively

$$
\begin{aligned}
\eta^+ &= \frac{f^2}{\rho}\left(1 + \frac{L}{2f}\right) = \frac{L^2}{2\rho}\kappa\,(2\kappa + 1) , \\
\eta^- &= \frac{f^2}{\rho}\left(1 - \frac{L}{2f}\right) = \frac{L^2}{2\rho}\kappa\,(2\kappa - 1) ,
\end{aligned} \tag{6.67}
$$

where $\kappa = f/L$.

Note that in this approximation the bending magnet is as long as the length of half the FODO cell since the quadrupoles are assumed to be thin lenses and no drift spaces have been included between the quadrupoles and the bending magnet. The bending radius ρ, therefore, is equal to the average bending radius in the FODO lattice. From the known values of the dispersion function at the beginning of the FODO lattice we can calculate this function anywhere else in this periodic cell. Similar to the discussion in Sect. 6.1 we chose an optimum reference lattice, where

$$\kappa_\mathrm{o} = \sqrt{2}, \tag{6.68}$$

and

$$\eta_\mathrm{o}^+ = \frac{L^2}{2\rho}\left(4 + \sqrt{2}\right),$$
$$\eta_\mathrm{o}^- = \frac{L^2}{2\rho}\left(4 - \sqrt{2}\right). \tag{6.69}$$

In Fig. 6.10 the values of the dispersion functions, normalized to those for the optimum FODO lattice in the middle of the FODO quadrupoles, are plotted versus the FODO cell parameter κ. From Fig. 6.10 we note a diminishing dispersion function in a FODO cell as the betatron phase per cell or the focusing is increased. This result will be important later for the design of storage rings for specific applications requiring either large or small beam emittance. The procedure to determine the dispersion functions in a FODO cell is straightforward and can easily be generalized to real FODO lattices with finite quadrupole length and shorter bending magnets although it may be desirable to perform the matrix multiplications on a computer. For exploratory designs of accelerators structures, however, the thin lens approximation is a powerful and fairly accurate design tool.

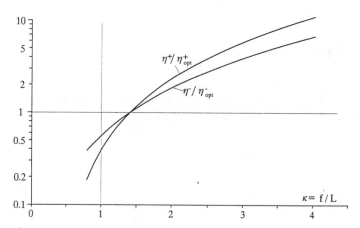

Fig. 6.10. Scaling of the dispersion function in a FODO lattice

6.4.2 General Solution for the Periodic Dispersion

In the previous section the dispersion function for a periodic and symmetric FODO lattice was derived. Many periodic lattice structures, however, are neither symmetric nor are they pure FODO structures and therefore we need to derive the periodic dispersion function in a more general form. To do this we include in the equation of motion also the linear energy error term from (4.86)

$$u'' + K(s) \cdot u = \frac{1}{\rho}(s) \cdot \delta. \tag{6.70}$$

For particles having the ideal energy, $\delta = 0$, the right hand side vanishes and the solutions are composed of betatron oscillations and the trivial solution

$$u_o(s) \equiv 0. \tag{6.71}$$

This trivial solution of (6.70) is clearly periodic and represents what is called in beam transport systems the *reference path* and in circular accelerators the *equilibrium orbit* or *closed orbit* about which particles perform betatron oscillations. The expression for the ideal equilibrium orbit is this simple since we decided to use a curvilinear coordinate system which follows the design orbit (6.71) as determined by the ideal placement of the bending magnets and quadrupoles.

For off momentum particles, $(\delta \neq 0)$, the closed orbit is displaced from the ideal closed orbit. Ignoring for a moment the s-dependence of K and ρ, this systematic displacement of the orbit is of the order of

$$\Delta u = \frac{1}{K \cdot \rho} \cdot \delta \tag{6.72}$$

as suggested by (6.70). In a real circular accelerator we expect a similar although s-dependent displacement of the equilibrium orbit for off momentum particles. For each particle energy only one equilibrium orbit exists in a given closed lattice. If there were two solutions u_1 and u_2 of (6.70) we could write for the difference

$$(u_1 - u_2)'' + K(s) \cdot (u_1 - u_2) = 0, \tag{6.73}$$

which is the differential equation for betatron oscillations. Different orbits for the same energy, therefore, differ only by betatron oscillations which are already included in the general solution as the homogeneous part of the differential equation (6.70). Therefore, in a particular circular lattice only one unique equilibrium orbit or closed orbit exists for each energy.

Chromatic transformation matrices have been derived in Sect. 5.6. If we apply these matrices to a circular lattice and calculate the total transformation matrix around the whole ring, we will be able to determine a self-consistent solution for equilibrium orbits. Before we calculate the periodic equilibrium orbits we note that the solutions of (6.70) are proportional to the momentum deviation δ. We therefore define the generalized periodic dispersion function as the equilibrium orbit for $\delta = 1$ which we called in the previous section the η-function. The transformation matrix for a periodic lattice of length L_{p} is

$$\boldsymbol{M}(s + L_{\mathrm{p}} \,|\, s) = \begin{bmatrix} C(s + L_{\mathrm{p}}) & S(s + L_{\mathrm{p}}) & D(s + L_{\mathrm{p}}) \\ C'(s + L_{\mathrm{p}}) & S'(s + L_{\mathrm{p}}) & D'(s + L_{\mathrm{p}}) \\ 0 & 0 & 1 \end{bmatrix} \tag{6.74}$$

and we get for the η-function with $\eta(s + L_{\mathrm{p}}) = \eta(s)$, $\eta'(s + L_{\mathrm{p}}) = \eta'(s)$ and (6.74)

$$\begin{aligned} \eta(s) &= C(s + L_{\mathrm{p}}) \cdot \eta(s) + S(s + L_{\mathrm{p}}) \cdot \eta'(s) + D(s + L_{\mathrm{p}}), \\ \eta'(s) &= C'(s + L_{\mathrm{p}}) \cdot \eta(s) + S'(s + L_{\mathrm{p}}) \cdot \eta'(s) + D'(s + L_{\mathrm{p}}). \end{aligned} \tag{6.75}$$

These two equations can be solved for $\eta(s)$ and $\eta'(s)$, the periodic dispersion function at the point s. The equilibrium orbit for any off momentum particle can be derived from this solution by multiplying with δ

$$u_{\delta}(s) = \eta(s) \cdot \delta. \tag{6.76}$$

In a more formal way the periodic solution for the dispersion function can be derived from (6.75) in a form without the arguments for increased clarity

$$\begin{aligned} (C - 1) \cdot \eta + S \cdot \eta' + D &= 0 \\ C' \cdot \eta + (S' - 1) \cdot \eta' + D' &= 0, \end{aligned} \tag{6.77}$$

which in vector notation is

$$(\boldsymbol{M}_{\eta} - \boldsymbol{I}) \cdot \boldsymbol{\eta} = 0, \tag{6.78}$$

where $\boldsymbol{M}_{\eta}$ is formed from (6.74) and $\boldsymbol{\eta} = (\eta, \eta', 1)$. The periodic dispersion function is therefore the eigenvector of the eigenvalue equation (6.78).

A particularly simple result is obtained if the point s is chosen at a symmetry point, where $\eta'_{\mathrm{sym}} = 0$. In this case the dispersion function at the symmetry point is

$$\eta_{\mathrm{sym}} = \frac{D}{1 - C} \qquad \text{and} \qquad \eta'_{\mathrm{sym}} = 0. \tag{6.79}$$

Once the values of the η-functions are known at one point it is straightforward to obtain the values at any other point in the periodic lattice by matrix multiplication. We may also try to derive an analytical solution for the periodic dispersion from the differential equation

$$\eta'' + K \cdot \eta = \frac{1}{\rho}. \tag{6.80}$$

The solution is again the composition of the solutions for the homogeneous and the inhomogeneous differential equation. Before we try a solution, however, we will transform (6.80) into normalized coordinates $w_\eta = \eta/\sqrt{\beta}$ and $d\varphi = ds/(\nu\beta)$. In these coordinates (6.80) becomes

$$\frac{d^2}{d\varphi^2} \cdot w_\eta + \nu^2 \cdot w_\eta = \nu^2 \cdot \beta^{3/2} \cdot \frac{1}{\rho} = \nu^2 \cdot F(\varphi). \tag{6.81}$$

An analytical solution to (6.81) has been derived in Sect. 4.8.3 and we have accordingly

$$w_\eta(\varphi) = w_{\eta o} \cdot \cos\nu\varphi + \frac{\dot{w}_{\eta o}}{\nu} \cdot \sin\nu\varphi$$

$$+ \nu \cdot \int_o^\varphi F(\tau) \cdot \sin\nu(\varphi - \tau) \cdot d\tau,$$

$$\frac{\dot{w}_\eta}{\nu}(\varphi) = -w_{\eta o} \cdot \sin\nu\varphi + \frac{\dot{w}_{\eta o}}{\nu} \cdot \cos\nu\varphi \tag{6.82}$$

$$+ \nu \cdot \int_o^\varphi F(\tau) \cdot \cos\nu(\varphi - \tau) \cdot d\tau,$$

where we have set $\dot{w} = \frac{d}{d\varphi} w(\varphi)$. To select a periodic solution we set

$$w_\eta(2\pi) = w_\eta(0) = w_{\eta o} \qquad \text{and} \qquad \dot{w}_\eta(2\pi) = \dot{w}_{\eta o}$$

and inserting these boundary conditions into the first equation (6.82) the general periodic solution for the normalized dispersion function becomes after some manipulations

$$w_\eta(\varphi) = \frac{\nu}{2\sin\pi\nu} \cdot \int_\varphi^{\varphi+2\pi} F(\tau) \cdot \cos[\nu(\varphi - \tau + \pi)] \cdot d\tau. \tag{6.83}$$

Now we return to the original variables, (η, s), and get the equation for the periodic dispersion or η-function

$$\eta(s) = \frac{\sqrt{\beta(s)}}{2\sin\pi\nu} \cdot \int_s^{s+L_p} \frac{\sqrt{\beta(\sigma)}}{\rho(\sigma)} \cdot \cos\nu[\varphi(s) - \varphi(\sigma) + \pi] \cdot d\sigma. \tag{6.84}$$

This solution shows clearly that the periodic dispersion function at any point s depends on all bending magnets in the ring. We also observe a fundamental resonance phenomenon which occurs should the tune of the ring approach an integer in which case finite equilibrium orbits for off momentum particles do not exist anymore. To get stable equilibrium orbits, the tune of the ring must not be chosen to be an integer or in accelerator terminology an *integer resonance* must be avoided

$$\nu \neq n,\tag{6.85}$$

where n is an integer.

This is consistent with the solution (6.79), where we learned that $C(s + L_p)$ must be different from unity. Since C is the matrix element for the total ring we have $C = \cos 2\pi\nu$ which obviously is equal to $+1$ only for integer values of the tune ν. While (6.84) is not particularly convenient to calculate the dispersion function, it clearly exhibits the resonance character and will be very useful later in some other context, for example, if we want to determine the effect of a single bending magnet.

Another way to solve the differential equation (6.81) will be considered to introduce a powerful mathematical method useful in periodic systems. We note that the perturbation term $F(s) = \beta^{3/2}(s)/\rho(s)$ is a periodic function with the period L_p or using normalized coordinates with the period 2π. The perturbation term can therefore be expanded into a Fourier series

$$\beta^{\frac{3}{2}}\frac{1}{\rho} = \sum F_n e^{in\varphi},\tag{6.86}$$

where

$$F_n = \frac{1}{2\pi} \oint \frac{\beta^{3/2}}{\rho} e^{-in\varphi} d\varphi\tag{6.87}$$

or if we go back to regular variables

$$F_n = \frac{1}{2\pi\nu} \oint \frac{\sqrt{\beta(\sigma)}}{\rho(\sigma)} \cdot e^{-in\varphi(\sigma)} \cdot d\sigma.\tag{6.88}$$

Similarly we may expand the periodic η-function into a Fourier series

$$w_\eta(\varphi) = \sum W_{\eta n} e^{in\varphi}.\tag{6.89}$$

Using both (6.86,89) in (6.81) we get

$$(-n^2 + \nu^2) \sum W_{\eta n} e^{-in\varphi} = \nu^2 \sum F_n e^{-in\varphi},\tag{6.90}$$

which we solve for the Fourier coefficients, $W_{\eta n}$, of the periodic dispersion function

$$W_{\eta n} = \frac{\nu^2 F_n}{\nu^2 - n^2}.\tag{6.91}$$

The periodic solution of the differential equation (6.81) is finally

$$w_\eta(\varphi) = \sum_{n=-\infty}^{+\infty} \frac{\nu^2 F_n \, e^{in\varphi}}{\nu^2 - n^2} . \tag{6.92}$$

It's obvious again, we must choose the tune $\nu \neq n$ to avoid an integer resonance. This solution is intrinsically periodic since φ is periodic and the relation to (6.83) can be established by replacing F_n by its definition (6.88). Using the property $F_{-n} = F_n$ and formula GR[1.445.6][1] results from we get for a symmetric lattice

$$\begin{aligned}
w_\eta(\varphi) &= \sum_{n=-\infty}^{+\infty} \frac{e^{in\varphi} \cdot \frac{\nu}{2\pi} \oint \frac{\sqrt{\beta(\sigma)}}{\rho(\sigma)} e^{-in\sigma} \, d\sigma}{\nu^2 - n^2} \\
&= \frac{\nu}{\pi} \oint \frac{\sqrt{\beta(\sigma)}}{\rho(\sigma)} \left[\frac{1}{2\nu^2} + \sum_{n=1}^{\infty} \frac{\cos n(\sigma - \varphi)}{\nu^2 - n^2} \right] \, d\sigma \\
&= \frac{1}{2 \sin \nu \pi} \oint \frac{\sqrt{\beta(\sigma)}}{\rho(\sigma)} \cos(\nu[\varphi - \sigma + \pi]) \cdot d\sigma ,
\end{aligned} \tag{6.93}$$

which is the same as (6.83) since $d\tau = \nu\beta d\sigma$ and $F(s) = \beta^{3/2}/\rho$. For an asymmetric lattice the proof is similar albeit somewhat more elaborate.

Solution (6.93) expresses the dispersion function as the combination of a constant and a sum of oscillatory terms. Evaluating the nonoscillatory part of the integral we find the average value of the dispersion function,

$$\langle \eta \rangle \approx \frac{\langle \beta \rangle}{\nu_o} . \tag{6.94}$$

This result by itself is of limited usefulness but can be used to obtain an estimate for the *momentum compaction factor* α_c defined analogous to (5.182) for a periodic lattice by

$$\alpha_c = \frac{1}{L_p} \oint \frac{\eta}{\rho} \cdot ds \approx \langle \eta/\rho \rangle . \tag{6.95}$$

A good approximation for the momentum compaction factor is therefore $\alpha_c = \langle \beta \rangle / (\rho \cdot \nu)$ and with (6.63) integrated only over the arcs of the ring

$$\alpha_c \approx \frac{1}{\nu^2} . \tag{6.96}$$

Thus we find the interesting result that the transition energy γ_t is approximately equal to the horizontal tune of a circular accelerator

$$\gamma_t \approx \nu_x . \tag{6.97}$$

---

[1] We will abbreviate in this way formulas from the Table of Integrals, Series and Products, I.S. Gradshteyn/I.M. Ryzhik, 4th edition

As a cautionary note for circular accelerators with long straight sections, only the tune of the arc sections should be used here since straight sections do not contribute to the momentum compaction factor but can add significantly to the tune.

6.5 Periodic Lattices in Circular Accelerators

Circular accelerators and arbitrarily long beam transport lines can be constructed from fundamental building blocks like FODO cells or other magnet sequences which are then repeated many times. Any cell or lattice unit for which periodic solution of the lattice functions can be found may be used as a basic building block for a periodic lattice. Such units need not be symmetric but a symmetric lattice segment is always periodic.

FODO cells as elementary building blocks for larger beam transport lattices may lack some design features necessary to meet the objectives of the whole facility. In a circular accelerator we need for example some component free spaces along the orbit to allow the installation of experimental detectors or other machine components like accelerating sections and injection magnets. A lattice made up of standard FODO cells with bending magnets would not provide such spaces.

The lattice of a circular accelerator in most cases therefore exhibits more complexity than that of a simple FODO cell. In general a circular accelerator is made up of a number of *superperiods* which may be further subdivided into segments with special features like dispersion suppression section, achromatic sections, insertions, matching sections or simple focusing and bending units like FODO cells. To illustrate basic lattice design concepts for circular accelerators we will discuss specific lattice solutions for a variety of applications.

6.5.1 Synchrotron Lattice

For a synchrotron whose sole function is to accelerate particles the problem of free space can be solved quite easily. Most existing synchrotrons are based on a FODO lattice recognizing its simplicity, beam dynamical stability and efficient use of space. To provide magnet free space we merely eliminate some of the bending magnets. As a consequence the whole ring lattice is composed of curved as well as straight FODO cells. The elimination of bending magnets must, however, be done thoughtfully since the dispersion function depends critically on the distribution of the bending magnets. Random elimination of bending magnets may lead to an uncontrollable dispersion function. Often it is desirable to have the dispersion function vanish or at least be small in magnet free straight sections to simplify injection and avoid possible instabilities if rf cavities are placed, where the dispersion

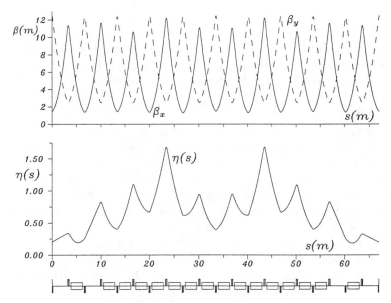

Fig. 6.11. Typical FODO lattice for a separated function synchrotron

function is finite. The general approach to this design goal is to use regular FODO cells for the arcs followed by a dispersion matching section, where the dispersion function is brought to zero or at least to a small value leading finally to a number of bending magnet free, straight FODO cells. As an example such a lattice is shown in Fig. 6.11 for a 3.5 GeV synchrotron.

Figure 6.11 shows one quadrant of the whole ring and we clearly recognize three different lattice segments including seven arc FODO half cells, two half cells to match the dispersion function and one half cell for installation of other machine components. Such a quadrant is mirror reflected at one or the other end to form one of two *superperiods* of the ring lattice. In this example the ring consists of two superperiods although another ring could be composed by a different number of superperiods. A specific property of the lattice shown in Fig. 6.11 is that as far as the focusing is concerned, the whole ring is made up of equal FODO cells including only two quadrupole families, QF and QD. The betatron functions are periodic and are not significantly affected by the presence or omission of bending magnets which are assumed to have negligible edge focusing. By eliminating bending magnets in an otherwise unperturbed FODO lattice, we obtain magnet free spaces equal to the length of the bending magnets which are used for the installation of accelerating components, injection magnets, and beam monitoring equipment.

211

6.5.2 Phase Space Matching

Periodic lattices like FODO channels exhibit unique periodic solutions for the betatron and dispersion functions. In realistic accelerator designs, however, we will not be able to restrict the lattice to periodic cells only. We will find the need for a variety of lattice modifications which necessarily require locally other than periodic solutions. Within a lattice of a circular accelerator, for example, we encountered the need to provide some magnet free spaces, where the dispersion function vanishes. In colliding beam storage rings it is desirable to provide for a very low value of the betatron function at the beam collision point to maximize the *luminosity*. These and other lattice requirements necessitate a deviation from the periodic cell structure.

Beam transport lines are in most cases not based on periodic focusing. If such transport lines carry beam to be injected into a circular accelerator or must carry beam from such an accelerator to some other point, we must consider proper *matching conditions* at locations, where lattices of different machines or beam transport systems meet [6.10, 11]. Joining arbitrary lattices may result in an inadequate over lap of the phase ellipse for the incoming beam with the acceptance of the downstream lattice as shown in Fig. 6.12a.

For a perfect match of two lattices all lattice functions must be the same at the joining point as shown in Fig. 6.12b

$$(\beta_x, \alpha_x, \beta_y, \alpha_y, \eta, \eta')_1 = (\beta_x, \alpha_x, \beta_y, \alpha_y, \eta, \eta')_2 . \tag{6.98}$$

In this case the phase ellipse at the end of lattice $_1$ is similar to the *acceptance phase ellipse* at the entrance of lattice $_2$. Equality of both ellipses occurs only if the acceptance in both lattices is the same. To avoid dilution of particles in phase space perfect matching is desired in proton and ion beam transport systems and accelerators. For electrons this is not required because electron beams approach through synchrotron radiation and damping the appropriate phase ellipse. The main goal of matching an electron beam is to assure that the emittance of the incoming beam is fully accepted by

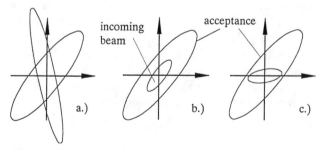

Fig. 6.12. Matching conditions in phase space; a) mismatch, b) perfect match, c) efficient match

the downstream lattice as shown in Fig. 6.12c. Perfect matching of all lattice functions and acceptances with beam emittance, however, provides the most economic solution since no unused acceptance exist. Matching of the dispersion function (η, η') in addition also assures that phase ellipses for off momentum particles match as well.

Matching in circular accelerators is much more restrictive than that between independent lattices. In circular accelerators a variety of lattice segments for different specific functions must be tied together to form a periodic magnet structure. To preserve the periodic lattice functions we must match them exactly between different lattice segments. Failure of perfect matching between lattice segments can lead to periodic solutions of lattice functions which are vastly different from design goals or do not exist at all.

In general there are six lattice functions to be matched requiring six variables or quadrupoles in the focusing structure of the upstream lattice to produce a perfect match. Matching quadrupoles must not be too close together in order to provide some independent matching power for individual quadrupoles. As an example, the betatron functions can be modified most effectively if a quadrupole is used at a location, where the betatron function is large and not separated from the matching point by multiples of π in betatron phase. Most independent matching conditions for both the horizontal and vertical betatron functions are created if matching quadrupoles are located, where one betatron function is much larger than the other allowing almost independent matching.

It is impossible to perform such general matching tasks by analytic methods and a number of numerical codes are available to solve such problems. Frequently used matching codes are *TRANSPORT* [6.11], *SYNCH* [6.12], *COMFORT* [6.13], or *MAD* [6.14]. These programs are an indispensable tool for lattice design and allow the fitting of any number of lattice functions to desired values including boundary conditions to be met along the matching section.

6.5.3 Dispersion Matching

A very simple, although not perfect, method to reduce the dispersion function in magnet free straight sections is to eliminate one or more bending magnets close to but not at the end of the arc and preferably following a focusing quadrupole, QF. In this arrangement of magnets the dispersion function reaches a smaller values compared to those in regular FODO cells with a slope that becomes mostly compensated by the dispersion generated in the last bending magnet. The match is not perfect but the dispersion function is significantly reduced, where this is desirable, and magnet free sections are created in the lattice. This method requires no change in the quadrupole or bending magnet strength and is therefore also operationally very simple as demonstrated in the example of a synchrotron lattice shown in Fig. 6.11. We note the less than perfect matching of the dispersion func-

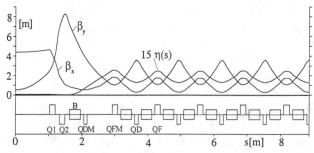

Fig. 6.13. Lattice for a 1.2 GeV low emittance damping ring

tion which causes a beating of an originally periodic dispersion function. In the magnet free straight sections, however, the dispersion function is considerably reduced compared to the values in the regular FODO cells.

More sophisticated matching methods must be employed, where a perfect match of the dispersion function is required. Matching of the dispersion to zero requires the adjustment of the two parameters, $\eta = 0$ and $\eta' = 0$, at the beginning of the straight section. This can always be achieved by controlling some of the upstream quadrupoles. Compared to a simple two parameter FODO lattice (Fig. 6.11) this variation requires a more complicated control system and additional power supplies to specially control the matching quadrupoles. This dispersion matching process disturbs the betatron functions which must be separately controlled and matched by other quadrupoles in dispersion free sections. Such a matching method is utilized in a number of storage rings with a special example shown in Fig. 6.13[6.16].

Here we note the perfect matching of the dispersion function as well as the associated perturbation of the betatron function requiring additional matching. Quadrupoles QFM and QDM are adjusted such that $\eta = 0$ and $\eta' = 0$ in the straight section. In principle this could be done even without eliminating a bending magnet, but the strength of the dispersion matching quadrupoles would significantly deviate from that of the regular FODO channel quadrupoles and cause a large distortion of the betatron function in the straight section. To preserve a symmetric lattice, the betatron function must be matched with the quadrupoles Q1 and Q2 to get $\alpha_x = 0$ and $\alpha_y = 0$ at the symmetry points of the lattice.

Dispersion Suppressor: A rather elegant method of dispersion matching has been developed by Keil [6.16]. Noting that dispersion matching requires two parameters he chooses to vary the last bending magnets at the end of the arcs rather than quadrupoles. The great advantage of this method is to leave the betatron functions and the tunes undisturbed at least as long as we may ignore the end field focusing of the bending magnets which is justified in large high energy accelerators. This *dispersion suppressor* consists of four FODO half cells following directly the regular FODO cells at a focusing

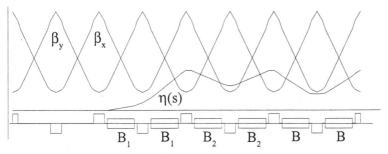

Fig. 6.14. Dispersion suppressor lattice

quadrupole QF as shown in Fig. 6.14. The strength of the bending magnets are altered into two types with a total bending angle of all four magnets to be equal to two regular bending magnets.

The matching conditions can be derived analytically from the transformation matrix for the full dispersion suppressor as a function of the individual magnet parameters. An algebraic manipulation program has been used to derive a result that is surprisingly simple. If θ is the bending angle for regular FODO cell bending magnets and ψ the betatron phase for a regular FODO half cell the bending angles θ_1 and θ_2 are determined by [6.16]:

$$\theta_1 = \theta \cdot \left(1 - \frac{1}{4 \cdot \sin^2 \psi}\right) \tag{6.99}$$

and

$$\theta_2 = \theta \cdot \left(\frac{1}{4 \cdot \sin^2 \psi}\right), \tag{6.100}$$

where

$$\theta = \theta_1 + \theta_2 . \tag{6.101}$$

This elegant method requires several FODO cells to match the dispersion function and is therefore most appropriately used in large systems. Where a compact lattice is important, matching by quadrupoles as discussed earlier might be more space efficient.

6.5.4 Magnet Free Insertions

An important part of practical lattice design is to provide *magnet free spaces* which are needed for the installation of other essential accelerator components or experimental facilities. Methods to provide limited magnet free spaces by eliminating bending magnets in FODO lattices have been discussed earlier. Often, however, much larger magnet free spaces are required and procedures to provide such sections need to be formulated.

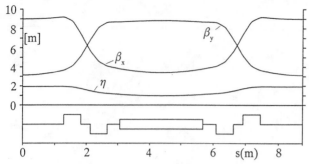

Fig. 6.15. Lattice of the SPEAR storage ring [6.18]

The most simple and straight forward approach is to use a set of quadrupoles and focus the lattice functions β_x, β_y and η into a magnet free section such that the derivatives α_x, α_y and η' vanish in the center of this section. This method is commonly applied to interaction areas in colliding beam storage rings to provide optimum beam conditions for maximum luminosity at the collision point. A typical example is shown in Fig. 6.13.

A more general design approach to provide magnet free spaces in a periodic lattice is exercised in the storage ring *ADONE* [6.19] shown in Fig. 6.16 or the storage ring *SPEAR* [6.18] as shown in Fig. 6.15. In the ADONE lattice the quadrupoles of a FODO lattice are moved together to form doublets and alternate free spaces are filled with bending magnets or left free for the installations of other components.

Another scheme to provide magnet free spaces is exercised in the SPEAR lattice, (Fig. 6.15), where the FODO structure remains unaltered except that the FODO cells have been separated in the middle of the QF quadrupoles. A separation in the middle of the QD quadrupoles would have worked as well. Since the middle of FODO quadrupoles are symmetry points a modest separation can be made with minimal perturbation to the betatron functions and no perturbation to the dispersion function since $\eta' = 0$ in the middle of FODO quadrupoles.

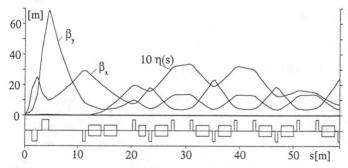

Fig. 6.16. Lattice of the ADONE storage ring [6.19]

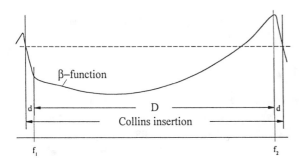

Fig. 6.17. Collins insertion

Collin's Insertion: A simple magnet free insertion for dispersion free segments of the lattice has been proposed by Collins [6.19]. The proposed insertion consists basically of a focusing and a defocusing quadrupole with a long drift space in between as shown in Fig. 6.17. In thin lens approximation we have the transformation matrix for the insertion

$$
\mathcal{M}_{\text{ins}} = \begin{bmatrix} 1 & d \\ 0 & 1 \end{bmatrix} \cdot \begin{bmatrix} 1 & 0 \\ \frac{1}{f} & 1 \end{bmatrix} \cdot \begin{bmatrix} 1 & D \\ 0 & 1 \end{bmatrix} \cdot \begin{bmatrix} 1 & 0 \\ -\frac{1}{f} & 1 \end{bmatrix} \cdot \begin{bmatrix} 1 & d \\ 0 & 1 \end{bmatrix} . \tag{6.102}
$$

This insertion matrix must be equated with the transformation matrix for this same insertion expressed in terms of lattice functions at the joining point with the regular lattice

$$
\mathcal{M}_{\text{ins}} = \begin{bmatrix} \cos \psi + \alpha \cdot \sin \psi & \beta \cdot \sin \psi \\ -\frac{1+\alpha^2}{\beta} \cdot \sin \psi & \cos \psi - \alpha \cdot \sin \psi \end{bmatrix} . \tag{6.103}
$$

Both matrices provide three independent equations to be solved for the drift lengths d and D and for the focal length f of the quadrupoles. After multiplications of all matrices we equate matrix elements and get

$$
D = \frac{\alpha^2}{\gamma}, \qquad d = \frac{1}{\gamma}, \qquad \text{and} \qquad f = -\frac{\alpha}{\gamma} . \tag{6.104}
$$

These relations are valid for both planes only if $\alpha_x = -\alpha_y$. Generally this is not the case for arbitrary lattices but for a weak focusing FODO lattice this condition is well met. We note that this design provides an insertion with a length D which is proportional to the value of the betatron functions at the insertion point and requires that $\alpha \neq 0$.

Of course any arbitrary insertion with a unity transformation matrix, $\mathcal{I}$, in both planes is a valid solution as well. Such solutions can in principle always be enforced by matching with a sufficient number of quadrupoles. If the dispersion function and its derivative is zero such an insertion may also have a transformation matrix of $-\mathcal{I}$. This property of insertions is widely used in computer designs of insertions when fitting routines are available to numerically adjust quadrupole strength such that desired lattice features are met including the matching of the lattice functions to the insertion point.

A special version of such a solution is the low beta insertion for colliding beam facilities.

6.5.5 Low Beta Insertions

In *colliding beam facilities* long magnet free straight sections are required to allow the installation of high energy particle detectors. In the center of these sections, where two counter rotating particle beams collide, the betatron functions must reach very small values forming a narrow beam waist. This requirement allows to minimize the destructive *beam-beam effect* when two beams collide and thereby maximize the luminosity of the colliding beam facility [6.20].

An example for the incorporation of such a *low beta insertion* is shown in Fig. 6.18 representing one [6.21] of many variations of a low beta insertion in colliding beam facilities. The special challenge in this matching problem is to provide a very small value for the betatron functions at the collision point. To balance the asymmetry of the focusing in the closest quadrupoles the betatron functions in both planes are generally not made equally small but the vertical betatron function is chosen smaller than the horizontal to maximize the luminosity. The length of the magnet free straight section is determined by the maximum value for the betatron function that can be accepted in the first vertically focusing quadrupole. The limit may be determined by just the physical aperture available or technically possible in these *insertion quadrupoles* or by the chromaticity and ability to correct and control chromatic and geometric aberrations.

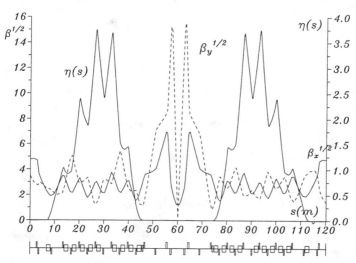

Fig. 6.18. Lattice functions of a colliding beam storage ring [6.21]. Shown is half the circumference with the collision point, low beta and vanishing dispersion in the center.

The maximum value of the betatron function at the entrance to the first quadrupole, the minimum value at the collision point, and the magnet free section are correlated by the equation for the betatron function in a drift space. Assuming the *collision point* to be a symmetry point the betatron functions develop from there like

$$\beta(s) = \beta^* + \frac{L_{\text{ins}}^2}{\beta^*}, \qquad (6.105)$$

where β^* is the value of the betatron function at the symmetry point and $2L_{\text{ins}}$ the full length of the insertion between the innermost quadrupoles.

6.5.6 Example of a Colliding Beam Storage Ring

In electron or hadron colliding beam storage rings many of the previously discussed design features are incorporated. Basically such facilities employ a lattice which consists of a number of identical superperiods, where each superperiod includes a collision point or *interaction region*, a transition section for matching of lattice functions and an arc section which in most cases is made up of a number of FODO cells. The collision points feature a minimum value of the betatron functions to maximize the collision rate or *luminosity* requiring a matching section to match the betatron functions to those of the FODO cells in the arcs. In addition the transition section also serves to match the finite dispersion from the arcs to the desired values in the interaction region.

In Fig. 6.19 the lattice of the Positron Electron Project, PEP, is shown for one half of six symmetric superperiods. We will use this lattice in this and Vol. II as a reference to discuss beam dynamics issues, beam stability characteristics, and to allow comparison with measurements. Some salient parameters for the PEP colliding beam facility are compiled in Table 6.2.

The interaction region was designed to provide 20 m of magnet free space for the installation of an experimental detector and the minimum value of the vertical betatron functions at the collision point was designed to be $\beta_y^* = 5$ cm. This interaction region continues into the transition section with betatron matching in the first part and betatron and dispersion matching close to the arcs. At the symmetry points of the superperiod a short magnet free section is included for installation of select beam manipulation and monitoring equipment. The lattice functions in the FODO section are not perfectly matched for economic reasons to minimize the number of independent power supplies.

From the lattice functions in Fig. 6.19 we note the low beta insertion and matching of the dispersion function to zero in the interaction region. We also note very large values of the betatron functions in the interaction region quadrupoles as a consequence of the low beta at the collision point and the long distance to the first focusing quadrupole. The long magnet free distance between the first quadrupole doublet and the beginning of the transition

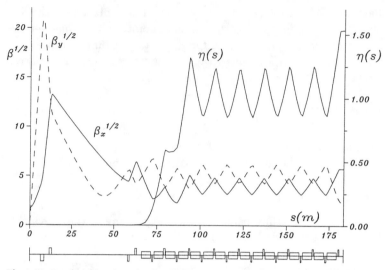

Fig. 6.19. Lattice functions in the PEP storage ring for one half of six symmetric super-periods. The collision point and low beta section is at $s = 0$ and the arc sections consist of FODO cells.

Table 6.2. PEP lattice parameters

energy, E(GeV)	15.0	beam current, I(mA)	100
circumference, C(m)	2200	superperiodicity,	6
beam emittance, ϵ_x(mm mrad)	0.125	energy spread, σ_E/E_o	0.0010
tunes, ν_x/ν_y	21.25/18.19	beta function at IP, $\beta^*_{x,y}$(m)	3.00/0.11
nat. chromaticity, ξ_{ox}/ξ_{oy}	-36.21/-99.47	momentum compaction factor,	0.00257
energy loss/turn, U_o(MeV)	26.98	radiation power, P_s(MW)	2.698
accelerating voltage, V_{rf}(MV)	39.43	synchrotron tune, ν_s	0.0451

FODO parameters:

cell length, L(m)	14.4	phase/cell, ψ_x/ψ_y(deg)	56.016/31.925
bending radius, ρ(m)	165.5	acceptance, A_x/A_y(mm-mrad)	29.88/11.01

section is useful for the installation of accelerator equipment, especially for accelerating rf cavities, but is mainly a necessary lattice feature. The transition from rather small betatron functions in the arc FODO lattice to large values in the insertion quadrupoles cannot be accomplished without an appropriate length of drift space to let the betatron functions grow. As this drift space is reduced the strength of the matching quadrupoles becomes very strong and quickly neither theoretical nor technical solution to the matching problem is possible. The focusing from the interaction region quadrupoles must be such that at the beginning of the matching section, here at Q3, not only the betatron functions reach values comparable to those in the arc section but the rate of change of the betatron functions, $\alpha = -\beta'/2$, must be comparable as well to those in the arcs.

This feature of lattice matching constitutes a severe limitation on the flexibility for small rings and short beam transport systems to incorporate insertions with specific properties of the lattice functions. Even if at low energies such insertions might be technically possible the strong focusing and large values of the betatron functions and divergencies can cause severe limitations in beam stability due to aberrations. This should not prevent the accelerator designer from trying to meet a particular design need, but it is prudent to address beam stability problems as early as a linear lattice design has been developed.

Problems

Use thin lens approximation unless otherwise noted.

Problem 6.1. Calculate for the real quadrupole length the quadrupole strength required to produce the quoted betatron phase advances per FODO cell in Table 6.1. Compare with thin lens quadrupole strengths.

Problem 6.2. Calculate the values of the betatron functions in the center of the quadrupoles for #1 and #2 FODO cells in Table 6.1 and compare with the actual thick lens betatron functions in Figs. 6.6,7. Discuss the difference.

Problem 6.3. Specify a FODO cell to be used as the basic lattice unit for a 50 GeV synchrotron or storage ring. The quadrupole bore radius shall be 3 cm. Adjust parameters such that a beam with an emittance in both planes of $\epsilon = 5$ mm mrad and an energy spread of $\sigma_\varepsilon/E_0 = 0.01$ would fit within the quadrupole aperture. Ignore wall thickness of vacuum chamber.

a) Considering the magnetic field limitations of conventional magnets, adjust bending radius, focal length and if necessary cell length to stay within realistic limits for conventional magnets.

b) What is the dipole field and the pole tip field of the quadrupoles? Adjust the total number of cells such that there is an even number of FODO cells and the tunes are far away from an integer or half integer resonance?

Problem 6.4. Plot the betatron and dispersion function for the lattice from problem 6.3. Could the maximum beam sizes be reduced by another choice of the phase advance per cell? Optimize phase advance such that beam sizes are close to the minimum and the tunes are not on a resonance.

Problem 6.5. The lattice of problem 6.3 is to be expanded to include dispersion free cells. Incorporate into the lattice two symmetric dispersion suppressors based on the ring FODO lattice following the scheme shown in Fig.

6.14. Adjust the bending magnet strength to retain a total bending angle of 2π in the ring. Incorporate the two dispersion suppressors symmetrically into the ring and make a schematic sketch of the lattice.

Problem 6.6. In the dispersion free region of problem 6.5 introduce a symmetric Collins insertion to provide a long magnet free section of the ring. Determine the parameters of the insertion magnets and drift spaces. Use thin lens approximation to calculate a few values of the betatron functions in the Collins insertions and plot betatron and dispersion functions through the Collins insertion.

Problem 6.7. For the complete ring lattice of problems 6.3,5,6 make a parameter list including such parameters as circumference, revolution time, magnet parameters, number of cells, tunes (use simple numerical integration to calculate the phase advance in the Collins insertion), max. beam sizes, etc.

Problem 6.8. Consider three cells of a symmetric FODO lattice $(1/2)QF_1 - QD_1 - QF_2 - QD_2 - QF_3 - QD_3 - (1/2)QF_4$ with a betatron phase advance ψ_F per cell. Further assume there are special coils in the quadrupoles to produce dipole fields which can be used to deflect the beam.

a) Construct a symmetric beam bump which starts at QF_1, ends at QF_4 and reaches an amplitude A_k in the center of QD_2. How many trim coils need to be activated?

b) Derive the relative kick angles required to construct the beam bump and calculate the beam displacement in each quadrupole. Is A_k the maximum amplitude of the beam bump? Why? Why not?

c) What are the criteria for either A_k being the maximum displacement or not? For which phase ψ_F would the dipole fields be minimum? Is there a more economic solution for a symmetric beam bump with an amplitude A_k in the center of Q_2?

Problem 6.9. Consider a proton synchrotron with a maximum energy of 400 GeV which is composed of a large number of FODO cells. Let the cell length be 50 m and the total circumference of the ring be 6000 m.

a) Determine the betatron phase advance per FODO cell close to 90 degrees per cell but such that the total ring tune is neither close to an integer nor close to a half integer value. What is the quadrupole strength and how long must iron dominated quadrupoles be in order not to exceed saturation limits if the bore radius is $R = 2.5$ cm?

b) Let the drift spaces between quadrupoles and bending magnets be as long as half a quadrupole. How much space is left for bending magnets and what field strength is required at the maximum particle energy?

c) What is the maximum energy achievable for fully ionized Si ions?

Problem 6.10. Produce a conceptual design for a proton synchrotron to be used to accelerate protons from a kinetic energy of 10 GeV/c to 150 GeV/c. The circular vacuum chamber aperture has a radius of 2 cm and is supposed to accommodate a beam with a maximum beam emittance of 40 mm mrad and a momentum spread of ±0.1%. The peak magnetic bending field at 150 GeV/c is 18 kG.

a) Choose a half cell length L which provides the desired transverse acceptance for the specified vacuum chamber aperture. What is the maximum value of the betatron and η-function? What is the quadrupole strength if the length is 5% of L and what is the deflection angle per bending magnet if the length is 75% of L. How many dipoles and quadrupoles are required?

b) If the copper cross sectional area in the dipole coils is 20 cm^2/coil, what is the total power dissipation in the ring at 150 GeV/c ($\rho_{Cu} = 2 \times 10^{-6}$ Ohm-cm)? What is the circumference and what are the tunes of the machine?

c) Is the injection energy above or below the transition energy? What is the revolution frequency at injection and at maximum energy?

Problem 6.11. How many protons would produce a circulating beam of 1 amp in the ring of problem 6.10? Calculate the total power stored in that beam at 150 GeV/c. By how many degrees could one liter of water be heated up by this energy? Assume the beam becomes suddenly missteered in the ring and hits the vacuum chamber. The proton beam has an emittance of $\epsilon_{x,y} = 30$ mm mrad at the injection energy of 10 GeV/c. Calculate the average beam width at 150 GeV/c along the lattice and assume this beam to hit because of a sudden missteering a straight piece of vacuum chamber at an angle of 10 mrad. The vacuum chamber is made of stainless steel and is 1 mm thick. If all available beam energy is absorbed in the steel by how much will the strip of steel hit by the beam heat up ? Will it melt? (specific heat $c_{Fe} = 0.11$ cal/g/°C, melting temperature $T_{Fe} = 1528$°C)

Problem 6.12. Consider a ring made from an even number of FODO cells. To provide component free space we cut the ring at a symmetry line through the middle of two quadrupoles on opposite sides of the ring and insert a drift space of length ℓ_d. Derive the transformation matrix for this ring and compare with that of the unperturbed ring. What is the tune change of the accelerator. The betatron functions will be modified. Derive the new value of the horizontal betatron function at the symmetry point in units of the unperturbed betatron function. Is there a difference to whether the free section is inserted in the middle of a focusing or defocusing quadrupole? How does the η-function change?

Problem 6.13. Sometimes two FODO channels of different parameters must be matched. Show that a lattice section can be designed with a phase advance of $\Delta\psi_x = \Delta\psi_y = \pi/2$ which will provide the desired matching of the betatron functions from the symmetry point of one FODO channel to the symmetry point of the other channel. Such a matching section is also called a *quarter wavelength transformer*. Does this transformer also work for curved FODO channels, where the dispersion is finite?

Problem 6.14. The fact that a Collins straight section can be inserted into any transport line without creating perturbations outside the insertion makes these insertions also a periodic lattice. A series of Collins straight sections can be considered as a periodic lattice composed of quadrupole doublets and long drift spaces in between. Construct a circular accelerator by inserting bending magnets into the drift spaces d and adjusting the drift spaces to $D = 5$ m. What is the phase advance per period? Calculate the periodic η-function and make a sketch with lattice and lattice functions for one period.

Problem 6.15. Consider a regular FODO lattice as shown in Fig. 6.11, where some bending magnets are eliminated to provide magnet free spaces and to reduce the η-function in the straight section. How does the minimum value of the η-function scale with the phase per FODO cell. Show if conditions exist to match the η-function perfectly in the straight section of this lattice?

Problem 6.16. The quadrupole lattice of the synchrotron in Fig. 6.11 forms a pure FODO lattice. Yet the horizontal betatron function shows some beating perturbation while the vertical betatron function is periodic. What is the source of perturbation for the horizontal betatron function? An even stronger perturbation is apparent for the dispersion function. Explain why the dispersion function is perturbed.

7 Perturbations in Beam Dynamics

The study of beam dynamics under ideal conditions is the first basic step toward the design of a beam transport system. In the previous sections we have followed this path and have allowed only the particle energy to deviate from its ideal value. In a real particle beam line or accelerator we may, however, not assume ideal and linear conditions. More sophisticated beam transport systems require the incorporation of nonlinear sextupole fields to correct for chromatic aberrations. Deviations from the desired field configurations can be caused by transverse or longitudinal misplacements of magnets with respect to the ideal beam path. Of similar concern are errors in the magnetic field strength, undesirable field effects caused in the field configurations at magnet ends, or higher order multipole fields resulting from design, construction, and assembly tolerances. Still other sources of errors may be *beam-beam perturbations*, insertion devices in beam transport systems or accelerating sections which are not part of the magnetic lattice configurations. Such systems may be magnetic detectors for high energy physics experiments, wiggler and undulator magnets for the production of synchrotron radiation, a gas jet or immaterial field sources like that of a free electron laser interacting with the particle beam to name just a few examples. The impact of such errors is magnified in strong focusing beam transport systems as has been recognized soon after the invention of the strong focusing principle. Early overviews and references can be found for example in [7.1– 12].

A horizontal bending magnet has been characterized as a magnet with only a vertical field component. This is true as long as this magnet is perfectly aligned, in most cases perfectly level. Small rotations about the magnet axis result in the appearance of horizontal field components which must be taken into account for beam stability calculations.

We also assumed that the magnetic field in a quadrupole vanishes at the center of magnet axis. In the horizontal midplane of a quadrupole the vertical field component has been derived as $B_y = g\,x$. If this quadrupole is displaced horizontally with respect to the beam axis by a small amount δx we observe a dipole field $\delta B_y = g\,\delta x$ at the beam axis. Similarly, a horizontal dipole field component is created for a vertical displacement of the quadrupole. These dipole field components in most cases are unintentional and lead to an undesired deflection of the beam.

In addition, a quadrupole can be rotated by a small angle with respect to the reference coordinate system. As a result we observe the appearance of a small component of a "rotated quadrupole". A sextupole magnet, when displaced, introduces a dipole as well as a quadrupole field component on the beam axis. In general we find that any displaced higher order multipole introduces field errors on the beam axis in all lower order field configurations.

Although such misalignments and field errors are unintentional and undesired, we have to deal with their existence since there is no way to avoid such errors in a real environment. The particular effects of different types of errors on beam stability will be discussed. Tolerance limits on these errors as well as corrective measures must be established to avoid destruction of the particle beam. Common to all these perturbations from ideal conditions is that they can be considered small compared to forces of linear elements. We will therefore discuss mathematical perturbation methods that allow us to determine the effects of perturbations and to apply corrective measures for beam stability.

7.1 Magnet Alignment Errors

In this section field errors created by *magnet misalignments* like displacements or rotations from the ideal location will be derived quantitatively. Such magnet alignment errors, however, are not the only cause for field errors. External sources like the earth magnetic field, the fields of nearby electrical current carrying conductors, magnets connected to vacuum pumps or ferromagnetic material in the vicinity of beam transport magnets can cause similar field errors. For example electrical power cables connected to other magnets along the beam transport line can be hooked up such that the currents in all cables are compensated. This occurs automatically for cases, where the power cables to and from a magnet run close together. In circular accelerators one might, however, be tempted to run the cables around the ring only once to save the high material and installation costs. This, however, causes an uncompensated magnetic field in the vicinity of cables which may reach as far as the particle beam pipe. The economic solution is to seek electrical current compensation among all magnet currents by running electrical currents in different directions around the ring. Careful design of the beam transport system can in most cases minimize the impact of such field perturbations while at the same time meeting economic goals.

Incidental field errors cannot be derived in a formal way but must be evaluated individually by magnetic measurements. The main component of such fields, however, can be described in most cases by a superposition of a dipole and a gradient field. In the following paragraphs we will restrict ourselves to the effects of magnet field and alignment errors. Misalignment errors can be expressed by the transformation

$$\begin{bmatrix} \tilde{x} \\ \tilde{y} \end{bmatrix} = \begin{bmatrix} \cos \delta\varphi & \sin \delta\varphi \\ -\sin \delta\varphi & \cos \delta\varphi \end{bmatrix} \begin{bmatrix} x \\ y \end{bmatrix} - \begin{bmatrix} \delta x \\ \delta y \end{bmatrix}, \tag{7.1}$$

where (x, y) are coordinates with respect to the ideal path, $(\delta x, \delta y)$ displacement errors of the magnets from the ideal path, and $\delta\varphi$ a rotational error of the magnet with respect to the magnet axis and ideal coordinate system (x, y, z). The sign convention adopted here is such that a counterclockwise rotation of the magnet is represented by a positive value of φ. The coordinates $(\tilde{x}, \tilde{y})$ describe the particle position with respect to the magnet axis. Such a transformation can be applied to the vector potential, given by (4.45) which in turn then allows the calculation of the field perturbations by simple differentiation.

To demonstrate the types of *field errors* generated by magnet misalignments we express the transformation (7.1) in polar coordinates which is more convenient to apply to the nth order vector potential $V_n(r_m, \varphi_m)$. We use the coordinate system (r_m, φ_m) which is centered and fixed in the displaced magnet. The transformation $r_m = r - \delta r$ and $\varphi_m = \varphi - \delta\varphi$ relates the magnet system to the coordinate system of the beam (r, φ) which is the reference path. The magnet potential expressed with respect to the beam center is

$$V_n(r, \varphi) = -\frac{cp}{e} \frac{1}{n!} A_n (r - \delta r)^n e^{in(\varphi - \delta\varphi)}, \tag{7.2}$$

where we apply the expansion $(r - \delta r)^n = \sum_{j=0}^{n} \binom{n}{j} \delta r^{n-j} r^j$. Since $\delta\varphi \ll 1$ we have $e^{-in\delta\varphi} \approx 1 - in\delta\varphi = 1 - n\delta\varphi e^{i\frac{\pi}{2}}$ and after some manipulation and keeping only linear terms in the displacements the vector potential is

$$\begin{aligned} V_n(r, \varphi) \approx &-\frac{cp}{e} \frac{A_n}{n!} \left[r^n e^{in\varphi} - in\delta\varphi \, r^n e^{in\varphi} \right. \\ &\left. + \sum_{j=0}^{n-1} \binom{n}{j} \delta r^{n-j} r^j e^{in\varphi} \right] + \mathcal{O}(2). \end{aligned} \tag{7.3}$$

The effects of *magnet misalignments* become obvious. The first term in the square bracket shows that to first order the original n-th order fields have not been changed due to magnet misalignments. The second term demonstrates the appearance of a "rotated" multipole component of the same n-th order with the strength $n\delta\varphi$ being linearly proportional to the rotational misalignment. Transverse misalignments lead to the third term which is the sum over a series of lower order field components. Since $\delta r^{n-j} r^j e^{in\varphi} = (\delta r \, e^{i\varphi})^{n-j} r^j e^{ij\varphi} = (\delta x + i\delta y)^{n-j} r^j e^{ij\varphi}$ we find the perturbation terms to be of the order $0 \le j \le n - 1$. A displaced *octupole magnet*, for example, generates at the beam axis all lower order field components, like sextupole, quadrupole and dipole field.

An upright quadrupole with the potential $V = -gxy$ after rotation by the angle $\delta\varphi$ is transformed into a composition of an upright and a rotated quadrupole

$$V(x, y, s) = -\cos(2\,\delta\varphi)\, g\, x\, y + \tfrac{1}{2} \sin(2\delta\varphi)\, g\, (x^2 - y^2), \qquad (7.4)$$

where $\delta\varphi < 0$ for a clockwise rotation. Similar transformations are true for other multipoles.

Multipole errors in magnets are not the only cause for perturbations. For beams with large divergence or large cross section *kinematic perturbation* terms may have to be included. Such terms are neglected in *paraxial beam optics* discussed here, but will be derived in detail in Vol. II.

7.2 Dipole Field Perturbations

Dipole fields are the lowest order magnetic fields and therefore also the lowest order field errors. Such *dipole field errors* deflect the beam from its ideal path and we are interested to quantify this perturbation and to develop compensating methods to minimize the distortions of the beam path. In an open beam transport line the effect of dipole field errors on the beam path can be calculated within the matrix formalism.

A dipole field error at point s_k deflects the beam by an angle θ. If $\mathbf{M}(s_m|s_k)$ is the transformation matrix of the beam line between the point s_k, where the kick occurs, and the point s_m, where we observe the beam position, we find a displacement of the beam center by

$$\Delta u = M_{12}\, \theta, \qquad (7.5)$$

where M_{12} is the element of the transformation matrix in the first row and the second column. Due to the linearity of the equation of motion, effects of many kicks caused by dipole errors can be calculated by summation of individual beam center displacements at the observation point s_m for each kick. The displacement of a beam at the location s_m due to many dipole field errors is then given by

$$\Delta u(s_m) = \sum_k M_{12}(s_m|s_k)\, \theta_k, \qquad (7.6)$$

where θ_k are kicks due to dipole errors at locations $s_k < s_m$ and $M_{12}(s_m|s_k)$ the M_{12}-matrix element of the transformation matrix from the perturbation s_k to the monitor s_m.

Generally we do not know the strength and location of errors. Statistical methods are applied therefore to estimate the expectation value for beam perturbation and displacement. With $M_{12}(s_m|s_k) = \sqrt{\beta_m \beta_k} \sin(\psi_m - \psi_k)$ we calculate the root-mean-square of (7.6) noting that the phases ψ_k are random and cross terms involving different phases cancel. With $\langle \theta_k^2 \rangle = \sigma_\theta^2$ and $\langle \Delta u^2 \rangle = \sigma_u^2$ we get finally from (7.6) the expectation value of the path distortion at s_m due to statistical errors with a standard value σ_θ

$$\sigma_u = \sqrt{\beta_m \langle \beta \rangle} \sqrt{N_\theta}\, \sigma_\theta, \qquad (7.7)$$

where $\langle \beta \rangle$ is the average betatron function at the location of errors and N_θ the number of dipole field errors.

7.2.1 Existence of Equilibrium Orbits

Particles orbiting around a circular accelerator perform in general betatron oscillations about the equilibrium orbit and we will discuss properties of this equilibrium orbit. Of fundamental interest is of course that such equilibrium orbits exist at all. We will not try to find conditions for the existence of equilibrium orbits in arbitrary electric and magnetic fields but restrict this discussion to fields with midplane symmetry as they are used in particle beam systems. The existence of equilibrium orbits can easily be verified for particles like electrons and positrons because these particles radiate energy in form of *synchrotron radiation* as they orbit around the ring.

This radiation constitutes an energy loss and we need accelerating fields to compensate these losses. The loss of momentum due to the emission of photons is parallel and opposite to the direction of the particle momentum. Since the particles perform betatron oscillations, the momentum loss can be split into a transverse and longitudinal component. The accelerating fields, on the other hand, provide a compensation of the lost momentum only in the longitudinal direction. In total the emission of synchrotron radiation eventually leads to a continuous loss of transverse momentum and the betatron oscillations of radiating particles are therefore damped. A particle orbiting with an arbitrary betatron oscillation must eventually damp down to a final *equilibrium orbit*.

In this section we find out the parameters and characteristics of this orbit and use the damping process to find the eventual equilibrium orbit in the presence of arbitrary dipole perturbations. To do this we follow an orbiting particle starting with the parameters $x = 0$ and $x' = 0$. This choice of initial parameters will not affect the generality of the argument since any other value of initial parameters is damped independently because of the linear superposition of the betatron oscillations.

As a particle orbits in a circular accelerator it will encounter a number of kicks from *dipole field errors* or field errors due to a deviation of the particle energy from the ideal energy. After one turn the particle position is the result of the superposition of all kicks the particle has encountered in that turn. Since each kick leads to a particle oscillation given by

$$x(s) = \sqrt{\beta(s)\,\beta_\theta}\,\theta\,\sin[\varphi(s) - \varphi_\theta]$$

we find for the superposition of all kicks in one turn

$$x(s) = \sqrt{\beta(s)} \sum_i \sqrt{\beta_i}\,\theta_i\,\sin[\varphi(s) - \varphi_i]\,, \tag{7.8}$$

where the index i indicates the location of the kicks. We ask ourselves now what is the oscillation amplitude after many turns. For that we add up the kicks from all past turns and include damping effects expressed by the factor $e^{-kT_o/\tau}$ on the particle oscillation amplitude, where T_o is the revolution time, kT_o is the time passed since the kick occurred k turns ago, and τ the *damping time*. The contribution to the *betatron oscillation* due to the kicks k turns ago is then given by

$$\Delta x_k = \sqrt{\beta(s)}\, e^{-kT_o/\tau} \sum_i \sqrt{\beta_i}\, \theta_i \, \sin[2\pi\nu k + \varphi(s) - \varphi_i]. \tag{7.9}$$

Adding the contributions from all past turns we get the position $x(s)$ of the particle

$$x(s) = \sum_{k=0}^{\infty} \sqrt{\beta(s)}\, e^{-kT_o/\tau} \sum_i \sqrt{\beta_i}\, \theta_i \, \sin[2\pi\nu k + \varphi(s) - \varphi_i]. \tag{7.10}$$

After some rearranging (7.10) becomes

$$x(s) = C_\theta \sum_{k=0}^{\infty} e^{-kT_o/\tau} \sin(2\pi\nu k) + S_\theta \sum_{k=0}^{\infty} e^{-kT_o/\tau} \cos(2\pi\nu k), \tag{7.11}$$

where

$$\begin{aligned} C_\theta &= \sum_i \sqrt{\beta(s)\,\beta_i}\, \theta_i \, \cos[\varphi(s) - \varphi_i], \\ S_\theta &= \sum_i \sqrt{\beta(s)\,\beta_i}\, \theta_i \, \sin[\varphi(s) - \varphi_i]. \end{aligned} \tag{7.12}$$

With the definition $q = e^{-T_o/\tau}$ we use the mathematical identities

$$\sum_{k=0}^{\infty} e^{-kT_o/\tau} \sin(2\pi\nu k) = \frac{q \sin 2\pi\nu}{1 - 2q \cos 2\pi\nu + q^2} \tag{7.13}$$

and

$$\sum_{k=0}^{\infty} e^{-kT_o/\tau} \cos(2\pi\nu k) = \frac{1 - q \cos 2\pi\nu}{1 - 2q \cos 2\pi\nu + q^2} \tag{7.14}$$

and get finally instead of (7.11)

$$x(s) = \frac{C_\theta\, q \sin 2\pi\nu + S_\theta\,(1 - q \cos 2\pi\nu)}{1 - 2p \cos 2\pi\nu + q^2}. \tag{7.15}$$

The revolution time is generally much shorter than the damping time $T_o \ll \tau$ and therefore $q \approx 1$. In this approximation we get after some manipulation and using (7.12)

$$x(s) = \frac{\sqrt{\beta(s)}}{2 \sin \pi\nu} \sum_i \sqrt{\beta_i}\, \theta_i \, \cos \nu[\varphi(s) - \varphi_i + \pi]. \tag{7.16}$$

Equation (7.16) describes the particle orbit reached by particles after some damping times. The solution does not include anymore any reference to earlier turns and kicks except those in one turn and the solution therefore is a steady state solution defined as the *equilibrium orbit*.

The cause and nature of the kicks θ_i is undefined and can be any perturbation, random or systematic. A particular set of such errors are systematic errors in the deflection angle for particles with a momentum error δ for which we set $\theta_i = \delta \phi_i$, where $\phi_i = \ell_i/\rho_i$ is the deflection angle of the bending magnet i. These errors are equivalent to those that led to the dispersion or η function. Indeed setting $\eta(s) = x(s)/\delta$ in (7.16) we get the solution (6.84) for the η function. The trajectories $x(s) = \eta(s)\delta$ therefore are the equilibrium orbits for particles with a relative momentum deviation $\delta = \Delta p/p_o$ from the ideal momentum p_o.

In the next subsection we will discuss the effect of random dipole field errors θ_i on the beam orbit. These kicks, since constant in time, are still periodic with the periodicity of the circumference and lead to a distorted orbit which turns out to be again equal to the equilibrium orbit found here.

To derive the existence of equilibrium orbits we have made use of the damping of particle oscillations. Since this damping derives from the energy loss of particles due to synchrotron radiation we have proof only for equilibrium orbits for radiating particles like electrons and positrons. Because of the equivalence of this result with the formal solution of the equilibrium orbit as the only periodic solution of the inhomogeneous differential equation and since nonradiating charged particles like protons follow the *Lorentz force* like electrons, we conclude that the concept of equilibrium orbits is true also for protons although they may never reach that orbit but continuously oscillate about it. At very high energies, however, even protons start to radiate significantly and reach the equilibrium orbit in a finite time.

7.2.2 Closed Orbit Distortion

The solution (7.16) for the equilibrium orbit can be derived also directly by solving the equation of motion. Under the influence of dipole errors the equation of motion is

$$u'' + K(s)\,u = p_o(s),\tag{7.17}$$

where the dipole perturbation $p_o(s)$ is independent of coordinates (x, y) and energy error δ. This differential equation has been solved earlier in Sects. 4.8.3 and 6.4.2, where a dipole field perturbation was introduced as an energy error of the particle. Therefore, we can immediately write down the solution for an arbitrary beam line for which the principal solutions $C(s)$ and $S(s)$ are known

$$u(s) = C(s)\,u_o + S(s)\,u_o' + P(s)\tag{7.18}$$

with

$$P(s) = \int\limits_0^s p(\sigma) \left[S(s)\,C(\sigma) - S(\sigma)\,C(s)\right] \mathrm{d}\sigma\,. \tag{7.19}$$

The result (7.18) can be interpreted as a composition of betatron oscillations with initial values $(u_\mathrm{o}, u_\mathrm{o}')$ and a superimposed perturbation $P(s)$ which defines the equilibrium trajectory for the betatron oscillations. In (7.19) we have assumed that there is no distortion at the beginning of the beam line, $P(0) = 0$. If there were already a perturbation of the reference trajectory from a previous beam line, we define a new equilibrium orbit by *linear superposition* of new perturbations to the continuation of the perturbed path from the previous beam line section. The particle position $(u_\mathrm{o}, u_\mathrm{o}')$ is composed of the betatron oscillation $(u_{o\beta}, u_{o\beta}')$ and the perturbation of the reference path $(u_\mathrm{oc}, u_\mathrm{oc}')$. With $u_\mathrm{o} = u_{o\beta} + u_\mathrm{oc}$ and $u_\mathrm{o}' = u_{o\beta}' + u_\mathrm{oc}'$ we get

$$u(s) = \left[u_{o\beta}\,C(s) + u_{o\beta}'\,S(s)\right] + \left[u_\mathrm{oc}\,C(s) + u_\mathrm{oc}'\,S(s)\right] + P(s). \tag{7.20}$$

In a circular accelerator we look for a self-consistent periodic solution. Because the differential equation (7.17) is identical to that for the dispersion function the solution must be similar to (6.84) and is called the *closed orbit*, *reference orbit* or *equilibrium orbit* given by

$$u_\mathrm{c}(s) = \frac{\sqrt{\beta(s)}}{2\sin \pi\nu} \oint\limits_s^{s+C} p_\mathrm{o}(\sigma)\,\sqrt{\beta(\sigma)}\,\cos\bigl(\nu\left[\varphi(s) - \varphi(\sigma) + \pi\right]\bigr)\,\mathrm{d}\sigma\,, \tag{7.21}$$

where C is the circumference of the accelerator. We cannot anymore rely on a superperiodicity of length L_p as in Sect. 6.4 since the perturbations $p(\sigma)$ due to misalignment or field errors are statistically distributed over the whole ring. Again the integer resonance character discussed earlier for the dispersion function is obvious, indicating there is no stable orbit if the tune of the circular accelerator is an integer. The influence of the integer resonance is noticeable even when the tune is not quite an integer. From (7.21) we find a perturbation $p(s)$ to have an increasing effect the closer the tune is to an integer value. The similarity of the closed orbit and the *dispersion function* in a circular accelerator is deeper than merely mathematical. The dispersion function defines closed orbits for energy deviating particles approaching the ideal orbit as $\delta \to 0$.

In trying to establish expressions for dipole errors due to field or alignment errors, we note that the bending fields do not appear explicitly anymore in the equations of motions because of the specific selection of the curvilinear coordinate system and it is, therefore, not obvious in which form dipole field errors would appear in the equation of motion (7.17). In (4.86) we note, however, a dipole field perturbation due to a particle with a momentum error δ. This chromatic term $\kappa_{xo}\,\delta$ is the dipole field error as seen by a particle with the momentum $cp_\mathrm{o}\,(1 + \delta)$. For particles with the ideal energy we may therefore replace in (4.86) the chromatic term $\kappa\,\delta$ by the field error $-\Delta\kappa$. All other perturbations are obtained by variations of magnet positions $(\Delta x, \Delta y)$ or magnet strengths. Up to second order the horizontal dipole *perturbation terms* due to magnet field and alignment errors are from (4.86)

$$
\begin{aligned}
p_{ox}(s) = {} & -\Delta\kappa_{xo} - (\kappa_{xo}^2 + k_o)\,\Delta x + (2\kappa_{xo}\,\Delta\kappa_{xo} + \Delta k)\,\Delta x \\
& - \tfrac{1}{2}m\,(\Delta x^2 - \Delta y^2 + 2x_c\,\Delta x - 2y_c\,\Delta y) + \mathcal{O}(3),
\end{aligned}
\tag{7.22}
$$

where we used $x = x_\beta + x_c + \Delta x$ and $y = y_\beta + y_c + \Delta y$ with (x_β, y_β) the betatron oscillations and (x_c, y_c) the closed orbit deviation in the magnet. In the presence of multipole magnets the perturbation depends on the displacement of the beam with respect to the center of multipole magnets.

There is a similar expression for vertical dipole perturbation terms and we get from (4.88) ignoring vertical bending magnets, $\kappa_{yo} = 0$, but not vertical dipole errors, $\Delta\kappa_{xo} \neq 0$,

$$
p_{oy}(s) = -\Delta\kappa_{yo} + k_o\,\Delta y + m\,(x_c\,\Delta y + y_c\,\Delta x) + \mathcal{O}(3).
\tag{7.23}
$$

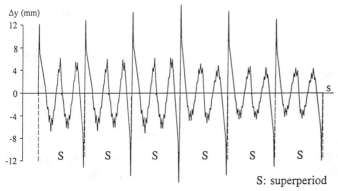

Fig. 7.1. Simulation of the closed orbit distortion in the sixfold symmetric PEP lattice due to statistical misalignments of quadrupoles by an amount $\langle\Delta x\rangle_{\mathrm{rms}} = \langle\Delta y\rangle_{\mathrm{rms}} = 0.05$ mm

A vertical closed orbit distortion is shown in Fig. 7.1 for the PEP storage ring with parameters compiled in Table 6.2 and Fig. 6.19. Here a gaussian distribution of horizontal and vertical alignment errors with an rms error of 0.05 mm in all quadrupoles has been simulated. In spite of the statistical distribution of errors a strong oscillatory character of the orbit is apparent and counting oscillations we find 18 oscillations which is equal to the vertical tune of PEP as we would expect from the denominator of (7.21). We also note large values of the orbit distortion adjacent to the interaction points (dashed lines), where the betatron function becomes large, again in agreement with expectations from (7.21) since $u_c \propto \sqrt{\beta}$. A more regular representation of the same orbit distortion can be obtained if we plot the normalized closed orbit $u_c(s)/\sqrt{\beta(s)}$ as a function of the betatron phase

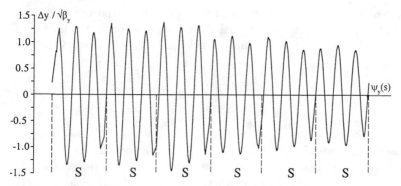

Fig. 7.2. Closed orbit distortion of Fig. 7.1 in normalized coordinates as a function of the betatron phase φ

$\psi(s)$ shown in Fig. 7.2. In this representation the strong harmonic close to the tune becomes evident while the statistical distribution of the perturbations shows mostly in the amplitude of the normalized orbit distortion.

For the sake of simplicity terms of third or higher order as well as terms associated with nonlinear magnets have been neglected in both equations (7.22,23). It is straightforward to treat these terms individually as they seem to become important in particular beam lines. Perturbations caused by sextupoles depend on the orbit itself and to get a self-consistent periodic solution of the distorted orbit iteration methods must be employed.

Solutions for equilibrium orbits can be obtained by inserting the perturbation (7.22) or (7.23) into (7.21). First we will concentrate on a situation, where we have only one perturbing kick in the whole lattice, assuming the perturbation to occur at $s = s_k$ and to produce a kick $\theta_k = \int p(\sigma)\,d\sigma$ in the particle trajectory. The orbit distortion at a location $s < s_k$ in the lattice is from (7.21)

$$u_o(s) = \tfrac{1}{2}\sqrt{\beta(s)\,\beta(s_k)}\,\theta_k\,\frac{\cos\nu\left[\pi - \varphi(s_k) + \varphi(s)\right]}{\sin\pi\nu}. \tag{7.24}$$

If on the other hand we look for the orbit distortion downstream from the perturbation $s > s_k$ the integration must start at s, follow the ring to $s = C$ and then further to $s = s+C$. The kick, therefore, occurs at the place $C + s_k$ with the phase $\varphi(C) + \varphi(s_k) = 2\pi + \varphi(s_u)$ and the orbit is given by

$$u_o(s) = \tfrac{1}{2}\sqrt{\beta(s)\,\beta(s_k)}\,\theta_k\,\frac{\cos\nu\left[\pi - \varphi(s) + \varphi(s_k)\right]}{\sin\pi\nu}. \tag{7.25}$$

This mathematical distinction of cases $s < s_k$ and $s > s_k$ is a consequence of the integration starting at s and ending at $s+C$ and is necessary to account for the discontinuity of the slope of the equilibrium orbit at the location of the kick. At the point $s = s_k$ obviously both equations are the same. In Fig. 7.3 the normalized distortion of the ideal orbit due to a single dipole kick

234

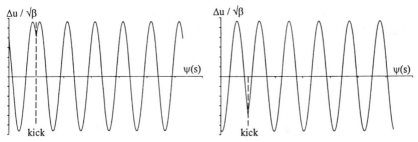

Fig. 7.3. Distorted orbit due to a single dipole kick for a tune just above an integer (left) and for a tune below an integer (right)

is shown. In a linear lattice this distortion is independent of the orbit and adds in *linear superposition*. If, however, sextupoles or other coupling or nonlinear magnets are included in the lattice, the distortion due to a single or multiple kick depends on the orbit itself and self-consistent solutions can be obtained only by iterations.

In cases where a single kick occurs at a symmetry point of a circular accelerator we expect the distorted orbit to also be symmetric about the kick. This is expressed in the asymmetric phase terms of both equations. Indeed since $\varphi(s_k) - \varphi(s) = \Delta\varphi$ for $s_k > s$ and $\varphi(s) - \varphi(s_k) = \Delta\varphi$ for $s > s_k$ the orbit distortion extends symmetrically in either direction from the location of the kick.

The solution for the perturbed equilibrium orbit is specially simple at the place, where the kick occurs. With $\varphi(s) = \varphi(s_k)$ the orbit distortion is

$$u_k = \tfrac{1}{2}\,\beta_k\,\theta_k\,\cot\pi\nu\,. \tag{7.26}$$

In situations where a short bending magnet like an orbit correction magnet and a beam position monitor are at the same place or at least close together we may use these devices to measure the betatron function at that place s_k by measuring the tune ν of the ring and the change in orbit u_k due to a kick θ_k. Equation (7.26) can then be solved for the betatron function β_k at the location s_k. This procedure can obviously be applied in both planes to experimentally determine β_x as well as β_y.

Statistical Distribution of Dipole Errors: In a real circular accelerator we have generally a large number of field and misalignment errors of unknown location and magnitude. If the accelerator is functional we may measure the distorted orbit with the help of beam position monitors and apply an orbit correction as discussed later in this section. During the design stage, however, we need to know the sensitivity of the ring design to such errors in order to determine *alignment tolerances* and the degree of correction required. In the absence of detailed knowledge about errors we use statistical methods to determine the most probable equilibrium orbit. All magnets are designed, fabricated, and aligned within statistical tolerances, which are

235

determined such that the distorted orbit allows the beam to stay within the vacuum pipe without loss. An expectation value for the orbit distortion can be derived by calculating the root-mean-square of (7.21)

$$
u_o^2(s) = \frac{\beta(s)}{4\sin^2 \pi\nu} \oint\limits_{s}^{s+C} \oint\limits_{s}^{s+C} p_o(\sigma)\, p_o(\tau)\, \sqrt{\beta(\sigma)}\, \sqrt{\beta(\tau)}
$$

$$
\cos\left[\nu(\varphi_s - \varphi_\sigma + \pi)\right] \cos\left[\nu(\varphi_s - \varphi_\tau + \pi)\right] d\sigma\, d\tau\,,
\tag{7.27}
$$

where we have set for simplicity $\varphi_s = \varphi(s)$ etc. This double integral can be evaluated by expanding the cosine functions to separate the phases φ_σ and φ_τ. We get terms like $\cos\nu\varphi_\sigma \cos\nu\varphi_\tau$ and $\sin\nu\varphi_\sigma \sin\nu\varphi_\tau$ or mixed terms. All these terms tend to cancel except when $\sigma = \tau$ since both the perturbations and their locations are statistically distributed in phase. Only for $\sigma = \tau$ will we get quadratic terms that contribute to a finite expectation value for the orbit distortion

$$
\langle p_o^2(\tau) \left[\cos^2 \nu(\varphi_s + \pi) \cos^2 \nu\varphi_\tau + \sin^2 \nu(\varphi_s + \pi) \sin^2 \nu\varphi_\tau\right]\rangle
$$
$$
= \langle p_o^2(\tau)\rangle[\cos^2 \nu(\varphi_s + \pi)\langle \cos^2 \nu\varphi_\tau\rangle + \sin^2 \nu(\varphi_s + \pi)\langle \sin^2 \nu\varphi_\tau\rangle]
$$
$$
= \langle p_o^2(\tau)\rangle \tfrac{1}{2}\,,
$$

and get with this for (7.27)

$$
\langle u_o^2(s)\rangle = \frac{\beta(s)}{8\sin^2 \pi\nu} \sum_i \langle\, p_o^2(\sigma_i)\, \beta(\sigma_i)\, \ell_i^2\,\rangle\,,
\tag{7.28}
$$

where the integrals have been replaced by a single sum over all perturbing fields of length ℓ_i. This can be done since we assume that the betatron phase does not change much over the length of individual perturbations. Equation (7.28) gives the expectation value for the orbit distortion at the point s and since the errors are statistically distributed we get from the *central limit theorem* a gaussian distribution of the orbit distortions with the standard deviation $\sigma_u^2(s) = \langle u_o^2(s)\rangle$ from (7.28). In other words if an accelerator is constructed with tolerances $\langle p_o^2(\sigma_i)\rangle$ there is a 68% probability that the orbit distortions are of the order $\sqrt{\langle u_o^2(s)\rangle}$ as calculated from (7.28) and a 98% probability that they are not more than twice that large.

As an example, we consider a uniform beam transport line, where all quadrupoles have the same strength and the betatron functions are periodic like in a FODO channel. This example seems to be very special since hardly any practical beam line has these properties, but it is still a useful example and may be used to simulate more general beam lines for a quick estimate of *alignment tolerances*. Assuming a gaussian distribution of *quadrupole misalignments* with a standard deviation σ_u and quadrupole strength k, we find the perturbations to be $p = k\sigma_u$ and the expected orbit distortion is

$$
\sqrt{\langle u_o^2(s)\rangle} = \sqrt{\beta(s)}\, A\, \sigma_u\,,
\tag{7.29}
$$

where A is called the error *amplification factor* defined by

$$A^2 = \frac{N}{8\sin^2\pi\nu}\langle(k\ell)^2\beta\rangle \approx \frac{N}{8\sin^2\pi\nu}\frac{\bar{\beta}}{f^2}, \qquad (7.30)$$

where $\langle(k\ell)^2\beta\rangle$ is taken as the average value for the expression in all N misaligned quadrupoles, f is the focal length of the quadrupoles, and $\bar{\beta}$ the average betatron function.

The expectation value for the maximum value of the orbit distortion $\langle\hat{u}_o^2(s)\rangle$ is larger. In (7.28) we have averaged the trigonometric functions

$$\langle\cos^2\nu\varphi(\tau)\rangle = \langle\sin^2\nu\varphi(\tau)\rangle = \tfrac{1}{2}$$

and therefore

$$\langle\hat{u}_o^2\rangle = 2\langle u_o^2(s)\rangle. \qquad (7.31)$$

These statistical methods obviously require a large number of misalignments to become accurate. While this is not always the case for shorter beam lines it is still useful to perform such calculations. In cases where the statistical significance is really poor one may use 10 or 20 sets of random perturbations and apply them to the beam line or ring lattice. This way a better evaluation of the distribution of possible perturbations is possible.

Clearly, the *tolerance requirements* increase as the average value of betatron functions, the quadrupole focusing, or the size of the accelerator or number of magnets N is increased. No finite orbit can be achieved if the tune is chosen to be an integer value. Most accelerators work at tunes which are about one quarter away from the next integer to maximize the trigonometric denominator $|\sin\pi\nu| \approx 1$. From a practical standpoint we may wonder what compromise to aim for between a large aperture and tight tolerances. It is good practice to avoid perturbations as reasonable as possible and then, if necessary, enlarge the magnet aperture to accommodate distortions which are too difficult to avoid. As a practical measure it is possible to restrict the uncorrected orbit distortion in most cases to 5 – 10 mm and provide magnet apertures that will accommodate this.

What happens if the expected orbit distortions are larger than the vacuum aperture which is virtually sure to happen at least during initial commissioning of more sensitive accelerators? In this case one relies on fluorescent screens or electronic monitoring devices located along the beam line which are sensitive enough to detect even small beam intensities passing by only once. By empirically employing corrector magnets the beam can be guided from monitor to monitor thus establishing a path and eventually a closed orbit. Once all monitors receive a signal, more sophisticated and computerized orbit control mechanism may be employed.

7.2.3 Closed Orbit Correction

Due to magnetic field and alignment errors a distorted equilibrium orbit is generated as discussed in the previous section. Specifically for distinct localized dipole field errors at position s_j we got

$$u_o(s) = \frac{\sqrt{\beta(s)}}{2\sin\pi\nu} \sum_j \theta_j \sqrt{\beta_j} \cos\nu[\varphi(s) - \varphi_j + \pi].$$ (7.32)

Since *orbit distortions* reduce the available aperture for betatron oscillations and can change other beam parameters it is customary in accelerator design to include a special set of magnets for the correction of distorted orbits. These orbit correction magnets produce *orbit kicks* and have, therefore, the same effect on the orbit as dipole errors; however, now the location and the strength of the kicks are known. Before we try to correct an orbit it must have been detected with the help of *beam position monitors*. The position of the beam in these monitors is the only direct information we have about the distorted orbit. From the set of measured orbit distortions u_i at the m monitors i we form a vector

$$\mathbf{u}_m = (u_1, u_2, u_3, \ldots, u_m)$$ (7.33)

and use the correctors to produce additional "orbit distortions" at the monitors through carefully selected kicks θ_k in *orbit correction magnets* which are also called *trim magnets*. For n corrector magnets the change in the orbit at the monitor i is

$$\Delta u_i = \frac{\sqrt{\beta_i}}{2\sin\pi\nu} \sum_{k=1}^{n} \theta_k \sqrt{\beta_k} \cos\nu(\varphi_i - \varphi_k + \pi),$$ (7.34)

where the index k refers to the corrector at $s = s_k$. The orbit changes at the beam position monitors due to the corrector kicks can be expressed in a matrix equation

$$\Delta\mathbf{u}_m = \boldsymbol{M}\,\boldsymbol{\theta}_n,$$ (7.35)

where $\Delta\mathbf{u}_m$ is the vector formed from the orbit changes at all m monitors, $\boldsymbol{\theta}_n$ the vector formed by all kicks in the n correction magnets, and $\boldsymbol{M}$ the transformation matrix $\boldsymbol{M} = (\mathrm{M}_{ik})$ with

$$\mathrm{M}_{ik} = \frac{\sqrt{\beta_i\beta_k}}{2\sin\pi\nu} \cos\nu(\varphi_i - \varphi_k + \pi).$$ (7.36)

The distorted orbit can be corrected at least at the position monitors with corrector kicks θ_k chosen such that $\Delta\mathbf{u}_m = -\mathbf{u}_m$ or

$$\boldsymbol{\theta}_n = -\boldsymbol{M}^{-1}\mathbf{u}_m.$$ (7.37)

Obviously, this equation can be solved exactly if $n = m$ and also for $n > m$ if not all correctors are used. Additional conditions could be imposed in the latter case like minimizing the corrector strength.

While an orbit correction according to (7.37) is possible it is not always the optimum way to do it. A perfectly corrected orbit at the monitors still leaves finite distortions between the monitors. To avoid large orbit distortions between monitors sufficiently many monitors and correctors must be distributed along the beam line. A more sophisticated *orbit correction scheme* would only try to minimize the sum of the squares of the orbit distortions at the monitors

$$(\mathbf{u}_m - \Delta \mathbf{u}_m)^2_{\min} = (\mathbf{u}_m - \mathcal{M}\, \boldsymbol{\theta}_n)^2_{\min}, \tag{7.38}$$

thus avoiding extreme corrector settings due to an unnecessary requirement for perfect correction at monitor locations.

This can be achieved for any number of monitors m and correctors n although the quality of the resulting orbit depends greatly on the actual number of correctors and monitors. To estimate the number of correctors and monitors needed we remember the similarity of dispersion function and orbit distortion. Both are derived from similar differential equations. The solution for the distorted orbit, therefore, can also be expressed by Fourier harmonics similar to (6.92). With F_n being the Fourier harmonics of $-\beta^{3/2}(s)\Delta\frac{1}{\rho}(s)$ we get for the distorted orbit

$$u_o(s) = \sqrt{\beta(s)}\, \nu^2 \sum_{n=-\infty}^{+\infty} \frac{F_n\, e^{in\varphi}}{\nu^2 - n^2}, \tag{7.39}$$

which exhibits a resonance for $\nu = k$. The harmonic spectrum of the uncorrected orbit $u_o(s)$ has therefore also a strong harmonic content for $k \approx \nu$. To obtain an efficient orbit correction both the beam position monitor and corrector distribution around the accelerator must have a strong harmonic close to the tune ν of the accelerator. It is, therefore, most efficient to distribute monitors and correctors uniformly with respect to the betatron phase $\varphi(s)$ rather than uniform with s and use about four of each per betatron wave length.

With sufficiently many correctors and monitors the orbit can be corrected in different ways. One could excite all correctors in such a way as to compensate individual harmonics in the distorted orbit as derived from beam position measurement. Another simple and efficient way is to look for the first corrector that most efficiently reduces the orbit errors then for the second most efficient and so on. This latter method is most efficient since the practicality of other methods can be greatly influenced by errors of the position measurements as well as field errors in the correctors. The single most effective corrector method can be employed repeatedly to obtain an acceptable orbit. Of similar practical effectiveness is the method of beam bumps. Here a set of three to four correctors are chosen and powered in such a way as to produce a beam bump opposite to an orbit distortion in

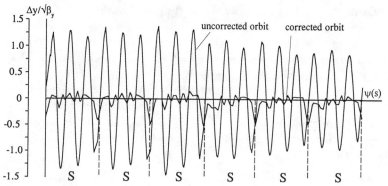

Fig. 7.4. Orbit of Fig. 7.2 before and after correction.

that area. This method, therefore, is a local orbit correction scheme while the others are global schemes.

As a practical example, we show the vertical orbit in the storage ring PEP before and after correction, Fig. 7.4, in normalized units. The orbit distortions are significantly reduced and the strong harmonic close to the betatron frequency has all but disappeared. Even in normalized units the orbit distortions have now a statistical appearance and further correction would require many more correctors. The peaks at the six interaction points of the lattice, indicated by dashed lines, are actually small orbit distortions and appear large only due to the normalization to a very small value of the betatron function at the interaction point, see Fig. 6.19.

7.3 Quadrupole Field Perturbations

The dipole perturbation terms cause a shift in the beam path or closed orbit without affecting the focusing properties of the beam line. The next higher perturbation terms which depend linearly on the transverse particle offset from the ideal orbit will affect the focusing characteristics because these perturbations act just like quadrupoles. Similar to the alignment tolerances we will derive tolerances on the linear perturbations. Quadrupole fields determine the betatron function as well as the phase advance or tune in a circular accelerator. We expect therefore that linear field errors will modify these parameters and we will derive the effect of gradient errors on lattice functions and tune.

7.3.1 Betatron Tune Shift

Gradient field errors have a first order effect on the betatron phase and tune. Specifically in circular accelerators we have to be concerned about the tune not to deviate too much from stable values to avoid beam loss. The effect of a linear perturbation on the tune can be easily derived in matrix formulation for one single perturbation. For simplicity we choose a symmetry point in the lattice of a circular accelerator and insert on either side of this point a thin lens perturbation with the transformation matrix

$$\mathcal{M}_{\mathrm{P}} = \begin{bmatrix} 1 & 0 \\ -1/f & 1 \end{bmatrix}, \tag{7.40}$$

where $f^{-1} = -\frac{1}{2} \int p_1(s)\, ds$ and $p_1(s)$ is the total perturbation. Combining this with the transformation of an ideal ring (5.135),

$$\mathcal{M}_{\mathrm{o}} = \begin{bmatrix} C(s) & S(s) \\ C'(s) & S'(s) \end{bmatrix} = \begin{bmatrix} \cos\psi_{\mathrm{o}} & \beta_{\mathrm{o}} \sin\psi_{\mathrm{o}} \\ -\frac{1}{\beta_{\mathrm{o}}} \sin\psi_{\mathrm{o}} & \cos\psi_{\mathrm{o}} \end{bmatrix}$$

we get for the trace of the total transformation matrix $\mathcal{M} = \mathcal{M}_{\mathrm{p}} \mathcal{M}_{\mathrm{o}} \mathcal{M}_{\mathrm{p}}$

$$\mathrm{Tr}\mathcal{M} = 2\cos\psi_{\mathrm{o}} - \frac{\beta_{\mathrm{o}}}{f} \sin\psi_{\mathrm{o}}, \tag{7.41}$$

where β_{o} is the unperturbed betatron function at the location of the perturbation and $\psi_{\mathrm{o}} = 2\pi\nu_{\mathrm{o}}$ the unperturbed phase advance per turn. The trace of the perturbed ring is $\mathrm{Tr}\mathcal{M} = 2\cos\psi$ and we have therefore

$$\cos\psi = \cos\psi_{\mathrm{o}} - \frac{1}{2} \frac{\beta_{\mathrm{o}}}{f} \sin\psi_{\mathrm{o}}. \tag{7.42}$$

With $\psi = 2\pi\nu = 2\pi\nu_{\mathrm{o}} + 2\pi\delta\nu$ and $\cos\psi = \cos\psi_{\mathrm{o}} \cos 2\pi\delta\nu - \sin\psi_{\mathrm{o}} \sin 2\pi\delta\nu$ we get for small perturbations the tune shift

$$\delta\nu = \frac{1}{4\pi} \frac{\beta_{\mathrm{o}}}{f} = -\frac{\beta_{\mathrm{o}}}{4\pi} \int p_1(\sigma)\, d\sigma. \tag{7.43}$$

For more than a single gradient error one would simply add the individual contribution from each error to find the total tune shift. The same result can be obtained from the perturbed equation of motion

$$u'' + K(s)\, u = p_1(s)\, u. \tag{7.44}$$

To show this we introduce normalized coordinates $w = u/\sqrt{\beta}$ and $\varphi = \int ds/\nu\beta$ and (7.44) becomes

$$\ddot{w} + \nu_{\mathrm{o}}^2 w = \nu_{\mathrm{o}}^2 \beta^2(s)\, p_1(s)\, w. \tag{7.45}$$

Since both the betatron function $\beta(s)$ and perturbations $p_1(s)$ are periodic we may Fourier expand the coefficient of $\nu_{\mathrm{o}}w$ on the r.h.s. and get for the lowest, nonoscillating harmonic

$$F_{\mathrm{o}} = \frac{1}{2\pi} \int_{0}^{2\pi} \nu_{\mathrm{o}} \beta^2 p_1 \, d\varphi = \frac{1}{2\pi} \oint \beta(s)\, p_1(s)\, ds. \tag{7.46}$$

241

Inserting this into (7.45) and collecting terms linear in w we get

$$\ddot{w} + (\nu_0^2 - \nu_0 F_0)\,w = 0 \tag{7.47}$$

and the new tune $\nu = \nu_0 + \delta\nu$ is determined by

$$\nu^2 = \nu_0^2 - \nu_0 F_0 \approx \nu_0^2 + 2\nu_0\,\delta\nu. \tag{7.48}$$

Solving for $\delta\nu$ gives the linear tune perturbation

$$\delta\nu = -\frac{F_0}{2} = -\frac{1}{4\pi}\oint \beta(s)\,p_1(s)\,ds \tag{7.49}$$

in complete agreement with the result obtained in (7.43). The produced by a linear perturbation has great diagnostic importance. By varying the strength of an individual quadrupole and measuring the tune shift it is possible to derive the value of the betatron function in this quadrupole.

The effect of linear perturbations contributes in first approximation only to a static tune shift. In higher approximation, however, we note specific effects which can affect beam stability and therefore must be addressed in more detail. To derive these effects we solve (7.45) with the help of a Green's function as discussed in Sect. 4.8.3 and obtain the perturbation

$$P(\varphi) = \int_0^\varphi \nu_0^2\,\beta^2(\chi)\,p_1(\chi)\,w(\chi)\,\frac{1}{\nu_0}\sin\nu_0(\varphi - \chi)\,d\chi, \tag{7.50}$$

where we have made use of (5.150) for the principal solutions. We select a particular, unperturbed trajectory, $w(\chi) = w_0\cos\nu\chi$ with $\dot{w}_0 = 0$ and get the perturbed particle trajectory

$$w(\nu\varphi) = w_0\cos\nu_0\varphi + w_0\nu_0\int_0^\varphi \beta^2\,p_1\cos\nu_0\chi\,\sin\nu_0(\varphi - \chi)\,d\chi, \tag{7.51}$$

where $\beta = \beta(\chi)$ and $p_1 = p_1(\chi)$. If, on the other hand, we consider the perturbations to be a part of the lattice, we would describe the same trajectory by

$$w(\psi) = w_0\cos\nu\psi. \tag{7.52}$$

Both solutions must be equal. Specifically the phase advance per turn must be the same and we get from (7.51, 52) after one turn $\varphi = 2\pi$ for the perturbed tune $\nu = \nu_0 + \delta\nu$

$$\cos 2\pi(\nu_0 + \delta\nu) = \cos 2\pi\nu_0$$
$$+ \nu_0\int_0^{2\pi}\beta^2(\chi)\,p_1(\chi)\,\cos(\nu_0\chi)\,\sin\nu_0(2\pi - \chi)\,d\chi, \tag{7.53}$$

which can be solved for the tune shift dν. Obviously the approximation breaks down for large values of the perturbation as soon as the r.h.s. becomes larger than unity. For small perturbations, however, we expand the trigonometric functions and get

$$
\delta\nu = -\frac{1}{4\pi} \oint \beta(s)\,p_1(s)\,\mathrm{d}s
$$
$$
-\frac{1}{4\pi \sin 2\pi\nu_0} \oint \beta(s)\,p_1(s)\,\sin 2\nu_0[\pi - \chi(s)]\,\mathrm{d}s\,,
\tag{7.54}
$$

where $\mathrm{d}s = \nu_0\,\beta(s)\,\mathrm{d}\chi$.

The first term is the average *tune shift* which has been derived before, while the second term is of oscillatory nature averaging to zero over many turns if the tune of the circular accelerator is not equal to a half integer or multiples thereof. We have found hereby a second resonance condition to be avoided which occurs for half integer values of the tunes

$$
\nu_0 \neq \tfrac{1}{2}n\,.
\tag{7.55}
$$

This resonance is called a *half integer resonance* and causes divergent solutions for the lattice functions.

7.3.2 Resonances and Stop Band Width

Calculating the tune shift from (7.53) we noticed that there is no solution if the perturbation is too large such that the absolute value of the r.h.s. becomes larger than unity. In this case the tune becomes imaginary leading to ever increasing betatron oscillation amplitudes and beam loss. This resonance condition occurs not only at a half integer resonance but also in a finite vicinity, where $1 - \cos 2\pi\nu_0$ is smaller than the perturbation term and the r.h.s of (7.53) is larger than unity. The region of instability is called the stop band and the width of unstable tune values is called the *stop band width* which can be calculated by using a higher approximation for the perturbed solution. Following the arguments of Courant and Snyder [7.12] we note that the perturbation (7.50) depends on the betatron oscillation $w(\varphi)$ itself and we now use in the perturbation integral the first order approximation (7.51) rather than the unperturbed solution to calculate the perturbation (7.50). Then instead of (7.53) we get

$$
\cos 2\pi(\nu_0 + \delta\nu) - \cos 2\pi\nu_0 =
$$
$$
+ \nu_0 \int_0^{2\pi} \beta^2(\chi)\,p_1(\chi)\,\cos(\nu_0\chi)\,\sin\nu_0(2\pi - \chi)\,\mathrm{d}\chi
$$
$$
+ \nu_0^2 \int_0^{2\pi} \beta^2(\chi)\,p_1(\chi)\,\sin\nu_0(2\pi - \chi)
\tag{7.56}
$$

$$\cdot \int\limits_0^\chi \beta^2(\zeta) \, p_1(\zeta) \, \cos \nu_o \zeta \, \sin \nu_o(\chi - \zeta) \, d\zeta \, d\chi \, .$$

This expression can be used to calculate the stop band width due to gradient field errors which we will do for the *integer resonance* $\nu_o = n + \delta\nu$ and for the *half integer resonance* $\nu_o = n + 1/2 + \delta\nu$, where n is an integer and $\delta\nu$ the deviation of the tune from these resonances. To evaluate the first integral I_1, on the r.h.s., we make use of the relation

$$\cos \nu_o \chi \, \sin \nu_o(2\pi - \chi) = \tfrac{1}{2} \sin 2\pi\nu_o + \tfrac{1}{2} \sin[2\nu_o(\pi - \chi)]$$

and get with $\oint \beta(s) \, p_1(s) \, ds = \nu_o \int\limits_0^{2\pi} \beta^2(\chi) \, p_1(\chi) \, d\chi = 2\pi \, F_o$ from (7.46)

$$I_1 = \pi F_o \sin 2\pi\nu_o + \tfrac{1}{2}\nu_o \int\limits_0^{2\pi} \beta^2(\chi) \, p_1(\chi) \, \sin 2\nu_o(\pi - \chi) \, d\chi \, .$$

The second term of the integral I_1 has oscillatory character and averages to zero over many turns. We have therefore

$$I_1 = \pi F_o \sin 2\pi\nu_o \approx \begin{cases} 2\pi^2 \, F_o \, \delta\nu & \text{for } \nu_o = n + \delta\nu \\ -2\pi^2 \, F_o \, \delta\nu & \text{for } \nu_o = n + \tfrac{1}{2} + \delta\nu. \end{cases} \qquad (7.57)$$

The second integral I_2 in (7.56) can best be evaluated while expressing the trigonometric functions in their exponential form. Terms like $e^{\pm i\nu(2\pi - 2\sigma)}$ or $e^{\pm i\nu(2\pi - 2\tau)}$ vanish on average over many turns. With

$$\int_0^{2\pi} d\chi \int_0^\chi d\zeta = \tfrac{1}{2} \int_0^{2\pi} d\chi \int_0^{2\pi} d\zeta$$

we get for the second integral

$$I_2 = -\frac{\nu_o^2}{16} \int\limits_0^{2\pi} \beta^2(\chi) \, p_1(\chi) \int\limits_0^{2\pi} \beta^2(\zeta) \, p_1(\zeta)$$

$$\cdot \left\{ \left(e^{i2\pi\nu_o} + e^{-i2\pi\nu_o}\right) - \left[e^{i2\nu_o(\pi - \chi + \zeta)} + e^{-i2\nu_o(\pi - \chi + \zeta)}\right] \right\} d\zeta \, d\chi \, .$$

Close to the integer resonance $\nu_o = n + \delta\nu$ we get

$$I_{2,n} = -\frac{\nu_o^2}{16} \int\limits_0^{2\pi} \beta^2(\chi) \, p_1(\chi) \int\limits_0^{2\pi} \beta^2(\zeta) \, p_1(\zeta)$$

$$\cdot \left\{ \left(e^{i2\pi\delta\nu} + e^{-i2\pi\delta\nu}\right) - \left[e^{i2n(\zeta - \chi)} + e^{-i2n(\zeta - \chi)}\right] \right\} d\zeta \, d\chi \qquad (7.58)$$

and in the vicinity of the half integer resonance $\nu_o = n + 1/2 + \delta\nu$

$$I_{2,n+1/2} = -\frac{\nu_o^2}{16} \int_o^{2\pi} \beta^2(\chi)\, p_1(\chi) \int_o^{2\pi} \beta^2(\zeta)\, p_1(\zeta)$$

$$\{-(e^{i2\pi\delta\nu} + e^{-i2\pi\delta\nu})$$

$$+ [e^{i2(n+1/2)(\zeta-\chi)} + e^{-i2(n+1/2)(\zeta-\chi)}]\}\, d\zeta\, d\chi. \tag{7.59}$$

The integrals are now easy to express in terms of Fourier harmonics of $\nu_o \beta^2(\varphi)\, p_1(\varphi)$, where the amplitudes of the harmonics F_p with $p > 1$ are given by

$$|F_p|^2 = F_p F_p^* = \frac{\nu_o^2}{\pi^2} \int_o^{2\pi} \beta^2(\chi)\, p_1(\chi)\, e^{-ip\chi}\, d\chi \int_o^{2\pi} \beta^2(\zeta)\, p_1(\zeta)\, e^{ip\zeta}\, d\zeta.$$

For F_o we have the well-known result of $2F_o = F_p(p = 0)$. With this and ignoring terms quadratic in $\delta\nu$ we get for (7,58,59)

$$I_{2,n} \approx \frac{\pi^2}{8}\left(F_{2n}^2 - 4F_o^2 \cos 2\pi\delta\nu\right) \tag{7.60}$$

and

$$I_{2,n+1/2} \approx -\frac{\pi^2}{8}\left(F_{2n+1}^2 - 4F_o^2 \cos 2\pi\delta\nu\right), \tag{7.61}$$

respectively. At this point we may collect the results and get on the l.h.s. of (7.56) for $\nu_o = n + \delta\nu$

$$\cos 2\pi(\nu_o + \delta\nu) - \cos 2\pi\nu_o = \cos 2\pi(\nu_o + \delta\nu) - 1 + 2\pi^2\,\delta\nu^2.$$

This must be equated with the r.h.s. which is the sum of both integrals I_1 and I_2 and we have with $F_o^2 \cos 2\pi\delta\nu \approx 1 - \mathcal{O}(\delta^4\nu)$

$$\cos 2\pi(\nu_o + \delta\nu) - 1 = -2\pi^2\,\delta\nu^2 + 2\pi^2\,F_o\,\delta\nu + \frac{\pi^2}{8}(F_{2n}^2 - 4F_o^2). \tag{7.62}$$

The boundaries of the *stop band* on either side of the integer resonance $\nu_o \approx n$ can be calculated from the condition that $\cos 2\pi(\nu_o + \delta\nu) \leq 1$ which has two solutions $\delta\nu_{1,2}$. From (7.62) we get therefore

$$\delta\nu^2 - F_o\,\delta\nu = \tfrac{1}{16}\left(|F_{2n}|^2 - 4F_o^2\right)$$

and solving for $\delta\nu$

$$\delta\nu_{1,2} = \tfrac{1}{2}F_o \pm \tfrac{1}{4}|F_{2n}| \tag{7.63}$$

the stop band width is finally

$$\Delta\nu = \delta\nu_1 - \delta\nu_2 = \tfrac{1}{2}|F_{2n}| = \frac{1}{2\pi}\oint \beta(s)\, p_1(s)\, e^{-i2n\zeta(s)}\, ds. \tag{7.64}$$

The stop band width close to the integer tune $\nu \approx n$ is determined by the second harmonic of the Fourier spectrum for the perturbation. The vicinity of the resonance for which no stable betatron oscillations exist increases with the value of the gradient field error and with the value of the betatron function at the location of the field error. For the half integer resonance $\nu_o \approx n + 1/2$, the stop band width has a similar form

$$\Delta\nu_{1/2} = \tfrac{1}{2}\,|\,F_{2n+1}\,| = \frac{1}{2\pi}\int\limits_0^{2\pi} \beta(s)\,p_1(s)\,e^{-i(2n+1)\zeta(s)}\,ds\,. \tag{7.65}$$

The lowest order Fourier harmonic $n = 0$ determines the static tune shift while the resonance width depends on higher harmonics. The existence of finite stop bands is not restricted to linear perturbation terms only. Nonlinear, higher order perturbation terms lead to higher order resonances and associated stop bands. In such cases one would replace in (7.45) the linear perturbation $\beta^{1/2}\,p_1(s)\,w$ by the nth order nonlinear perturbation $\beta^{n/2}\,p_n(s)\,w^n$ and basically go through the same derivation. In Vol. II we will use a different way to describe resonance characteristics caused by higher order perturbations. At this point we note only that perturbations of order n are weighted by the $n/2$ power of the betatron function at the location of the perturbation and increased care must be exercised, where large values of the betatron functions cannot be avoided. Undesired fields at such locations must be minimized.

7.3.3 Perturbation of Betatron Functions

The existence of linear perturbation terms causes not only the tunes but also betatron functions to vary around the ring or along a beam line. This variation can be derived by observing the perturbation of a particular trajectory like for example the sine-like solution given by

$$S_o(s_o|s) = \sqrt{\beta(s)}\,\sqrt{\beta_o}\,\sin\nu_o[\varphi(s) - \varphi_o]\,. \tag{7.66}$$

The sine-like function or trajectory in the presence of linear perturbation terms is by the principle of linear superposition the combination of the unperturbed solution (5.135) and perturbation (4.107)

$$\begin{aligned}
S(s_o|s) = &\sqrt{\beta(s)}\,\sqrt{\beta_o}\,\sin\nu_o\varphi(s) \\
&+\sqrt{\beta(s)}\int_{s_o}^s p_1(\sigma)\,\sqrt{\beta(\sigma)}\,S_o(s_o|\sigma)\,\sin\nu_o[\varphi(s) - \varphi(\sigma)]\,d\sigma\,.
\end{aligned} \tag{7.67}$$

Following the sinusoidal trajectory for the whole ring circumference or length of a superperiod L_p, we have with $s = s_o + L_p$, $\beta(s_o + L_p) = \beta(s_o) = \beta_o$ and $\varphi(s_o + L_p) = 2\pi + \varphi_o$

$$S(s_o|s_o + L_p) = \beta_o \sin 2\pi\nu_o + \beta_o \oint_{s_o}^{s_o+L_p} \beta(\sigma)\, p_1(\sigma)$$

$$\sin\nu_o[\varphi(\sigma) - \varphi_o] \sin\nu_o[2\pi + \varphi_o - \varphi(\sigma)]\, d\sigma\,. \tag{7.68}$$

We compute the difference due to the perturbation from the unperturbed trajectory (7.66) at $s = s_o + L_p$ and get

$$\Delta S = S(s_o|s_o + L_p) - S_o(s_o|s_o + L_p) \tag{7.69}$$

$$= \beta_o \int_{s_o}^{s_o+L_p} \beta(\sigma)\, p_1(\sigma)\, \sin[\nu_o(\varphi_\sigma - \varphi_o)] \sin[\nu_o(2\pi + \varphi_o - \varphi_\sigma)]\, d\sigma\,,$$

where we abbreviated $\varphi(s_o) = \varphi_o$ etc. The variation of the sine like function can be derived also from the variation of the M_{12} element of the transformation matrix for the whole ring

$$\Delta S = \Delta(\beta \sin 2\pi\nu) = \Delta\beta \sin 2\pi\nu_o + \beta_o\, 2\pi\, \Delta\nu \cos 2\pi\nu_o\,. \tag{7.70}$$

We use (7.49) for the tune shift $\delta\nu = -\tfrac{1}{2} F_o$, equate (7.70) with (7.69) and solve for $\Delta\beta/\beta$. After some manipulations, where we replace the trigonometric functions by their exponential expressions, the variation of the betatron function becomes at $\varphi_o = \varphi(s)$

$$\frac{\Delta\beta(s)}{\beta(s)} = \frac{1}{2\sin 2\pi\nu_o} \oint \beta(\sigma)\, p_1(\sigma)\, \cos 2\nu_o[\varphi(s) - \varphi(\sigma) + \pi]\, d\sigma\,. \tag{7.71}$$

The perturbation of the betatron function shows clearly resonance character and a half integer tune must be avoided. We observe a close similarity with the solution (6.84) of the dispersion function or the closed orbit (7.16). Setting $d\sigma = \nu_o\beta(\sigma)\, d\varphi$ we find by comparison that the solution for the perturbed betatron function can be derived from a differential equation similar to a modified Eq. (6.81)

$$\frac{d^2}{d\varphi^2}\left(\frac{\Delta\beta}{\beta}\right) + (2\nu_o)^2\, \frac{\Delta\beta}{\beta} = (2\nu_o)^2\, \tfrac{1}{2}\, \beta^2(s)\, p_1(s)\,. \tag{7.72}$$

Expanding the periodic function $\nu_o\, \beta^2\, p_1 = \sum_q F_q\, e^{iq\varphi}$ we try the periodic ansatz

$$\frac{\Delta\beta}{\beta} = \sum_q B_q\, F_q\, e^{iq\varphi} \tag{7.73}$$

and get from (7.72)

$$\sum_q \left[-q^2 + (2\nu_o)^2\right] B_q\, F_q\, e^{iq\varphi} = 2\nu_o \sum_q F_q\, e^{iq\varphi}\,.$$

This can be true for all values of the phase φ only if the coefficients of the exponential functions vanish separately for each value of q or if

$$B_q = \frac{2\nu_o}{(2\nu_o)^2 - q^2} . \tag{7.74}$$

Inserting into the periodic ansatz (7.73) we get finally another form for the perturbation of the betatron function

$$\frac{\Delta\beta}{\beta} = \frac{\nu_o}{2} \sum_q \frac{F_q \, e^{iq\varphi}}{\nu_o^2 - (q/2)^2} . \tag{7.75}$$

Again we recognize the half inter resonance leading to an infinitely large perturbation of the betatron function. In the vicinity of the half integer resonance $\nu_o \approx n + 1/2 = q/2$ the betatron function can be expressed by the resonant term only

$$\frac{\Delta\beta}{\beta} \approx \tfrac{1}{2}|F_{2n+1}| \frac{\cos(2n+1)\varphi}{\nu_o - (n + 1/2)}$$

and with $|F_{2n+1}| = 2\Delta\nu_{1/2}$ from (7.65) we get again the perturbation of the betatron function (7.71). The *beat factor* for the variation of the betatron function is define by

$$BF = 1 + \left(\frac{\Delta\beta}{\beta_o}\right)_{max} = 1 + \frac{\Delta\nu_{2n+1}}{2\nu_o - (2n+1)} , \tag{7.76}$$

where $\Delta\nu_{2n+1}$ is the half integer stop band width. The beating of the betatron function is proportional to the stop band width and therefore depends greatly on the value of the betatron function at the location of the perturbation. Even if the tune is chosen safely away from the next resonance, a linear perturbation at a large betatron function may still cause an unacceptable beat factor. It is generally prudent to design lattices in such a way as to avoid large values of the betatron functions. As a practical note, any value of the betatron function which is significantly larger than the quadrupole distances should be considered large. For many beam transport problems this is easier said than done. Therefore, where large betatron functions cannot be avoided or must be included to meet special design goals, results of perturbation theory warn us to apply special care for beam line component design, alignment, and to minimize undesirable stray fields.

7.4 Resonance Theory

Perturbation terms in the equation of motion can lead to a special class of beam instability called *resonances* which can occur if perturbations act on a particle in synchronism with its oscillatory motion. While such a situation is conceivable in a very long beam transport line composed of many periodic sections, the appearance of resonances is generally restricted to circular accelerators. There perturbations occur periodically at every turn and we may Fourier analyze the perturbation with respect to the revolution frequency. If any of the harmonics of the perturbation terms coincides with the eigenfrequency of the particles a resonance can occur and the particles may get lost. Such resonances caused by field imperfections of the magnet lattice are also called *structural resonances* or *lattice resonances*. We have already come across two such resonances, the integer and the half integer resonances.

The characteristics of these two resonances is that the equilibrium orbit and the overall focusing is not defined. Any small dipole or quadrupole error would therefore lead to particle loss as would any deviation of the particle position and energy from ideal values. Since these resonances are caused by linear field errors we call them also linear resonances. Higher order resonances are caused in a similar way by nonlinear fields which are considered to be field errors with respect to the ideal linear lattice even though we may choose to specifically include such magnets, like sextupoles or octupoles, into the lattice to compensate for particular beam dynamics problems.

7.4.1 Resonance Conditions

In this section the characteristics of resonances in circular accelerators will be derived in a more general way starting from the equation of motion in normalized coordinates (5.143, 144) with only the nth order multipole perturbation term. This is no restriction of generality since in linear approximation each multipole perturbation having its own resonance structure will be superimposed to that of other multipole perturbations. On the other hand, the treatment of only a single multipole perturbation will reveal much clearer the nature of the resonance. The equation of motion in normalized horizontal coordinates for an nth order perturbation is

$$\ddot{w} + \nu_o^2 \, w = \bar{p}_n(\psi) \, w^{n-1} \tag{7.77}$$

and a similar equation holds for vertical oscillations $v(\psi)$.

The perturbations can be any of the terms in the equations of motion (4.86) and (4.88) , however, we will consider primarily perturbation terms which occur most frequently in circular accelerators and are due to rotated quadrupole or nonlinear multipole fields. The general treatment of resonances for other perturbations is not fundamentally different and is left to the interested reader. From Chap. 4 we extract the dominant perturbation terms in normalized coordinates and compile them in Table 7.1 ordered by horizontal perturbations of order n and vertical perturbations of order r.

The perturbations $\bar{p}_n(\varphi)$ are periodic in φ and can be expanded into a Fourier series

Table 7.1. Perturbation terms

order n r	$\bar{p}_{nx}(\varphi)\, w^{n-1}\, v^{r-1}$	$\bar{p}_{ny}(\varphi)\, v^{n-1}\, w^{r-1}$
1 2	$-\nu_{xo}^2\, \beta_x^{3/2}\, \beta_y^{1/2}\, \underline{k}\, v$	
2 1		$-\nu_{yo}^2\, \beta_y^{3/2}\, \beta_x^{1/2}\, \underline{k}\, w$
3 1	$-\nu_{xo}^2\, \beta_x^{5/2}\, \tfrac{1}{2} m\, w^2$	
2 2		$-\nu_{yo}^2\, \beta_x^{1/2}\, \beta_y^2\, m w v$
1 3	$+\nu_{xo}^2\, \beta_x^{3/2}\, \beta_y\, \tfrac{1}{2} m\, v^2$	
4 1	$-\nu_{xo}^2\, \beta_x^3\, \tfrac{1}{6} r\, w^3$	
3 2		$+\nu_{yo}^2\, \beta_x\, \beta_y^2\, \tfrac{1}{2} r\, w^2 v$
2 3	$+\nu_{xo}^2\, \beta_x^2\, \beta_y\, \tfrac{1}{2} r\, w v^2$	
1 4		$-\nu_{yo}^2\, \beta_y^3\, \tfrac{1}{6} r\, v^3$

$$\bar{p}_n(\varphi) = \sum_m \bar{p}_{nm}\, e^{im\varphi}. \tag{7.78}$$

Since the perturbation is supposed to be small we will insert the unperturbed oscillation w_o on the right-hand side of (7.77). The general form of the unperturbed betatron oscillation can be written like

$$w_o(\varphi) = a\, e^{i\nu_o\varphi} + b\, e^{-i\nu_o\varphi}, \tag{7.79}$$

where a and b are arbitrary constants and we may now express the amplitude factor in the perturbation term $w^{n-1}(\varphi)$ by a sum of exponential terms which we use on the right hand side of (7.77) as a first approximation

$$w^{n-1}(\varphi) \approx w_o^{n-1}(\varphi) = \sum_{|q|\le n-1} W_q\, e^{iq\nu_o\varphi}. \tag{7.80}$$

We use both (7.78,80) in (7.77) and get for the equation of motion

$$\ddot{w} + \nu_o^2 w = \sum_{q,m} W_q\, \bar{p}_{nm}\, e^{i(m+q\nu_o)\varphi}. \tag{7.81}$$

The solution of this equation includes resonant terms whenever there is a perturbation term with a frequency equal to the eigenfrequency ν_o of the oscillator. The *resonance condition* is therefore

$$m + q\nu_o = \nu_o \quad \text{with} \quad |q| \le n-1. \tag{7.82}$$

From earlier discussions we expect to find the *integer resonance* caused by dipole errors $n = 1$. In this case the index q can only be $q = 0$ and we get from (7.82)

$$\nu_o = m \tag{7.83}$$

which is indeed the condition for an integer resonance.

Magnetic *gradient field errors*, $n = 2$, can cause both a *half integer-resonance* as well as an integer resonance. The index q can have the values $q = 0$ and $q = \pm 1$. Note however that not all coefficients W_q necessarily are nonzero. In this particular case, the coefficient for $q = 0$ is indeed zero as becomes obvious by inspection of (7.79). The *resonance conditions* for these second order resonances are

$$
\begin{aligned}
m + \nu_o &= \nu_o \rightarrow m = 0 & \text{tune shift}, \\
m - \nu_o &= \nu_o \rightarrow m = 2\nu_o & \text{integer and half integer resonance}, \quad (7.84) \\
m &= \nu_o & \text{no resonance because } W_o = 0.
\end{aligned}
$$

Among the resonance conditions (7.84) we notice that for $m = 0$ the effect of the perturbation on the particle motion exists independent of the particular choice of the tune ν_o. The perturbation includes a nonvanishing average value $\bar{p}_{2o}$ which in this particular case represents the average gradient error of the perturbation. Like any other gradient field in the lattice, this gradient error also contributes to the tune and therefore causes a tune shift. From (7.77) we find the new tune to be determined by $\nu^2 = \nu_o^2 - \bar{p}_{2o}$ and the tune shift is $\Delta\nu \approx -\bar{p}_{2o}/(2\nu_o)$ in agreement with our earlier result in Sect. 7.3.

Third order resonances $n = 3$ can be driven by sextupole fields and the index q can have values

$$q = -2, -1, 0, +1, +2. \tag{7.85}$$

Here we note that $W_1 = W_{-1} = 0$ and therefore no resonances occur for $q = \pm 1$. The *resonance conditions* for sextupole field perturbations are then

$$
\begin{aligned}
m - 2\nu_o &= \nu_o \rightarrow m = 3\nu_o & \text{third order resonance}, \\
m &= \nu_o \rightarrow m = \nu_o & \text{integer resonance}, \quad (7.86) \\
m + 2\nu_o &= \nu_o \rightarrow m = -\nu_o & \text{integer resonance}.
\end{aligned}
$$

Sextupole fields can drive *third order resonances* at tunes of

$$\nu_o = p + \tfrac{1}{3} \qquad \text{or} \qquad \nu_o = p - \tfrac{1}{3}, \tag{7.87}$$

where p is an integer. Finally we derive resonance conditions for octupole fields, $n = 4$, for

$$q = -3, -2, -1, 0, +1, +2, +3 \tag{7.87}$$

and again some values of q do not lead to a resonance since the amplitude coefficient W_q is zero. For octupole terms this is the case for $q = 0$ and $q = \pm 2$. The remaining resonance terms are then

$$m - 3\nu_o = \nu_o \rightarrow m = 4\nu_o \qquad \text{1/4 integer resonance},$$
$$m - \nu_o = \nu_o \rightarrow m = 2\nu_o \qquad \text{1/2 integer resonance},$$
$$m + \nu_o = \nu_o \rightarrow m = 0 \qquad \text{tune spread}, \tag{7.89}$$
$$m + 3\nu_o = \nu_o \rightarrow m = -2\nu_o \qquad \text{1/2 integer resonance}.$$

The resonance condition for $m = 0$ leads to a shift in the oscillation frequency. Different from gradient errors, however, we find the tune shift generated by octupole fields to be amplitude dependent $\nu^2 = \nu_o^2 - \bar{p}_{4o} W_o^2$. The amplitude dependence of the tune shift causes an asymmetric tune spread to higher or lower values depending on the sign of the perturbation term $\bar{p}_{4o}$ while the magnitude of the shift is determined by the oscillation amplitude of the particle.

The general *resonance condition* for betatron oscillations in one plane can be expressed by

$$|m| = (|q| \pm 1)\nu_o, \tag{7.90}$$

where the value $|q| + 1$ is the *order of resonance*. The index m is equal to the order of the perturbation Fourier harmonics and we call therefore these resonances structural or *lattice resonances* to distinguish them from resonances caused, for example, by externally driven oscillating fields.

The maximum *order of resonances* in this approximation depends on the order of nonlinear fields present. An nth order multipole field can drive all resonances up to nth order with driving amplitudes that depend on the actual multipole field strength and locations within the lattice. The term resonance is used very generally to include also effects which do not necessarily lead to a loss of the beam. Such "resonances" are characterized by $m = 0$ and are independent of the tune. In the case of gradient errors this condition was shown to lead to a stable shift in tune for the whole beam. Unless this tune shift moves the beam onto another resonance the beam stability is not affected. Similarly, octupole fields introduce a spread of tunes in the beam proportional to the square of the oscillation amplitude. Again no loss of particles occurs unless the tune spread reaches into the *stop band* of a resonance. By induction we conclude that all even perturbation terms, where n is an even integer, lead to some form of tune shift or spread. No such tune shifts occur for uneven perturbations in the approximation used here. Specifically we note that dipoles, sextupoles or decapoles etc. do not lead to a tune shift for weak perturbations. In Vol. II, however, we will discuss the Hamiltonian resonance theory and find, for example, that strong sextupole perturbations can indeed cause a tune spread.

In this derivation of resonance parameters we have expanded the perturbations into Fourier series and have assumed the full circular accelerator lattice as the expansion period. In general, however, a circular accelerator is composed of one or more equal *superperiods*. For a circular lattice composed of N superperiods the Fourier expansion has nonzero coefficients only every Nth harmonic and therefore the modified resonance conditions are

$$|j| N = (|q| \pm 1) \nu_\text{o}, \tag{7.91}$$

where j is an integer. A high *superperiodicity* actually eliminates many resonances and is therefore a desirable design feature for circular accelerator lattices. The integer and half integer resonances, however, will always be present independent of the superperiodicity because the equilibrium orbits and the betatron functions respectively are not defined. On the other hand, integer and half integer resonances driven by multipole perturbations may be eliminated in a high periodicity lattice with the overall effect of a reduced *stop band width*. It should be noted here, that the reduction of the number of resonances works only within the applied approximation. "Forbidden" resonances may be driven through field and alignment errors which compromise the high lattice periodicity or by strong nonlinearities and coupling creating resonant driving terms in higher order approximation. Nevertheless, the forbidden resonances are weaker in a lattice of high periodicity compared to a low periodicity lattice.

7.4.2 Coupling Resonances

Betatron motion in a circular accelerator occurs in both the horizontal and vertical plane. Perturbations can be present which depend on the betatron oscillation amplitude in both planes. Such terms are called *coupling terms*. The lowest order coupling term is caused by a rotated quadrupole or by the rotational misalignment of regular quadrupoles. In general we have the equation of motion

$$\ddot{w} + \nu_{ox}^2 w = \bar{p}_{nr}(\varphi) w^{n-1} v^{r-1}, \tag{7.92}$$

where n, r are integers and w describes betatron oscillations in one, the horizontal plane and v the betatron oscillation in the vertical plane.

Again we use the unperturbed solutions $w_\text{o}(\varphi)$ and $v_\text{o}(\varphi)$ of the equations of motion in the form (7.79) and express the higher order amplitude terms in the perturbation by the appropriate sums of trigonometric expressions:

$$\begin{aligned}
\bar{p}_{nr}(\varphi) &= \sum_m \bar{p}_{nrm} \, e^{im\varphi}, \\
w^{n-1}(\varphi) &= \sum_{|l| \le n-1} W_l \, e^{il\nu_{ox}\varphi}, \\
v^{r-1}(\varphi) &= \sum_{|q| \le r-1} V_q \, e^{iq\nu_{oy}\varphi}.
\end{aligned} \tag{7.93}$$

Insertion into (7.92) gives after some sorting

$$\ddot{w} + \nu_{ox}^2 w = \sum \bar{p}_{nrm} \, W_l \, V_q \, e^{i[(m+l\nu_{ox}+q\nu_{oy})\varphi]}, \tag{7.94}$$

where m, l and q are integers. The *resonance condition* is

$$m + l\,\nu_{ox} + q\,\nu_{oy} = \nu_{ox}\,, \tag{7.95}$$

and the quantity

$$|l| + |q| + 1 \tag{7.96}$$

designates the order of the *coupling resonances*. Again, for a superperiodicity N we replace m by $j\,N$, where j is an integer. As an example, we discuss a perturbation term caused by a rotated quadrupole for which the equation of motion is

$$\ddot{w} + \nu_o^2\,w = \bar{p}_{1,2}(\varphi)\,v\,. \tag{7.97}$$

In this case we have $n = 1$ and $r = 2$ and the *resonance condition* with $s = 0$ and $q = \pm 1$ is from (7.95)

$$m + q\,\nu_{oy} = \nu_{ox}\,. \tag{7.98}$$

Resonance occurs for

$$|m| = \nu_{ox} + \nu_{oy} \qquad \text{and} \qquad |m| = \nu_{ox} - \nu_{oy}\,. \tag{7.99}$$

There is no *coupling resonance* for $q = 0$ since $v_o = 0$. The resonances identified in (7.99) are called *linear coupling resonances* or a linear *sum resonance* and a linear *difference resonance* respectively. In circular accelerator design we therefore adjust the tunes such that a sum resonance is avoided.

Delaying proof for a later discussion in Vol. II we note at this point that the sum resonance can lead to a loss of beam while the *difference resonance* does not cause a loss of beam but rather leads to an exchange of horizontal and vertical betatron oscillations.

7.4.3 Resonance Diagram

The resonance condition (7.95) has been derived for horizontal motion only, but a similar equation can be derived for the vertical motion. Both resonance conditions can be written in a more symmetric way

$$k\,\nu_{ox} + l\,\nu_{oy} = i\,N\,, \tag{7.100}$$

where k, l, i are integers and $|k| + |l|$ is the order of the resonance. Plotting all straight lines from (7.100) for different values of k, l, i in a (ν_y, ν_x) diagram produces what is called a *resonance diagram*. In Fig. 7.5 an example of a resonance diagram for $N = 1$ is shown displaying all resonances up to 3rd order with $|k| + |l| \leq 3$.

The *operating points* for a circular accelerator are chosen to be clear of any of these resonances. It should be noted here that the resonance lines are not mathematically thin lines in the resonance diagram but rather exhibit some "thickness" which is called the *stop band width*. This stop band width depends on the strength of the resonance as was discussed in Sect. 7.3.2.

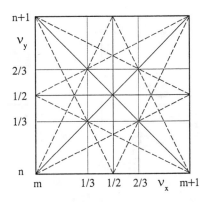

Fig. 7.5. Resonance diagram for a ring with superperiodicity one, $N = 1$

Not all resonances are of the same strength and generally get weaker with increasing order. While a particle beam would not survive on an integer or a half integer resonance all other resonances are basically survivable. Only in particular cases, where strong multipole field perturbations cause a *higher order resonance*, may we observe beam loss. This is very likely to be the case for third order resonances in rings, where strong sextupole magnets are employed to correct for *chromatic aberrations*.

The beneficial effect of a high superperiodicity or symmetry N in a circular accelerator becomes apparent in such a resonance diagram because the density of resonance lines is reduced by the factor N and the area of stability between resonances to operate the accelerator becomes proportionately larger. In Fig. 7.6 the resonance diagram for a ring with superperiodicity four $N = 4$ is shown and the reduced number of resonances is obvious. Wherever possible a high symmetry in the design of a circular accelerator should be attempted. Conversely breaking a high order of symmetry can lead to a reduction in stability if not otherwise compensated.

7.5 Chromatic Effects in a Circular Accelerator

Energy independent perturbations as discussed in previous sections can have a profound impact on the stability of particle beams in the form of perturbations of the betatron function or through resonances. Any beam transport line must be designed and optimized with these effects in mind since it is impossible to fabricate ideal magnets and align them perfectly. Although such field and alignment errors can have a destructive effect on a beam, it is the detailed understanding of these effects that allow us to minimize or even avoid such problems by a careful design which is within proven technology.

To complete the study of perturbations, we note that a realistic particle beam is never quite monoenergetic and includes a finite distribution of particle energies. Bending as well as focusing is altered if the particle

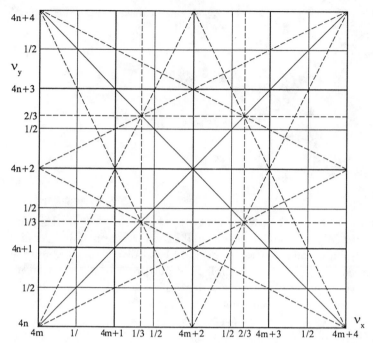

Fig. 7.6. Resonance diagram for a ring with superperiodicity four, $N = 4$

momentum is not the ideal momentum. We already derived the momentum dependent reference path in transport lines involving bending magnets. Beyond this basic momentum dependent effect we observe other chromatic aberrations which contribute in a significant way to the perturbations of lattice functions. The effect of chromatic aberrations due to a momentum error is the same as that of a corresponding magnet field error and to determine the stability characteristics of a beam, we must therefore include such *chromatic aberrations*.

7.5.1 Chromaticity

In beam transport systems perturbations of beam dynamics can occur even in the absence of magnet field and alignment errors. Deviations of particle energies from the ideal design energy cause perturbations in the solutions of the equations of motion. We have already derived the variation of the equilibrium orbit for different energies. Energy related or *chromatic effects* can be derived also for other lattice functions. Focusing errors due to an energy error cause such particles to be imaged at different focal points causing a blur of the beam spot. In a beam transport system, where the final beam spot size is of great importance as, for example, at the collision point of linear colliders, such a blur causes a severe degradation of the attainable

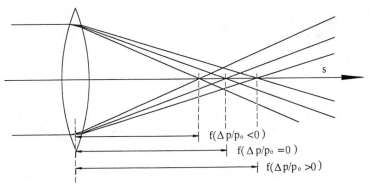

Fig. 7.7. Chromatic focusing errors

luminosity. In circular accelerators we have no such direct imaging task but note that the tune of the accelerator is determined by the overall focusing and tune errors occur when the focusing system is in error.

In this chapter we will specifically discuss effects of *energy errors* on tunes of a circular accelerator and means to compensate for such *chromatic aberrations*. The basic means of correction are applicable to either circular or open beam transport systems if, for the latter case, we only replace the tune by the phase advance of the transport line in units of 2π. The control of these chromatic effects in circular accelerators is important for two reasons, to avoid loss of particles due to tune shifts into resonances and to prevent beam loss due to an instability which we call the *head tail instability* to be discussed in more detail in Volume II.

The lowest order chromatic perturbation is caused by the variation of the focal length of the quadrupoles with energy (Fig. 7.7). This kind of error is well known from light optics, where a correction of this chromatic aberration can at least partially be corrected by using different kinds of glasses for the lenses in a composite focusing system.

In particle beam optics no equivalent approach is possible. To still correct for the chromatic perturbations we remember that particles with different energies can be separated by the introduction of a dispersion function. Once the particles are separated by energy we apply different focusing corrections depending on the energy of the particles. Higher energy particles are focused less than ideal energy particles and lower energy particles are over focused. For a correction of these focusing errors we need a magnet which is focusing for the higher energy particles and defocusing for the lower energy particles (Fig. 7.8). A sextupole has just that property.

The variation of tunes with energy is called the *chromaticity* and is defined by

$$\xi = \frac{\Delta\nu}{\Delta p/p_0}.$$

(7.101)

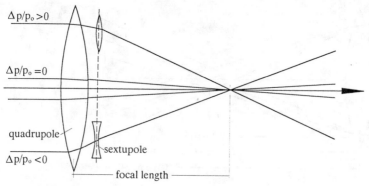

Fig. 7.8. Chromaticity correction with sextupoles

The chromaticity derives from second and higher order perturbations in (x, y, δ) and the relevant equations of motion are from (4.86) and (4.88)

$$x'' + k\,x = k\,x\,\delta - \tfrac{1}{2}m\,(x^2 - y^2),$$
$$y'' - k\,x = -k\,y\,\delta + m\,xy. \tag{7.102}$$

Setting $x = x_\beta + \eta_x\,\delta$ and $y = y_\beta$, assuming that $\eta_y \equiv 0$, we retain only terms depending on x_β or y_β to derive the chromatic tune shifts. In doing so we note three types of chromatic perturbation terms, those depending on the betatron motion only, those depending on the momentum error only, and terms depending on both. With these expansions (7.102) becomes

$$x''_\beta + k\,x_\beta = k\,x_\beta\,\delta - m\,\eta_x\,x_\beta\,\delta - \tfrac{1}{2}m\,(x^2_\beta - y^2_\beta),$$
$$y''_\beta - k\,y_\beta = -k\,y_\beta\,\delta + m\,\eta_x\,y_\beta\,\delta + m\,x_\beta\,y_\beta. \tag{7.103}$$

We ignore for the time being non chromatic terms of second order which will be discussed later as geometric aberrations and get

$$x''_\beta + k\,x_\beta = (k - m\,\eta_x)\,x_\beta\,\delta,$$
$$y''_\beta - k\,y_\beta = -(k - m\,\eta_x)\,y_\beta\,\delta. \tag{7.104}$$

The perturbation terms now are linear in the betatron amplitude and therefore have the character of a gradient error. From Sect. 7.3 we know that these types of errors lead to a tune shift which by comparison with (7.49) becomes in terms of a phase shift

$$\Delta\psi_x = -\frac{\delta}{2}\int_{s_o}^{s_o + L_{\mathrm{p}}} \beta_x\,(k - m\eta_x)\,\mathrm{d}s,$$
$$\Delta\psi_y = \frac{\delta}{2}\int_{s_o}^{s_o + L_{\mathrm{p}}} \beta_y\,(k - m\eta_x)\,\mathrm{d}s. \tag{7.105}$$

Equations (7.105) are applicable for both circular and open beam lines. Using the definition of the chromaticity for circular accelerators we have finally

$$\xi_x = -\frac{1}{4\pi} \oint \beta_x \left(k - m\,\eta_x\right) \mathrm{d}s \,,$$

$$\xi_y = \frac{1}{4\pi} \oint \beta_y \left(k - m\,\eta_x\right) \mathrm{d}s \,.$$

(7.106)

Similar to the definition of tunes the chromaticities are also an integral property of the circular accelerator lattice. Setting the sextupole strength m to zero one gets the *natural chromaticities* determined by focusing terms only

$$\xi_{xo} = -\frac{1}{4\pi} \oint \beta_x \, k \, \mathrm{d}s \,,$$

$$\xi_{yo} = \frac{1}{4\pi} \oint \beta_y \, k \, \mathrm{d}s \,.$$

(7.107)

The natural chromaticities are always negative which is to be expected since focusing is less effective for higher energy particles, $\delta > 0$, and therefore the number of betatron oscillations is reduced. For a thin lens symmetric FODO lattice the calculation of the chromaticity becomes very simple. With the betatron function β^+ at the center of a focusing quadrupole of strength $k^+ = k$ and β^- at the defocusing quadrupole of strength $k^- = k$ we have the chromaticity of one FODO half cell

$$\xi_{xo} = -\frac{1}{4\pi} \left(\beta^+ \int k^+ \,\mathrm{d}s + \beta^- \int k^- \,\mathrm{d}s\right) = -\frac{\beta^+ - \beta^-}{4\pi} \int k \, \mathrm{d}s \,.$$

With β^+ and β^- from (6.3,5) and $\int k \, \mathrm{d}s = f^{-1} = 1/(\kappa L)$, where κ is the FODO cell parameter (6.4) and L the length of a FODO half cell, we get the chromaticity per FODO cell in a more practical formulation

$$\xi_{xo} = -\frac{1}{2\pi} \frac{1}{\sqrt{\kappa^2 - 1}} = -\frac{1}{\pi} \tan\left(\psi_x/2\right) \,,$$

(7.108)

where ψ_x is the horizontal betatron phase for the full FODO cell. The same result can be obtained for the vertical plane.

The natural chromaticity for each 90° FODO cell is therefore equal to $1/\pi$. Although this value is rather small, the total chromaticity for the complete lattice of a storage ring or synchrotron, made up of many FODO cells, can become quite large. For the stability of a particle beam and the integrity of the imaging process by quadrupole focusing it is important that the natural chromaticity be corrected.

It is interesting at this point to discuss for a moment the chromatic effect if, for example, all bending magnets have a systematic field error with respect to other magnets. In an open beam transport line the beam would

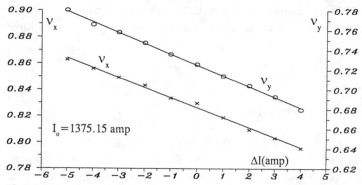

Fig. 7.9. Experimental determination of the natural chromaticity in a storage ring by measuring the tunes as a function of the excitation current $I = I_0 + \Delta I$ in the bending magnets.

follow an off momentum path as determined by the difference of the beam energy and the bending magnet "energy". Any chromatic aberration from quadrupoles as well as sextupoles would occur just as discussed.

In a circular accelerator the effect of systematic field errors might be different. We consider, for example, the case where we systematically change the strength of all bending magnets. In case of an electron storage ring, the particle beam would automatically stay at the ideal design orbit and the particle energy would be defined by the strength of the bending magnets. The strength of the quadrupoles and sextupole magnets, however, would now be systematically too high or too low with respect to the bending magnet field and particle energy. Quadrupoles introduce therefore a *chromatic tune shift* proportional to the natural chromaticity while the sextupoles are ineffective because the beam orbit is in the center. By changing the strength of the bending magnets in an electron circular accelerator and measuring the tunes one can determine experimentally the natural chromaticity of the ring. In Fig. 7.9 the measurement of the tunes as a function of the bending magnet current is shown for the storage ring SPEAR. From the slope of the graphs we derive the natural chromaticities of the SPEAR storage ring as $\xi_x = -11.4$ and $\xi_y = -11.7$.

In a proton accelerator the beam energy must be changed through acceleration or deceleration together with a change of the bending magnet strength to keep the beam on the reference orbit.

7.5.2 Chromaticity Correction

Equations (7.106) clearly suggest the usefulness of sextupole magnets for chromatic correction. Sextupoles must be placed along the orbit of a circular accelerator or along a beam transport line at locations, where the dispersion function does not vanish, $\eta_x \neq 0$. A single sextupole is sufficient, in principle, to correct the chromaticity for the whole ring or transport line but its strength may exceed technical limits or cause problems of geometric aberrations to the beam stability. This is due to the nonlinear nature of sextupole fields which causes dynamic instability for large amplitudes for which the sextupole field is no more a perturbation. The largest betatron oscillation amplitude which is still stable in the presence of nonlinear fields is called the *dynamic aperture*. To maximize the dynamic aperture it is prudent to distribute chromaticity correcting sextupoles most evenly along the beam line or circular accelerator.

To correct both the horizontal and the vertical chromaticity two different groups of sextupoles are required. For a moment we assume that there be only two sextupoles. To calculate the required strength of these sextupoles for *chromaticity correction* we use thin lens approximation and by replacing integrals in (7.106) by a sum the corrected chromaticities are

$$
\xi_x = \xi_{xo} + \frac{1}{4\pi} \left(m_1 \, \eta_{x1} \, \beta_{x1} + m_2 \, \eta_{x2} \, \beta_{x2} \right) \ell_s = 0 ,
$$
$$
\xi_y = \xi_{yo} + \frac{1}{4\pi} \left(m_1 \, \eta_{x1} \, \beta_{y1} + m_2 \, \eta_{x2} \, \beta_{y2} \right) \ell_s = 0 .
\tag{7.109}
$$

Here we assume that two different sextupoles, each of length ℓ_s, are available at locations s_1 and s_2. Solving for the sextupole strengths we get from (7.109)

$$
m_1 \ell_s = -\frac{4\pi}{\eta_{x1}} \frac{\xi_{xo} \beta_{y2} - \xi_{yo} \beta_{x2}}{\beta_{x1} \beta_{y2} - \beta_{x2} \beta_{y1}} ,
$$
$$
m_2 \ell_s = -\frac{4\pi}{\eta_{x2}} \frac{\xi_{xo} \beta_{y1} - \xi_{yo} \beta_{x1}}{\beta_{x1} \beta_{y2} - \beta_{x2} \beta_{y1}} .
\tag{7.110}
$$

It is obvious that the dispersion function at the sextupoles should be large to minimize the sextupoles strength. It is also clear that the betatron functions must be different preferably with $\beta_x \gg \beta_y$ at one and $\beta_x \ll \beta_y$ at the other sextupole to avoid "fighting" between sextupoles leading to excessive strength requirements.

In general this approach based on only two sextupoles in a ring to correct chromaticities leads to very strong sextupoles causing both magnetic design problems and strong higher order aberrations. A more gentle correction uses two groups or families of sextupoles with individual magnets distributed more evenly around the circular accelerator and the total required sextupole strength is spread over all sextupoles. In cases of severe aberrations, as discussed in Vol. II, we will need to subdivide all sextupoles into more than two families for a more sophisticated correction of chromaticities. Instead of (7.109) we write for the general case of *chromaticity correction*

$$
\xi_x = \xi_{xo} + \frac{1}{4\pi} \sum_i m_i \, \eta_{xi} \, \beta_{xi} \, \ell_{si} ,
$$
$$
\xi_y = \xi_{yo} + \frac{1}{4\pi} \sum_i m_i \, \eta_{xi} \, \beta_{yi} \, \ell_{si} ,
\tag{7.111}
$$

where the sum is taken over all sextupoles. In the case of a two family correction scheme we still can solve for m_1 and m_2 by grouping the terms into two sums.

The chromaticity of a circular accelerator as defined in this section obviously does not take care of all chromatic perturbations. Since the function $(k - m\eta_x)$ in (7.104) is periodic, we can Fourier analyze it and note that the chromaticity only describes the effect of the nonoscillating lowest order Fourier component (7.107). All higher order components are treated as *chromatic aberrations*. In Vol. II we will discuss in more detail such higher order chromatic and geometric aberrations.

Problems

Problem 7.1. Consider a compact storage ring optimized for x-ray lithography. The circumference is 9 m, $\bar\beta_x = \bar\beta_y = 1.2$ m, and the beam emittances are $\epsilon_x = 10$ mm mrad and $\epsilon_y = 5$ mm mrad. The arcs consist of two $180°$ bending magnets with a bending radius of $\rho = 1.1$ m and a vertical aperture $g = 10$ cm. The bending magnet field drops off linearly in the fringe region which we assume to be one vertical aperture long. Determine the five strongest nonchromatic perturbation terms in the bending magnets and discuss their effect. Which terms contribute to orbit distortions and which to the focusing or tune. What is the average orbit distortion and tune shift? Are these effects the same for all particles in the beam?

Problem 7.2. Use parameters of example #4 in Table 6.1 for a FODO lattice and construct a full ring. Adjust the quadrupole strength such that both tunes are an integer plus a quarter. Calculate the rms alignment tolerance on the quadrupoles required to keep the beam within $\sigma_x = 0.1$ mm and $\sigma_y = 0.1$ mm of the ideal orbit. What is the amplification factor? Determine the rms deflection tolerance of the bending magnets to keep the beam within 0.1 mm of the ideal orbit. A rotation of the bending magnets about its axis creates vertical orbit distortions. If the magnets are aligned to a rotational tolerance of 0.17 mrad (this is about the limit of conventional alignment techniques) what is the expectation value for the vertical orbit distortion?

Problem 7.3. Repeat the calculation of problem 7.2 with the lattice example #1 in Table 6.1. The alignment tolerances are much relaxed with respect to the ring in problem 7.2. What are the main three contributions influencing the tolerance requirements? Make general recommendations to relax tolerances.

Problem 7.4. Consider statistical transverse alignment errors of the quadrupoles in the large hadron collider lattice example #4 in Table 6.1 of

$\langle \delta x \rangle_{\text{rms}} = 0.1$ mm. What is the rms path distortion at the end of one turn? Determine the allowable rotational alignment error of the bending magnets to produce a vertical path distortion of no more than that due to quadrupole misalignments. How precise must the bending magnet fields be to not contribute more path distortion than the quadrupole misalignments.

Problem 7.5. Consider a horizontal orbit distortion Δx in the center of a QF quadrupole for example #4 in Tab. 6.1. Use adjacent quadrupoles which we assume to contain special trim coils to produce dipole field and construct a symmetric beam bump with an amplitude of $\Delta x = 1$ cm. What are the required dipole deflection angles? If a horizontal beam bump of 1 cm is desired in the center of a QD quadrupole what would then the deflection angles be? Why is it harder to displace the beam in a QD quadrupole?

Problem 7.6. Imagine to inject a particle beam in the center of a QF quadrupole on axis but at an angle with respect to the ideal orbit of the ring of problem 7.2. What is the maximum injection angle error allowable for a pencil beam so that the path of the injected beam in the ring will not deviate from the ideal orbit by more than 2 cm.

Problem 7.7. Use the lattice of example #3 in Table 6.1 and introduce vertical rms misalignments of all quadrupoles by $\langle \delta y \rangle_{\text{rms}} = 0.1$ mm. Calculate the vertical rms dispersion function. Now also add rotational alignment errors of the bending magnets by $\langle \delta \alpha \rangle_{\text{rms}} = 0.17$ mrad and calculate again the vertical rms dispersion.

Problem 7.8. We insert into the path of a particle beam two bending magnets of equal but opposite strength. Such a deflection arrangement causes a parallel displacement d of the beam path. Show that in this case the contribution to the dispersion at the end of the second bending magnet is $\Delta D = -d$ and $\Delta D' = 0$.

Problem 7.9. For the ring in problem 7.2 or 7.3 calculate the rms tolerance on the quadrupole strength to avoid the integer or half integer resonance. What is the corresponding tolerance on the quadrupole length? To avoid gradient fields in bending magnets the pole profiles must be aligned parallel with respect to the horizontal midplane. What is the angular tolerance for parallelism of the poles?

Problem 7.10. Calculate the expectation value for the integer and half integer stop band width of the ring in problem 7.7. Gradient errors introduce a perturbation of the betatron functions. What is the probable perturbation of the betatron function for the case in problem 7.7?

Problem 7.11. Consider a FODO cell equal to examples #1, #2, and #4 in Table 6.1, adjust the phase advance per cell to equal values and calculate the natural chromaticities. Insert thin sextupoles into the center of the quadrupoles and adjust to zero chromaticities. How strong are the sextupoles? Why must the sextupoles for lattice #2 be so much stronger compared with lattice #4 even though the chromaticity per cell is about the same?

Problem 7.12. Consider the transformation of phase ellipses through one full FODO cell of the examples in problem 7.11. Let the emittance for the phase ellipses be $\epsilon = 10$ mm mrad. First transform the equation for the phase ellipse into a circle by setting $u = x$ and $v = \alpha x + \beta x'$. Transform the phase circle from the center of the QF through one full FODO cell to the center of the next QF ignoring any sextupole terms. Repeat this transformation but include now the sextupole in the first QF only as calculated in problem 7.11. Discuss the distortions of the phase circle for the three different FODO lattices.

Problem 7.13. Derive an expression for the probable variation of the betatron function in a regular FODO lattice due to a statistical orbit distortion σ_u in sextupoles. Assume the sextupoles to occupy the same location as the quadrupoles and to have been adjusted to compensate the natural chromaticity of the FODO lattice. What is the contribution of this perturbation to the natural chromaticity?

Problem 7.14. Derive an expression for the half integer stop band width due to orbit errors in chromaticity correcting sextupoles. Calculate the half integer stop band width for the ring of problem 7.7 as a function of the rms orbit distortion.

Problem 7.15. Derive an analytical expression for the nth order integer stop band width.

8 Charged Particle Acceleration

Accelerator physics is primarily the study of the interaction of charged particles with electromagnetic fields. In previous chapters we have concentrated the discussion on the interaction of transverse electrical and magnetic fields with charged particles and have derived appropriate formalisms to apply this interaction to the design of beam transport systems. The characteristics of these transverse fields is that they allow to guide charged particles along a prescribed path but do not contribute directly to the energy of the particles through acceleration. For particle acceleration we must generate fields with nonvanishing force components in the direction of the desired acceleration. Such fields are called *longitudinal fields* or *accelerating fields*. In a very general way we describe in this section the interaction of longitudinal electric fields with charged particles to derive the process of particle acceleration, its scaling laws, and its stability limits.

The usefulness and application of electric fields to accelerate charged particles depends greatly on the temporal variations of these fields. Accelerating fields can be static or pulsed or they may be electromagnetic fields oscillating at high frequencies. Conceptually, the most simple way to accelerate charged particles is through a static field applied to two electrodes as discussed in Chap. 2. In this case the total kinetic energy a particle can gain while traveling from one electrode to the other is equal to the product of the particle charge and the voltage between the electrodes. Electric breakdown phenomena, however, limit the maximum applicable voltage and thereby the maximum energy gain. Nonetheless, this method is intriguingly simple and efficient compared to other accelerating methods and therefore still plays a significant role among modern particle accelerators. Electrostatic acceleration schemes are specifically useful for low energy particles for which other methods of acceleration would be inefficient. Somewhat higher voltages and particle energies can be reached if the electric fields are applied in the form of very short pulses. Application of static high voltages to accelerate particles is limited to about 10 million volts due to high *voltage breakdown*.

For higher particle energies different acceleration methods must be used. The most common and efficient way to accelerate charged particles to high energies is to use *high frequency electromagnetic fields* in specially designed accelerating structures. Acceleration to high energies occurs while charged particles either pass once through many or many times through one or few accelerating structures each excited to electric field levels below the break down threshold. In this section we concentrate the discussion on charged particle acceleration by electromagnetic *radio frequency fields*.

8.1 Longitudinal Particle Motion

Application of radio frequency fields in short, *rf fields*, has become exceptionally effective for the acceleration of charged particles. Both, fields and particle motion can be synchronized in an effective way to allow the acceleration of charged particles in principle to arbitrary large energies were it not for other limitations.

The first idea and experiment for particle acceleration with radio frequency fields has been published by Ising [8.1] although he did not actually succeed to accelerate particles due to an inefficient approach to rf technology. Later Wideroe [8.2] introduced the concept of generating the accelerating fields in resonating *rf cavities* and was able to accelerate heavy ions. Original papers describing these and other early developments of particle acceleration by rf fields are collected in a monogram edited by Livingston [8.3].

To study the interaction of electromagnetic rf fields with charged particles we assume a plane electromagnetic wave of frequency ω propagating in the z-direction. A free electromagnetic wave does not have a longitudinal electric field component and therefore a special physical environment, called the *accelerating structure*, must be provided to generate accelerating field components in the direction of propagation. This aspect has been discussed in more detail in Chap. 2, where conditions necessary to generate longitudinal field components have been derived. To study particle dynamics in longitudinal fields we assume that we were able to generate longitudinal rf fields expressed by

$$\mathbf{E}(z,t) = \mathbf{E}_{o} \cdot e^{i(\omega t - kz)} = \mathbf{E}_{o} \cdot e^{i\psi}, \tag{8.1}$$

where the phase $\psi = (\omega t - kz)$ and k is a constant. The particle momentum changes at a rate equal to the electric force exerted on the particle by the rf field

$$\frac{d\mathbf{p}}{dt} = e \cdot \mathbf{E}(\psi) = \frac{d}{dt}(mc\gamma\beta). \tag{8.2}$$

Multiplying this with the particle velocity we get the rate of change of the kinetic energy, $dE_{kin} = c\beta \cdot dp$. Integration of (8.2) with respect to the time becomes unnecessarily complicated for general fields because of the simultaneous variation of the electric field and particle velocity with time. We therefore integrate (8.2) with respect to the longitudinal coordinate and obtain instead of the momentum gain the increase in the kinetic or total energy for the complete accelerating structure

$$\Delta E = (\gamma - \gamma_0)\, mc^2 = e \int \mathbf{E}(\psi)\, d\mathbf{z}\,, \tag{8.3}$$

where $\gamma_0\, mc^2$ is the energy of the particle before acceleration. Of course the trick to integrate the electric field through the *accelerating section* rather than over time following the particle is only a conceptual simplification and the time integration will have to be executed at some point. Generally this is done when the particular accelerating section, the fields, and the synchronization are known.

Travelling electromagnetic waves are used in linear accelerators and the accelerating structure is designed such that the phase velocity of the wave is equal to the velocity of the particles to be accelerated. In this case the particle travels along the structure in synchronism with the wave and is therefore accelerated or decelerated at a constant rate. Maximum acceleration is obtained if the particles ride on the crest of the wave.

In a *standing wave* accelerating section the electric field has the form

$$\mathbf{E}(z,t) = \mathbf{E}_0(z)\, e^{i\omega t + \delta}\,, \tag{8.4}$$

where δ is the phase at the moment the particle enters the accelerating section at $t = 0$. When we refer to an accelerating voltage, V, in a standing wave cavity we mean to say a particle traveling close to the speed of light through the cavity will gain a maximum kinetic energy of eV while passing the cavity center at the moment the field reaches its crest. Such a particle would enter the cavity some time before the field reaches a maximum and will exit when the field is decaying again. For slower particles the energy gain would be lower because of the longer *transit time*.

8.1.1 Longitudinal Phase Space Dynamics

Successful particle acceleration depends on stable and predictable interaction of charged particles and electromagnetic fields. Because oscillating rf fields are used, special criteria must be met to assure systematic particle acceleration rather than random interaction with rf fields producing little or no *acceleration*. The constructive interaction of particles and waves have been investigated in 1945 independently by Veksler [8.4] and McMillan [8.5] leading to the discovery of the fundamental principle of *phase focusing*. In this subsection we will derive the physics of phase focusing and apply it to the design of particle accelerators.

The degree of acceleration depends on the momentary phase ψ of the field as seen by the particle while travelling through or with an electromagnetic field. Straight superposition of an electromagnetic wave and charged particle motion will not necessarily lead to a net acceleration. In general, the particles are either too slow or too fast with respect to the *phase velocity* of the wave and the particle will, during the course of interaction with the electromagnetic wave, integrate over a range of phases and may gain little or no net energy from the electric fields. Therefore special boundary conditions for the *accelerating rf wave* must be met such that maximum or at least net acceleration can be achieved. This can be done by exciting and guiding the electromagnetic waves in specially designed accelerating structures designed such that the phase velocity of the electromagnetic wave is equal to the particle velocity. Only then can we choose a specific phase and integration of (8.3) becomes straightforward for particles travelling in the direction of propagation of the electromagnetic waves.

For practical reasons, specifically in circular accelerators, particle acceleration occurs in short, straight accelerating sections placed along the particle path. In this case no direct traveling wave exists between adjacent accelerating sections and specific *synchronicity conditions* must be met for the fields in different accelerating sections to contribute to particle acceleration as desired. For the purpose of developing a theory of stable particle acceleration we may imagine an rf wave traveling along the path of the particle with a phase velocity equal to the particle velocity and an amplitude which is zero everywhere except in discrete accelerating cavities.

To derive the synchronicity conditions we consider two accelerating sections separated by the distance L as shown in Fig. 8.1. Once the proper operating conditions are known for two sections a third section may be added by applying the same synchronicity condition between each pair of cavities. The successive accelerating sections need not necessarily be physically different sections but could be the same section or the same sections passed through by the particles at periodic time intervals. For example, the distance L between successive accelerating sections may be equal to the circumference of a circular accelerator.

For systematic acceleration the phase of the rf fields in each of the accelerating sections must reach specific values at the moment the particles arrive. If the phase of the fields in each of N accelerating sections is adjusted to be the same at the time of arrival of the particles, the total acceleration is N times the acceleration in each individual section. This phase is called the *synchronous phase* ψ_s defined by

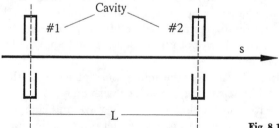

Fig. 8.1. Discrete accelerating sections

$$\psi_s = \omega t - kz = \text{const}, \tag{8.5}$$

where ω is the oscillating frequency of the electromagnetic field. The time derivative of (8.5) vanishes and the *synchronicity condition* is

$$\dot{\psi}_s = \omega - k\,\beta c = 0, \tag{8.6}$$

since $dz/dt = \beta c$. This condition can be met if we set

$$k = \frac{2\pi}{L} \tag{8.7}$$

and the frequency of the electromagnetic field is then from (8.6)

$$\omega_1 = k_1\,\beta c = \frac{2\pi}{L}\,\beta c = \frac{2\pi}{\Delta T}, \tag{8.8}$$

can be derived from (8.6), where ω_1 is the lowest frequency satisfying the synchronicity condition and ΔT is the time needed for particles with velocity βc to travel the distance L. This equation relates the time of travel between successive accelerating sections and the frequency of the *accelerating rf fields* in a conditional way to assure systematic particle acceleration and the relation (8.8) is therefore called the *synchronicity condition*.

However, any integer multiple of the frequency ω_1 satisfies the synchronicity condition as well and we may instead of (8.8) define permissible frequencies of the accelerating rf fields by

$$\omega_h = h\omega_1 = k_h\,\beta c = \frac{2\pi}{L}\,h\,\beta c = \frac{2\pi}{\Delta T}\,h, \tag{8.9}$$

where h is an integer called the *harmonic number* with $k_h = h\,k_1$.

The *synchronicity condition* must be fulfilled for any spatial arrangement of the accelerating structures. To illuminate the principle, we assume here a series of short, equidistant accelerating gaps or *accelerating sections* along the path of a particle. Let each of these gaps be excited by its own power source to produce an accelerating rf field. The synchronicity condition (8.8) is fulfilled if the rf frequency is the same in each of these gaps, although, it does not require each accelerating gap to reach the same rf phase at the arrival time of the particles. Each cavity in a set of accelerating cavities oscillating at the same frequency may be tuned to an arbitrary *rf phase* and the synchronicity condition still would be met. From a practical point of view, however, it is inefficient to choose arbitrary phases and it is more reasonable to adjust the phase in each cavity to the optimum phase desired.

The assumption that the rf frequency of all cavities be the same is unnecessarily restrictive considering that any harmonic of the fundamental frequency is acceptable. Therefore a set of accelerating cavities in a circular accelerator for example may include cavities resonating at frequencies differing by an integer multiple of the fundamental frequency ω_1.

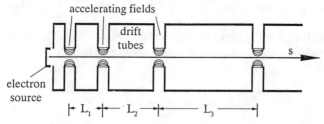

Fig. 8.2. Wideroe linac structure [8.2]

A straightforward application of the synchronicity condition can be found in the design of the *Wideroe linear accelerator structure* [8.2] as shown in Fig. 8.2. Here the fields are generated by an external rf source and applied to a series of metallic drift tubes. Accelerating fields build up at gaps between the tubes while the tubes themselves serve as a field screen for the particles during the time the electric fields is changing sign and would be decelerating. The length of the field free drift tubes is determined by the velocity of the particles and is $L = c\beta T_{rf}$ where T_{rf} is the period of the rf field. As the particle energy increases so does the velocity $c\beta$ and the length L of the tube must increase too. Only when the particles become highly relativistic will the distance between field free drift sections become a constant together with the velocity of the particles. Structures with varying drift lengths are generally found in low energy proton or ion accelerators based on the Alvarez structure [8.7], which is a technically more efficient version of the *Wideroe principle*.

For electrons it is much easier to reach relativistic energies where the velocity is sufficiently constant such that in general no longitudinal variation of the accelerating structure is needed. In circular accelerators we cannot adjust the distance between cavities or the circumference as the particle velocity β increases. The synchronicity condition therefore must be applied differently. From (8.9) we find the rf frequency to be related to the particle velocity and distances between cavities. Consequently we have the relation

$$\beta \lambda_{rf} h = L, \tag{8.10}$$

which requires that the distance between any pair of accelerating cavities be an integer multiple of $\beta\lambda_{rf}$. Since L and h are constants, this condition requires that the rf frequency be changed during acceleration proportional to the particle velocity β. Only for particles reaching relativistic energies, when $\beta \approx 1$, will the distance between cavities approach an integer multiple of the rf wave length and the circumference C must then meet the condition

$$C = \beta h \lambda_{rf}. \tag{8.11}$$

8.1.2 Equation of Motion in Phase Space

So far we have assumed that both the particle velocity β and the wave number k are constant. This is not a valid general assumption. For example, we cannot assume that the time of flight from one gap to the next is the same for all particles. For low energy particles we have a variation of the time of flight due to the variation of the particle velocities for different particle momenta. The *wave number* k or the distance between accelerating sections need not be the same for all particles either. A momentum dependent path length between accelerating sections exists if the lattice between such sections includes bending magnets. As a consequence the synchronicity condition must be modified to account for such chromatic effects.

Removing the restriction of a constant wave number k we obtain by a variation of (8.6)

$$\Delta \dot{\psi} = -\Delta(k\,\beta c) = -ck\,\Delta\beta - \beta c\,\frac{\partial k}{\partial p}\frac{\partial p}{\partial t}\,\Delta t\,, \qquad (8.12)$$

where

$$k = k_{\rm h} = h\,\frac{2\pi}{L_{\rm o}} = \frac{2\pi}{\lambda_{\rm rf}} = h\,\frac{\omega}{\beta c}\,, \qquad (8.13)$$

and $L_{\rm o}$ is the distance between accelerating gaps along the ideal path. The *synchronous phase* is kept constant $\psi_{\rm s} = {\rm const}$ or $\dot{\psi}_{\rm s} = 0$ and serves as the *reference phase* against which all deviations are measured.

The momentum dependence of the wave number comes from the fact that the path length L between accelerating gaps may be different from $L_{\rm o}$ for off momentum particles. The variation of the wave number with particle momentum is therefore

$$\frac{\partial k}{\partial p}\bigg|_{\rm o} = \frac{\partial k}{\partial L}\frac{\partial L}{\partial p}\bigg|_{\rm o} = -\frac{k_{\rm h}}{L_{\rm o}}\frac{\partial L}{\partial p}\bigg|_{\rm o} = -\frac{k_{\rm h}}{p_{\rm o}}\,\alpha_{\rm c}\,, \qquad (8.14)$$

and the factor $\alpha_{\rm c}$ is called the *momentum compaction factor* defined by

$$\alpha_{\rm c} = \frac{\Delta L/L_{\rm o}}{\Delta p/p_{\rm o}}\,. \qquad (8.15)$$

We evaluate the *momentum compaction factor* starting from (4.124) $L = \int_{\rm o}^{L_{\rm o}} \sigma'\,{\rm d}z$ and get while keeping only linear terms in the expression for σ' the path length $L = \int_{\rm o}^{L_{\rm o}}(1 + \kappa_x\,x)\,{\rm d}z$. For the transverse particle motion $x = x_\beta + \eta\,(\Delta p/p_{\rm o})$ and employing average values of the integrants the integral becomes

$$L = L_{\rm o} + \langle \kappa_x\,x_\beta \rangle\,L_{\rm o} + \langle \kappa_x\,\eta \rangle\,\frac{\Delta p}{p_{\rm o}}\,L_{\rm o}\,. \qquad (8.16)$$

Because of the oscillatory character of the betatron motion $\langle \kappa_x x_\beta \rangle = 0$ and the relative path length variation is $\Delta L/L_o = \langle \eta/\rho \rangle (\Delta p/p_o)$ or the momentum compaction factor becomes

$$\alpha_c = \left\langle \frac{\eta}{\rho} \right\rangle . \tag{8.17}$$

The momentum compaction factor increases only in curved sections where $\rho \neq 0$ and the path length is longer or shorter for higher energy particles depending on the dispersion function being positive or negative, respectively. For a linear accelerator the momentum compaction factor vanishes since the length of a straight line does not depend on the momentum. With $(\partial p/\partial t)\Delta t = \Delta p$ and $mc\gamma^3 \Delta \beta = \Delta p$ we get finally for (8.12) with (8.14) and after some manipulation

$$\dot{\psi} = -\beta c\, k_h\, (\gamma^{-2} - \alpha_c)\, \frac{\Delta cp}{cp_o} . \tag{8.18}$$

The term γ^{-2} in (8.18) appears together with the momentum compaction factor α_c and therefore has the same physical relevance. This term represents the variation of the particle velocity with energy. Therefore, even in a linear accelerator where $\alpha_c = 0$ the time of flight between accelerating gaps is energy dependent as long as particles are still nonrelativistic. The combination of both terms form the *momentum compaction* defined by

$$\eta_c = \gamma^{-2} - \alpha_c . \tag{8.19}$$

After differentiation of (8.18) with respect to the time, we get the equation of motion in the longitudinal direction describing the variation of the phase with respect to the synchronous phase, ψ_s for particles with a total momentum deviation Δp

$$\ddot{\psi} + \frac{\partial}{\partial t}\left(\beta c\, k_h\, \eta_c\, \frac{\Delta cp}{cp_o} \right) = 0 . \tag{8.20}$$

In most practical applications, parameters like the particle velocity β or the energy vary only slowly during acceleration compared to the rate of change of the phase and we consider them for the time being as constants. Later we will find that the slow variation of these parameters constitutes an adiabatic variation of external parameters for which *Ehrenfest's theorem* gives us a tool to determine the effect on beam parameters. The equation of motion in the potential of the rf field becomes in this approximation

$$\ddot{\psi} + \frac{\beta c\, k_h\, \eta_c}{cp_o} \frac{\partial}{\partial t} \Delta cp = 0 . \tag{8.21}$$

Integration of the electrical fields along the accelerating sections returns the kinetic energy gain per turn

$$\Delta E = e \int_L \mathbf{E}(\psi) \, d\mathbf{z} = e V(\psi), \tag{8.22}$$

where $V(\psi)$ is the total particle accelerating voltage seen by particles along the distance L. For particles with the ideal energy and following the ideal orbit the acceleration is $e V(\psi_s)$ where ψ_s is the synchronous phase.

Acceleration, however, is not the only source for the energy change of particles. There are also gains or losses from, for example, interaction with the *vacuum chamber environment*, external fields like *free electron lasers*, *synchrotron radiation* or anything else exerting longitudinal forces on the particle other than accelerating fields. We may separate all longitudinal forces into two classes, one for which the energy change depends only on the phase of the accelerating fields $V(\psi)$ and the other where the energy change depends only on the energy of the particle $U(E)$ itself. The total energy gain ΔE per unit time or per turn is the composition of both types of external effects

$$\Delta E = e V(\psi) - U(E), \tag{8.23}$$

where $U(E)$ is the energy dependent loss per turn due for example to synchrotron radiation.

Small Oscillation Amplitudes: For arbitrary variations of the *accelerating voltage* with phase we cannot further evaluate the equation of motion unless the discussion is restricted to small variations in the vicinity of the synchronous phase. While the ideal particle arrives at the accelerating cavities exactly at the synchronous phase ψ_s most other particles in a real beam arrive at slightly different phases. For small deviations φ from the synchronous phase

$$\varphi = \psi - \psi_s \tag{8.24}$$

we can expand the accelerating voltage into a Taylor series at $\psi = \psi_s$ and get for the average rate of change of the particle energy with respect to the energy of the synchronous particle from (8.22)

$$\frac{d}{dt} \Delta E = \frac{1}{T_o} \left[e V(\psi_s) + e \left. \frac{dV}{d\psi} \right|_{\psi_s} \varphi - U(E_o) - \left. \frac{dU}{dE} \right|_{E_o} \Delta E \right], \tag{8.25}$$

where the particle energy $E = E_o + \Delta E$ and T_o is the time of flight for the reference particle

$$T_o = \frac{L_o}{\beta c}. \tag{8.26}$$

At equilibrium $e V(\psi_s) = U(E_o) + \Delta E(\psi)$ where $\Delta E(\psi)$ is the energy loss that does not depend on the energy like higher order mode losses. We note that such losses lead only to a shift in the synchronous phase and we therefore ignore such losses here. In Volume II we will take up this discussion again in connection with the evaluation of the effects of *higher order mode losses*. Since $\beta \, \Delta cp = \Delta E$ we get with (8.25) and $\ddot{\varphi} = \ddot{\psi}$ from (8.21) the equation of motion or *phase equation*

$$\ddot{\varphi} + \frac{c \, k_h \, \eta_c}{cp_o \, T_o} \, e \, \frac{dV}{d\psi}\bigg|_{\psi_s} \varphi - \frac{c \, k_h \, \eta_c}{T_o} \, \frac{dU}{dE}\bigg|_{E_o} \frac{\Delta cp}{cp_o} = 0. \tag{8.27}$$

With (8.18) and $\psi = \psi_s + \varphi$ we get finally from (8.27) the differential equation for small phase oscillations

$$\ddot{\varphi} + 2 \, \alpha_s \, \dot{\varphi} + \Omega^2 \, \varphi = 0, \tag{8.28}$$

where the damping decrement is defined by

$$\alpha_s = -\frac{1}{2 T_o} \frac{dU}{dE}\bigg|_{E_o} \tag{8.29}$$

and the *synchrotron frequency* by

$$\Omega^2 = \frac{c \, k_h \, \eta_c}{cp_o \, T_o} \, e \, \frac{dV}{d\psi}\bigg|_{\psi_s}. \tag{8.30}$$

Particles orbiting in a circular accelerator perform *longitudinal oscillations* with the frequency Ω. These *phase oscillations* are damped or antidamped depending on the sign of the damping decrement. Damping occurs only if there is an energy loss which depends on the particle energy itself as in the case of synchrotron radiation. In most cases of accelerator physics we find the damping time to be much longer than the phase oscillation period and we may therefore discuss the phase equation while ignoring the damping terms. Whenever *damping* becomes of interest we will include this term again.

The phase equation as derived is valid only for small oscillation amplitudes because only the linear term has been used in the expansion for the rf voltage. For larger amplitudes this approximation cannot be made anymore and direct integration of the differential equation is necessary. The small amplitude approximation, however, is accurate to describe most of the fundamental features of phase oscillations. At large amplitudes the nonlinear terms will introduce a change in the phase oscillation frequency and finally a limit to stable oscillations to be discussed later in this chapter.

The phase equation has the form of the equation of motion for a damped harmonic oscillator and we will look for conditions leading to a positive frequency and stable phase oscillations. Because the phase equation was derived first for synchrotron accelerators the oscillations are also called *synchrotron oscillations* and are of fundamental importance for beam stability in all circular accelerators based on rf acceleration. For real values of the oscillation frequency we find that particles which deviate from the synchronous phase are subjected to a restoring force leading to harmonic oscillations about the equilibrium or synchronous phase. From the equation of motion (8.27) it becomes clear that the phase focusing is proportional to the derivative of the accelerating voltage rather than to the accelerating voltage itself and is also proportional to the *momentum compaction* η_c.

To gain further insight into the phase equation and determine stability criteria we must make an assumption for the waveform of the *accelerating voltage*. In most cases the rf accelerating fields are created in resonant cavities and therefore the accelerating voltage can be expressed by a sinusoidal waveform

$$V(\psi) = \widehat{V}_o \sin \psi \tag{8.31}$$

and expanded about the synchronous phase to get with $\psi = \psi_s + \varphi$

$$V(\psi_s + \varphi) = \widehat{V}_o \left(\sin \psi_s \cos \varphi + \sin \varphi \cos \psi_s \right).$$

Keeping only linear terms in φ the phase equation is

$$\ddot{\varphi} + \Omega^2 \varphi = 0, \tag{8.32}$$

where the *synchrotron oscillation frequency* becomes now

$$\Omega^2 = \frac{c \, k_h \, \eta_c}{c p_o \, T_o} \, e \widehat{V}_o \, \cos \psi_s . \tag{8.33}$$

A particle passing periodically every T_o seconds through localized and synchronized accelerating fields along its path performs synchrotron oscillations with the frequency Ω about the synchronous phase.

In circular accelerators we have frequently the situation that several rf cavities are required to provide the desired acceleration. The reference time T_o is most conveniently taken as the revolution time and the rf voltage $\widehat{V}_o$ is the total accelerating voltage seen by the particle while orbiting around the ring once. The *rf frequency* is an integer multiple of the *revolution frequency*,

$$f_{rf} = h \, f_{rev} , \tag{8.34}$$

where the integer h is the *harmonic number* and the revolution frequency is with the circumference C

$$f_{rev} = \frac{1}{T_o} = \frac{C}{c \beta} . \tag{8.35}$$

With $\omega_{rev} = 2\pi f_{rev}$ we get from (8.33) the synchrotron frequency in more practical units

$$\Omega^2 = \omega_{rev}^2 \, \frac{h \, \eta_c \, e \widehat{V}_o \, \cos \psi_s}{2\pi \, \beta \, c p_o} . \tag{8.36}$$

Similar to the betatron oscillation tunes we define the *synchrotron oscilla-tion tune* or short the *synchrotron tune* as the ratio

$$\nu_{\mathrm{s}} = \frac{\Omega}{\omega_{\mathrm{rev}}}.$$ (8.37)

For real values of the synchrotron oscillation frequency the *phase equation* assumes the simple form

$$\varphi = \widehat{\varphi} \cos(\Omega t + \chi_{\mathrm{i}}),$$ (8.38)

where χ_{i} is an arbitrary phase function for the particle i at time $t = 0$.

With $\dot{\psi} = \dot{\varphi}$ we find from (8.18,33) the relation between the momentum and phase deviation for real values of the synchrotron oscillation frequency

$$\delta = \frac{\Delta cp}{cp_{\mathrm{o}}} = -\frac{\dot{\varphi}}{h\omega_{\mathrm{rev}}\,\eta_{\mathrm{c}}} = \frac{\Omega\,\widehat{\varphi}}{h\,\omega_{\mathrm{rev}}\,\eta_{\mathrm{c}}} \sin(\Omega t + \chi_{\mathrm{i}}).$$ (8.39)

The particle momentum deviation being the conjugate variable to the phase also oscillates with the synchrotron frequency about the ideal momentum. Both the phase and momentum oscillations describe the particle motion in *longitudinal phase space* as shown in Figs. 8.3,4. At the time t_{o} when in (8.39) the phase $\Omega t_{\mathrm{o}} + \chi_{\mathrm{i}} = 0$ we expect the momentum deviation to be zero while the phase reaches the maximum value $\widehat{\varphi}$. Particles with a negative momentum compaction $\eta_{\mathrm{c}} < 0$ move clockwise in phase space about the reference point while a positive momentum compaction causes the particles to rotate counter clockwise.

The same process that has led to phase focusing will also provide the focusing of the particle momentum. Any particle with a momentum differ-ent from the ideal momentum will undergo oscillations at the synchrotron frequency which are from (8.39) described by

$$\delta = -\widehat{\delta} \sin(\Omega t + \chi_{\mathrm{i}}),$$ (8.40)

where the maximum momentum deviation is related to the maximum phase excursion $\widehat{\varphi}$ by

$$\widehat{\delta} = \left| \frac{\Omega}{h\,\omega_{\mathrm{rev}}\,\eta_{\mathrm{c}}} \right| \widehat{\varphi}.$$ (8.41)

By inverse deduction we may express the momentum equation similar to the phase equation (8.32) and get with $\Delta p/p_{\mathrm{o}} = \delta$ the differential equation for the momentum deviation

$$\frac{\mathrm{d}^2\delta}{\mathrm{d}t^2} + \Omega^2\,\delta = 0.$$ (8.42)

Similar to the transverse particle motion we may eliminate from (8.38,39) the argument of the trigonometric functions to obtain an invariant of the

276

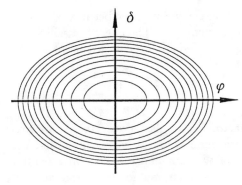

Fig. 8.3. Synchrotron oscillations in phase space for stable motion $\Omega^2 > 0$ and small amplitudes $\widehat{\varphi} \ll 1$

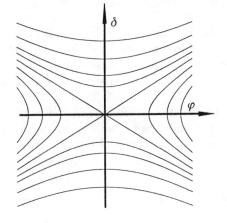

Fig. 8.4. Synchrotron oscillation in phase space for unstable motion $\Omega^2 < 0$

motion describing the particle trajectory in phase space in the form of closed ellipses or open hyperbolas

$$\frac{\delta^2}{\widehat{\delta}^2} \pm \frac{\varphi^2}{\widehat{\varphi}^2} = 1 \quad \text{with} \quad \widehat{\delta} = \frac{\Omega}{h\,\omega_{\mathrm{rev}}}\,\widehat{\varphi}, \tag{8.43}$$

where the sign is chosen to indicate stable or unstable motion depending on whether the synchrotron oscillation frequency Ω is real or imaginary respectively. The trajectories for both cases are shown in Figs. 8.3,4. Clearly the case of imaginary values of the synchrotron oscillation frequency leads to exponential growth in the *oscillation amplitude*.

8.1.3 Phase Stability

The *synchrotron oscillation frequency* must be real and the right-hand side of (8.33) must therefore be positive to obtain stable solutions for phase oscillations. All parameters in (8.33) are positively defined quantities except for the momentum compaction η_{c} and the phase factor $\cos\psi_{\mathrm{s}}$. For low particle energies the *momentum compaction* is in general positive because $\gamma^{-2} > \alpha_c$ but becomes negative for higher particle energies. The energy at which the momentum compaction changes sign is called the *transition energy* defined by

$$\gamma_{tr} = \frac{1}{\sqrt{\alpha_c}}.$$

(8.44)

Since the momentum compaction factor for circular accelerators is approximately equal to the inverse horizontal tune $\alpha_c \approx \nu_x^{-2}$ we conclude that the *transition energy* γ_{tr} is of the order of the tune and therefore in general a small number reaching up to the order of a hundred for very large accelerators. For electrons the transition energy is of the order of a few MeV and for protons in the GeV regime. In circular electron accelerators the injection energy always is selected to be well above the transition energy and no stability problems occur during acceleration since the transition energy is not crossed. Not so for protons. Proton linear accelerators with an energy of the order of 10 GeV or higher are very costly and therefore protons and ions in general must be injected into a circular accelerator below transition energy.

The *synchronous rf phase* must be selected depending on the particle energy being below or above the transition energy. Stable phase focusing can be obtained in either case if the rf synchronous phase is chosen as follows

$$
\begin{aligned}
0 < \psi_s < \frac{\pi}{2} \qquad &\text{for} \qquad \gamma < \gamma_{tr}, \\
\frac{\pi}{2} < \psi_s < \pi \qquad &\text{for} \qquad \gamma > \gamma_{tr}.
\end{aligned}
$$

(8.45)

In a proton accelerator with an injection energy below transition energy the rf phase must be changed very quickly when the transition energy is being crossed. Often the technical difficulty of this sudden change in the rf phase is ameliorated by the use of pulsed quadrupoles [8.8, 9], which is an efficient way of varying momentarily the momentum compaction factor by perturbing the dispersion function. A sudden change of a quadrupole strength can lower the transition energy below the actual energy of the particle. This helpful "perturbation" lasts for a small fraction of a second while the particles are still being accelerated and the rf phase is changed. By the time the quadrupole pulse terminates, the rf phase has been readjusted and the particle energy is now above the unperturbed transition energy.

In general we find that a stable phase oscillation for particles under the influence of accelerating fields can be obtained by properly selecting the synchronous phase ψ_s in conjunction with the sign of the momentum compaction such that

$$\Omega^2 > 0.$$

(8.46)

This is the principle of *phase focusing* [8.4, 5] and is a fundamental process to obtain stable particle beams in circular high-energy particle accelerators. An oscillating accelerating voltage together with a finite momentum compaction produces a stabilizing focusing force in the longitudinal degree of freedom just as transverse magnetic or electric fields can produce focusing forces for the two transverse degrees of freedom. With the focusing of transverse amplitudes we found a simultaneous focusing of its conjugate variable, the transverse momentum. The same occurs in the longitudinal phase where the particle energy or the energy deviation from the ideal energy is the conjugate variable to the time or phase of a particle. Both variables are related by (8.18) and a focusing force not only exists for the phase or longitudinal particle motion but also for the energy keeping the particle energy close to the ideal energy.

Focusing conditions have been derived for all six degrees of freedom where the source of focusing originates either from the magnet lattice for transverse motion or from a combination of accelerating fields and a magnetic lattice property for the energy and phase coordinate. The phase stability can be seen more clearly by observing the particle trajectories in phase space. Equation (8.32) describes the motion of a pendulum with the frequency Ω which, for small amplitudes $\sin\varphi \approx \varphi$ becomes equal to the equation of motion for a linear harmonic oscillator and can be derived from the Hamiltonian

$$\mathcal{H} = \tfrac{1}{2}\dot{\varphi}^2 + \tfrac{1}{2}\Omega^2\varphi^2 \,. \tag{8.47}$$

Small amplitude oscillations in phase space are shown in Figs. 8.3 and we note the confinement of the trajectories to the vicinity of the reference point. In case of unstable motion the trajectories quickly lead to unbound amplitudes in energy and phase, see Fig. 8.4.

Large Oscillation Amplitudes: For larger oscillation amplitudes we cannot anymore approximate the trigonometric function $\sin\varphi \approx \varphi$ by its argument. Following the previous derivation for the equation of motion (8.32) we get now

$$\ddot{\varphi} = -\Omega^2\sin\varphi \,, \tag{8.48}$$

which can be derived from the Hamiltonian

$$\mathcal{H} = \tfrac{1}{2}\dot{\varphi}^2 - \Omega^2\cos\varphi \tag{8.49}$$

being identical to that of a mechanical pendulum. As a consequence of our ability to describe synchrotron motion by a *Hamiltonian* and *canonical variables*, we expect the validity of the *Poincaré integral*

$$J_1 = \int_s \mathrm{d}\dot{\varphi}\,\mathrm{d}\varphi = \text{const} \tag{8.50}$$

under canonical transformations. Since the motion of particles during synchrotron oscillations can be described as a series of *canonical transformations* [8.10], we find the particle density in the $(\varphi, \dot{\varphi})$ phase space to be a constant of motion. The same result has been used in transverse phase space and the area occupied by this beam in phase space has been called the beam emittance. Similarly we define an emittance for the longitudinal phase space. Different choices of canonical variables can be defined as required to emphasize the physics under discussion. Specifically we find it often convenient to use the particle momentum instead of $\dot{\varphi}$ utilizing the relation (8.18).

Particle trajectories in phase space can be derived directly from the Hamiltonian by plotting solutions of (8.49) for different values of the "energy" $\mathcal{H}$ of the system. These trajectories, well known from the theory of harmonic oscillators, are shown in Fig. 8.5 for the case of a synchronous phase $\psi_s = \pi$.

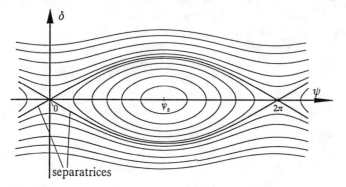

Fig. 8.5. Phase space diagrams for a synchronous phase $\psi_s = \pi$

The trajectories in Fig. 8.5 are of two distinct types. In one type the trajectories are completely local and describe oscillations about equilibrium points separated by 2π along the abscissa. For the other type the trajectories are not limited to a particular area in phase and the particle motion assumes the characteristics of *libration*. This phenomenon is similar to the two cases of possible motion of a mechanical pendulum or a swing. At small amplitudes we have periodic motion about the resting point of the swing. For increasing amplitudes, however, that oscillatory motion could become a libration when the swing continues to go over the top. The lines separating the regime of libration from the regimes of oscillation are called *separatrices*.

Particle motion is stable inside the separatrices due to the focusing properties of the potential well which in this representation is just the $\cos \varphi$-term in (8.49). The area within separatrices is commonly called an *rf bucket* describing a place where particles are in stable motion. In Fig. 8.6, the Hamil-

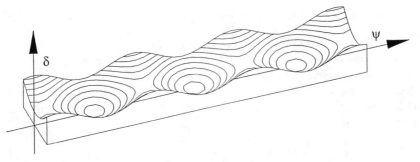

Fig. 8.6. Potential well for stationary rf buckets, ($\psi_s = \pi$)

tonian (8.49) is shown in a three-dimensional representation with contour lines representing the equipotential lines. The stable *potential wells, rf buckets*, within the *separatrices*, keeping the particles focused toward the equilibrium position, are clearly visible.

Inside the separatrices the average energy gain vanishes due to oscillatory phase motion of the particles. This is obvious from (8.31) which for $\psi_s = \pi$ becomes

$$V(\psi) = \widehat{V}_\mathrm{o} \sin\psi = \widehat{V}_\mathrm{o} \sin(\psi_s + \varphi) = \widehat{V}_\mathrm{o} \sin\varphi$$

averaging to zero since the average phase $\langle \varphi \rangle = 0$.

The area within such separatrices is called a *stationary rf bucket*. Such buckets, while not useful for particle accelerations, provide the necessary potential well to produce stable bunched particle beams in facilities where the particle energy need not be changed as for example in a proton or ion storage ring where bunched beams are desired. Whenever particles must receive energy from accelerating fields, may it be for straight acceleration or merely to compensate for energy losses like synchrotron radiation, the synchronous phase must be different from zero. As a matter of fact, due to the *principle of phase focusing*, particles within the regime of stability automatically oscillate about the appropriate *synchronous phase* independent of their initial parameters.

In the discussion of large amplitude oscillations we have tacitly assumed that the synchrotron oscillation frequency remains constant and equal to (8.33). From (8.30), however, we note that the frequency is proportional to the variation of the rf voltage with phase. Specifically we note in Fig. 8.5 that the trajectories in phase space are elliptical only for small amplitudes but are periodically distorted for larger amplitudes. This distortion leads to a spread of the synchrotron oscillation frequency.

8.1.4 Acceleration of Charged Particles

In the preceding paragraph we have arbitrarily assumed that the synchronous phase be zero $\psi_s = 0$ and as a result of this choice we obtained stationary, nonaccelerating rf buckets. No particle acceleration occurs since the particles pass through the cavities when the fields cross zero. Whenever particle acceleration is required a finite synchronous phase must be chosen. The average energy gain per revolution is then

$$\Delta E = V(\psi_s) = \widehat{V}_o \sin \psi_s. \tag{8.51}$$

The beam dynamics and stability becomes much different for $\psi_s \neq 0$. From (8.27-31) we get with $\psi = \psi_s + \varphi$ a phase equation more general than (8.48)

$$\ddot{\varphi} + \frac{\Omega^2}{\cos \psi_s} [\sin(\psi_s + \varphi) - \sin \psi_s] = 0, \tag{8.52}$$

or after expanding the trigonometric term into its components

$$\ddot{\varphi} + \frac{\Omega^2}{\cos \psi_s} (\sin \psi_s \cos \varphi + \sin \varphi \cos \psi_s - \sin \psi_s) = 0. \tag{8.53}$$

Equation (8.53) can also be derived directly from the *Hamiltonian*

$$\frac{1}{2} \dot{\varphi}^2 - \frac{\Omega^2}{\cos \psi_s} [\cos(\psi_s + \varphi) - \cos \psi_s + \varphi \sin \psi_s] = \mathcal{H}, \tag{8.54}$$

which is the Hamiltonian for the dynamics of phase motion.

The phase space trajectories or diagrams differ now considerably from those in Fig. 8.5 depending on the value of the synchronous phase ψ_s. In Fig. 8.7 *phase space diagrams* are shown for different values of the synchronous phase and a negative value for the momentum compaction η_c.

We note clearly the reduction in stable phase space area as the synchronous phase is increased or as the particle acceleration is increased. Outside the phase stable areas the particles follow unstable trajectories leading to continuous energy loss or gain depending on the sign of the momentum compaction. Equation (8.54) describes the particle motion in phase space for arbitrary values of the synchronous phase and we note that this equation reduces to (8.47) if we set $\psi_s = 0$. The energy gain for the *synchronous particle* at $\psi = \psi_s$ becomes from (8.20)

$$\Delta E = e \int \mathbf{E}(\psi_s) \, d\mathbf{z}. \tag{8.55}$$

We obtain a finite energy gain or loss whenever the synchronous phase in accelerating sections is different from an integer multiple of 180° assuming that all accelerating sections obey the synchronicity condition. The form of (8.55) actually is more general insofar as it integrates over all fields encoun-

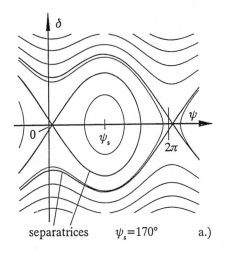

separatrices $\psi_s=170°$ a.)

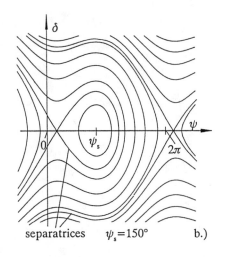

separatrices $\psi_s=150°$ b.)

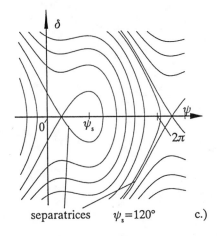

separatrices $\psi_s=120°$ c.)

Fig. 8.7. Phase space diagrams for different values of the synchronous phase and above transition energy $\gamma > \gamma_{tr}$

tered along the path of the particle. In case some accelerating sections are not synchronized, the integral collects all contributions as determined by the phase of the rf wave at the time the particle arrives at a particular section whether it be accelerating or decelerating. The *synchronicity condition* merely assures that the acceeration in all *accelerating sections* is the same for each turn.

The particle trajectories in phase space are determined by the Hamiltonian (8.54) which is similar to (8.49) except for the linear term in φ. Due to this term the potential well is now tilted, Fig. 8.8, compared to the stationary case, Fig. 8.6. We still have quadratic minima in the *potential well* function to provide stable *phase oscillations*, but particles which escape over the maxima of the potential well will be lost because they continuously loose or gain energy as can be seen by following such trajectories in Fig. 8.8. This is different from the case of stationary buckets where such a particle would

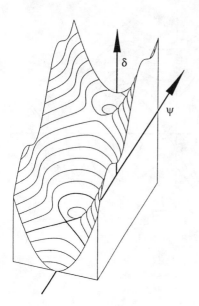

Fig. 8.8. Potential well for moving rf buckets $\psi_s \neq 0$

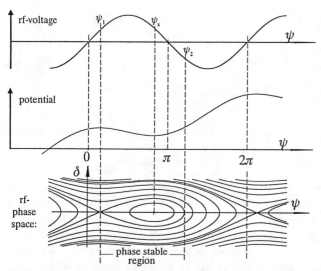

Fig. 8.9. Phase space focusing for moving rf buckets displaying the phase relationship of accelerating field, potential, and rf bucket

just wander from bucket to bucket while staying close to the ideal energy at the center of the buckets. Phase stable regions in case of finite values of the synchronous phase are called *moving rf buckets*.

The situation is best demonstrated by the three diagrams in Fig. 8.9 showing the accelerating field, the potential, and the *phase space diagram* as a function of the phase.

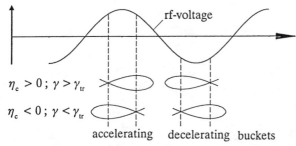

Fig. 8.10. Relationship between rf phase and orientation of moving rf buckets for accelerating as well as decelerating fields

In this particular case we have assumed that the particle energy is above transition energy and that the synchronous phase is such that $\cos \psi_s < 0$ to obtain stable synchrotron oscillations. The center of the bucket is located at the synchronous phase ψ_s and the longitudinal stability range is limited by the phases ψ_1 and ψ_2. In the next section we will derive analytical expressions for the *longitudinal stability limit* and use the results to determine the *momentum acceptance* of the bucket as well.

While both phases ψ_s as well as $\pi - \psi_s$ would supply the desired energy gain only one phase provides stability for the particles. The stable phase is easily chosen by noting that the synchrotron oscillation frequency Ω must be real and therefore $\eta_c \cos \psi_s > 0$. Depending on such operating conditions the *rf bucket* has different orientations as shown in Fig 8.10.

We still can choose whether the electric field should accelerate or decelerate the beam by choosing the sign of the field. For the decelerating case which, for example, is of interest for *free electron lasers*, the "fish" like buckets in the phase space diagram are mirror imaged.

8.2 Longitudinal Phase Space Parameters

We will here investigate in more detail specific properties and parameters of longitudinal phase space motion. From these parameters it will be possible to define stability criteria.

8.2.1 Separatrix Parameters

During the discussions of particle dynamics in longitudinal phase space we found specific trajectories in phase space, called separatrices which separate the *phase stable region* from the region where particles follow unstable trajectories leading away from the synchronous phase and from the ideal momentum. Within the phase stable region particles perform oscillations

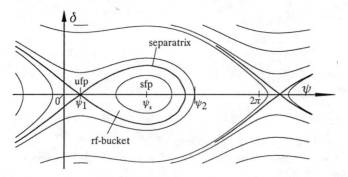

Fig. 8.11. Characteristic bucket and separatrix parameters

about the synchronous phase and the ideal momentum. This "focal point" in the phase diagram is called a *stable fixed point*, sfp. The *unstable fixed point*, ufp, is located where the two branches of the separatrix crosses. The location of *fixed points* can be derived from the two conditions:

$$\frac{\partial \mathcal{H}}{\partial \dot{\psi}} = 0 \quad \text{and} \quad \frac{\partial \mathcal{H}}{\partial \psi} = 0. \tag{8.56}$$

From the first condition we find with (8.54) that $\dot{\psi}_f = 0$ independent of any other parameter. All fixed points are therefore located along the ψ-axis of the *phase diagram* as shown in Fig. 8.11.

The second condition leads to the actual location of the fixed points, ψ_f on the ψ-axis and is with $\psi = \psi_s + \varphi$

$$\sin \psi_f - \sin \psi_s = 0. \tag{8.57}$$

This equation can be solved for $\psi_f = \psi_s$ and $\psi_f = \pi - \psi_s$ and the coordinates of the fixed points are

$$
\begin{aligned}
(\psi_{sf}, \dot{\psi}_{sf}) &= (\psi_s, 0) && \text{for the stable fixed point, sfp, and} \\
(\psi_{uf}, \dot{\psi}_{uf}) &= (\pi - \psi_s, 0) && \text{for the unstable fixed point, ufp.}
\end{aligned}
\tag{8.58}
$$

The distinction between a stable and unstable fixed point is made through the existence of a minimum or maximum in the potential at these points respectively. In Fig. 8.9 this distinction becomes obvious where we note the *stable fixed points* in the center of the potential minima and the unstable fixed points at the saddle points. The maximum stable phase elongation or *bunch length* is limited by the *separatrix* and the two extreme points ψ_1 and ψ_2 which we will determine in Sect. 8.2.3.

8.2.2 Momentum Acceptance

Particles on trajectories just inside the separatrix reach maximum deviations in phase and momentum from the ideal values in the course of performing synchrotron oscillations. A characteristic property of the separatrix therefore is the definition of the maximum phase or momentum deviation a particle may have and still undergo stable synchrotron oscillations. The value of the maximum momentum deviation is called the *momentum acceptance* of the accelerator. To determine the numerical value of the momentum acceptance we use the coordinates of the unstable fixed point (8.58) and calculate the value of the Hamiltonian for the separatrix which is from (8.54) with $\psi_{\mathrm{uf}} = \psi_{\mathrm{s}} + \varphi_{\mathrm{uf}} = \pi - \psi_{\mathrm{s}}$ and $\dot{\psi}_{\mathrm{uf}} = 0$

$$\mathcal{H}_f = \frac{\Omega^2}{\cos\psi_{\mathrm{s}}} \left[2\cos\psi_{\mathrm{s}} - (\pi - 2\psi_{\mathrm{s}})\sin\psi_{\mathrm{s}}\right]. \tag{8.59}$$

Following the separatrix from this unstable fixed point we eventually reach the location of maximum distance from the ideal momentum. Since $\dot{\varphi}$ is proportional to $\Delta p/p_{\mathrm{o}}$ we get the location of the maximum momentum acceptance through a differentiation of (8.54) with respect to φ

$$\dot{\varphi}\frac{\partial\dot{\varphi}}{\partial\varphi} - \Omega^2 \frac{\sin\psi_{\mathrm{s}} - \sin(\psi_{\mathrm{s}} + \varphi)}{\cos\psi_{\mathrm{s}}} = 0. \tag{8.60}$$

At the extreme points where the momentum reaches a maximum or minimum we have $\partial\dot{\varphi}/\partial\varphi = 0$ which occurs at the phase

$$\sin(\psi_{\mathrm{s}} + \varphi) = \sin\psi_{\mathrm{s}} \qquad \text{or} \quad \varphi = 0. \tag{8.61}$$

This is exactly the condition we found in (8.57) for the location of the stable fixed points and is independent of the value of the Hamiltonian. The maximum momentum deviation or momentum acceptance $\dot{\varphi}_{\mathrm{acc}}$ occurs therefore for all trajectories at the phase of the stable fixed points $\psi = \psi_{\mathrm{s}}$. We equate at this phase (8.59) with (8.54) to derive an expression for the maximum momentum acceptance

$$\tfrac{1}{2}\dot{\varphi}^2_{\mathrm{acc}} = \Omega^2 \left[2 - (\pi - 2\psi_{\mathrm{s}})\tan\psi_{\mathrm{s}}\right]. \tag{8.62}$$

In accelerator physics it is customary to define an *over voltage factor*. This factor is equal to the ratio of the maximum rf voltage in the cavities to the desired energy gain in the cavity U_{o}

$$q = \frac{eV_{\mathrm{o}}}{U_{\mathrm{o}}} = \frac{1}{\sin\psi_{\mathrm{s}}} \tag{8.63}$$

and can be used to replace trigonometric functions of the synchronous phase. To solve (8.62), we use the expression

$$\frac{\pi}{2} - \psi_{\mathrm{s}} = \arccos\frac{1}{q}$$

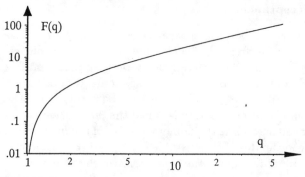

Fig. 8.12. Over voltage function F(q)

derived from the identity $\cos\left(\frac{\pi}{2} - \psi_s\right) = \sin\psi_s$, replace the *synchrotron oscillation frequency* Ω by its representation (8.36) and get with (8.18) the momentum acceptance for a moving bucket

$$\left(\frac{\Delta p}{p_o}\right)^2_{\mathrm{acc}} = \frac{eV_o \sin\psi_s}{\pi h \, |\eta_c| \, cp_o} \, 2 \left(\sqrt{q^2 - 1} - \arccos\frac{1}{q}\right). \tag{8.64}$$

The function

$$F(q) = 2 \left(\sqrt{q^2 - 1} - \arccos\frac{1}{q}\right) \tag{8.65}$$

is shown in Fig. 8.12 as a function of the over voltage factor q. The *synchronous phase* is always different from zero or π when charged particles are to be accelerated. In circular electron and very high-energy proton accelerators the synchronous phase must be nonzero to compensate for synchrotron radiation losses even if no net acceleration is desired. In low and medium energy circular proton or *heavy ion storage rings* no noticeable synchrotron radiation occurs and the synchronous phase is either $\psi_s = 0$ or π depending on the energy being below or above the transition energy. In either case $\sin\psi_s = 0$ which, however, does not necessarily lead to a vanishing momentum acceptance since the function $F(q)$ approaches the value $2q$ and the factor $\sin\psi_s \, F(q) \to 2$ in (8.64) while $q \to \infty$. Therefore stable buckets for protons and heavy ions can be produced with a finite energy acceptance. The maximum momentum acceptance for such *stationary buckets* is from (8.64)

$$\left(\frac{\Delta p}{p_o}\right)^2_{\mathrm{max,stat.}} = \frac{2 \, eV_o}{\pi h \, |\eta_c| \, \beta \, cp_o}. \tag{8.66}$$

Note that this expression for the maximum momentum acceptance appears to be numerically inconsistent with (8.41) for $\hat\varphi = \pi$ because (8.41) has been derived for small oscillations only $\hat\varphi \ll \pi$. From Fig. 8.11 we note that the aspect ratio of the *phase space ellipses* change between the bucket center and the vicinity of the separatrices. The linear proportionality between maximum momentum deviation and maximum phase of (8.41) becomes distorted for large values of $\hat{\varphi}$ such that the acceptance of the rf bucket is reduced by the factor $2/\pi$ from the value of (8.41).

The momentum acceptance is further reduced for *moving buckets* as the synchronous phase increases. In circular accelerators where the required energy gain for acceleration or compensation of *synchrotron radiation losses* is U_o per turn the momentum acceptance is

$$\left(\frac{\Delta p}{p_o}\right)^2_{\text{max,moving}} = \frac{U_o}{\pi\,h\,|\eta_c|\,\beta\,cp_o}\,F(q) = \frac{F(q)}{2\,q}\left(\frac{\Delta p}{p_o}\right)^2_{\text{max,static}} .(8.67)$$

The reduction $F(q)/2q$ in momentum acceptance is solely a function of the synchronous phase and is shown in Fig. 8.13 for the case $\gamma > \gamma_{\text{tr}}$.

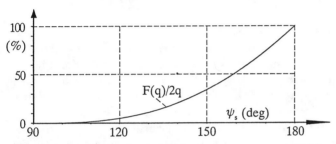

Fig. 8.13. Reduction factor of the momentum acceptance $F(q)/2q$ as a function of the synchronous phase

Overall the momentum acceptance depends on lattice and rf parameters and scales proportional to the square root of the rf voltage in the accelerating cavities. Strong transverse focusing decreases the momentum compaction thereby increasing the momentum acceptance while high values for the rf frequency diminishes the momentum acceptance. Very high frequency accelerating systems based, for example, on high intensity lasers to produce high accelerating fields are expected to have a rather small momentum acceptance and work therefore best with monoenergetic beams.

It is often customary to use other parameters than the momentum as the coordinates in longitudinal phase space. The most common parameter is the particle energy deviation $\Delta E/\omega_{\text{rf}}$ together with the phase. In these units we get for the stationary bucket instead of (8.66)

$$\left.\frac{\Delta E}{\omega_{\text{rf}}}\right|_{\text{max,stat.}} = \sqrt{\frac{2\,eV_o\,E_o\,\beta^2}{\pi\,h\,|\eta_c|\,\omega_{\text{rf}}^2}}, \tag{8.68}$$

which is measured in eV-sec. Independent of the conjugate coordinates used the momentum acceptance for moving rf-buckets can be measured in units of a stationary rf-bucket, where the proportionality factor depends only on the synchronous phase.

8.2.3 Bunch Length

During the course of synchrotron oscillations, particles oscillate between extreme values in momentum and phase with respect to the reference point and both modes of oscillation are out of phase by $90°$. All particles of a beam perform incoherent phase oscillations about a common reference point and generate thereby the appearance of a steady longitudinal distribution of particles which we call a *particle bunch*. The total *bunch length* is twice the maximum longitudinal excursion of particles from the bunch center defined by

$$\frac{\ell}{2} = \pm \frac{c}{h\,\omega_{\rm rev}}\,\widehat{\varphi} = \pm \frac{\lambda_{\rm rf}}{2\pi}\,\widehat{\varphi}, \tag{8.69}$$

where $\widehat{\varphi}$ is the maximum phase deviation.

In circular electron accelerators the rf parameters are generally chosen to generate a bucket which is much larger than the core of the beam. Statistical emission of synchrotron radiation photons generates a gaussian particle distribution in phase space and therefore the rf acceptance is adjusted to provide stability far into the tails of this distribution. To characterize the beam, however, only the core (one standard deviation) is used. In the case of bunch length or energy deviation we consider therefore only the situation for small oscillation amplitudes. In this approximation the bunch length becomes with (8.41)

$$\frac{\ell}{2} = \pm \frac{c\,|\eta_{\rm c}|}{\Omega}\,\frac{\Delta p}{p_{\rm o}}\bigg|_{\rm max} \tag{8.70}$$

or with (8.36)

$$\frac{\ell}{2} = \pm \frac{c\,\sqrt{2\pi}}{\omega_{\rm rev}}\,\sqrt{\frac{\eta_{\rm c}\,c p_{\rm o}}{h\,e\widehat{V}\,\cos\psi_{\rm s}}}\,\frac{\Delta p}{p_{\rm o}}\bigg|_{\rm max}. \tag{8.71}$$

The *bunch length* in a circular electron accelerator depends on a variety of rf and lattice parameters. It is inversely proportional to the square root of the rf voltage and frequency. A high frequency and rf voltage can be used to reduce the bunch length of which only the rf voltage remains a variable once the system is installed. Practical considerations, however, limit the range of bunch length adjustment this way. The *momentum compaction* is a lattice function and theoretically allows the bunch length to adjust to any small value. For high-energy electron rings $\eta_{\rm c} \approx -\alpha_{\rm c}$ and by arranging the focusing such that the *dispersion functions* changes sign, the momentum compaction factor of a ring can become zero or even negative. Rings for which $\eta_c = 0$ are called *isochronous rings* [8.11]. By adjusting the momentum compaction to zero phase focusing is lost similar to the situation going through transition in proton accelerators and total beam loss may occur. In this case, however, nonlinear, higher order effects become dominant which must be taken into consideration. If on the other hand the momentum compaction is adjusted to very small values, beam instability may be avoidable [8.12]. The benefit of an isochronous or *quasi-isochronous ring* would be that the bunch length in an electron storage ring could be made very small. This is important, for example, to either create short synchrotron radiation pulses or maximize the efficiency of a *free electron laser* by preserving the micro bunching at the laser wavelength as the electron beam orbits in the storage ring.

In a circular proton or ion accelerator we need not be concerned with the preservation of gaussian tails and therefore the whole rf bucket could be filled with such particles at high density. In this case the bunch length is limited by the extreme phases ψ_1 and ψ_2 of the separatrix. Because the longitudinal extend of the separatrix depends on the synchronous phase we expect the bunch length to depend also on the synchronous phase. One limit is given by the unstable fixed point at $\psi_1 = \pi - \psi_s$. The other limit must be derived from (8.54), where we replace $\mathcal{H}$ by the potential of the separatrix from Eq. (8.59). Setting $\dot{\varphi} = 0$ we get for the second limit of stable phases the transcendental equation

$$\cos\psi_{1,2} + \psi_{1,2}\sin\psi_s = (\pi - \psi_s)\sin\psi_s - \cos\psi_s. \tag{8.72}$$

This equation has two solutions mod2π of which ψ_1 is one solution and the other is ψ_2. Both solutions and their difference are shown in Fig. 8.14 as functions of the synchronous phase. The bunch length of proton beams is therefore determined only by the synchronous phase and is given by

$$\ell_p = \frac{\lambda_{rf}}{2\pi}(\psi_2 - \psi_1). \tag{8.73}$$

Different from the electron case we find the bunch length to be directly proportional to the rf wavelength. On the other hand there is no direct way

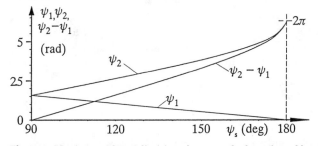

Fig. 8.14. Maximum phases limiting the extend of moving rf buckets

of compressing a proton bunch by raising or lowering the rf voltage. This difference stems from the fact that electrons radiate and adjust by damping to a changed rf bucket while nonradiating particles do not have this property. However, applying adiabatic rf voltage variation we may modify the bunch length as will be discussed in Sect. 8.2.5.

8.2.4 Longitudinal Beam Emittance

Separatrices distinguish between unstable and stable regions in the longitudinal phase space. The area of *stable phase space* in analogy to transverse phase space is called the *longitudinal beam emittance*; however, it should be noted that the definition of longitudinal emittance as used in the accelerator physics community generally includes the factor π in the numerical value of the emittance and is therefore equal to the real phase space area. To calculate the longitudinal emittance we evaluate the integral $\oint p\,dq$ where p and q are the conjugate variables describing the synchrotron oscillation.

Similar to transverse beam dynamics we distinguish again between beam acceptance and beam emittance. The acceptance is the maximum value for the beam emittance as determined by the transport line or accelerator components. In the longitudinal phase space the *acceptance* is the area enclosed by the separatrices. Of course, we ignore here other possible acceptance limitations which are not related to the parameters of the accelerating system. The equation for the separatrix can be derived by equating (8.54) with (8.59) which gives with (8.18, 36)

$$\left(\frac{\Delta cp}{cp_o}\right)^2 = \frac{eV_o}{\pi h |\eta_c| \beta \, cp_o} \left[\cos\varphi + 1 + (2\psi_s + \varphi - \pi)\sin\psi_s\right]. \qquad (8.74)$$

We define a longitudinal beam emittance by

$$\epsilon_\varphi = \int_S \frac{\Delta E}{\omega_{\rm rf}}\,d\varphi\,, \qquad (8.75)$$

where the integral is to be taken along a path S tightly enclosing the beam in phase space. Only for *stationary buckets*, where $\psi_s = n\,\pi$ can this integral be solved analytically. The maximum value of the beam emittance so defined is the acceptance of the system. Numerically the acceptance of a stationary bucket can be calculated by inserting (8.74) into (8.75) and integration along the enclosing separatrices resulting in

$$\epsilon_{\varphi,\rm acc} = 8\sqrt{\frac{2\,eV_o\,E_o\,\beta^2}{\pi\,h\,|\eta_c|\,\omega_{\rm rf}^2}}\,. \qquad (8.76)$$

Comparison with the momentum acceptance (8.75) shows the simple relation that the longitudinal acceptance is eight times the energy acceptance

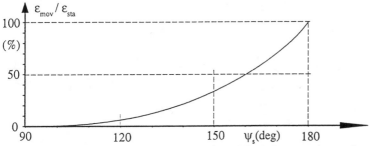

Fig. 8.15. Acceptance of moving rf buckets in units of the acceptance of a stationary rf bucket

$$\epsilon_{\varphi,\mathrm{acc}} = 8 \left.\frac{\Delta E}{\omega_{\mathrm{rf}}}\right|_{\mathrm{max,stat}}. \tag{8.77}$$

For *moving buckets* the integration (8.75)must be performed numerically between the limiting phases ψ_1 and ψ_2. The resulting acceptance in percentage of the acceptance for the stationary rf bucket is shown in Fig. 8.15 as a function of the synchronous phase angle.

The acceptance for $\psi_s < 180°$ is significantly reduced imposing some practical limits on the maximum rate of acceleration for a given maximum rf voltage. During the acceleration cycle, the magnetic fields in the lattice magnets are increased consistent with the available maximum rf voltage and by virtue of the *principle of phase focusing* the particles will keep close to the synchronous phase whenever the rate of energy increase is slow compared to the synchrotron oscillation frequency which is always the case. In high-energy electron synchrotrons or storage rings the required "acceleration" is no more a free parameter but is mainly determined by the energy loss due to synchrotron radiation and a stable beam can be obtained only if sufficient rf voltage is supplied to provide the necessary acceptance.

8.2.5 Phase Space Matching

In transverse phase space a need for matching exists while transferring a beam from one accelerator to another accelerator. Such matching conditions exist also in longitudinal phase space. In the absence of matching part of the beam may be lost due to lack of overlap with the rf bucket or severe phase space dilution may occur if a beam is injected unmatched into a too large rf bucket. In the case of electrons a mismatch generally has no detrimental effect on the beam unless part or all of the beam exceeds rf bucket limitations. Because of synchrotron radiation and concomitant damping, electrons always assume a gaussian distribution about the reference phase and ideal momentum. The only matching then requires that the rf bucket is large enough to enclose the gaussian distribution far into the tails up to 7 - 10 standard deviations.

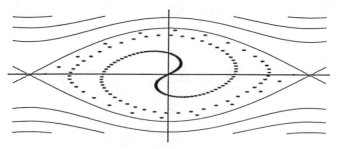

Fig. 8.16. Phase space filamentation

In proton and heavy ion accelerators such damping is absent and careful phase space matching during the transfer of particle beams from one accelerator to another is required to preserve beam stability and phase space density. A continuous monochromatic beam, for example, being injected into an accelerator with too large an rf bucket as shown in Fig. 8.16 will lead to a greatly diluted emittance.

This is due to the fact that the synchrotron oscillation is to some extend nonlinear and the frequency changes with the oscillation amplitude with the effect that for all practical purposes the beam eventually occupies all available phase space. This does not contradict *Liouville's theorem* since the microscopic phase space is preserved albeit fragmented and spread through *filamentation* over the whole bucket.

The situation is greatly altered if the rf voltage is reduced and adjusted to just cover the energy spread in the beam. Not all particles will be accepted, specifically those in the vicinity of the unstable fixed points, but all particles that are injected inside the rf bucket remain there and the *phase space density* for that part of the beam is not diluted. The *acceptance efficiency* is equal to the bucket overlap on the beam in phase space. A more sophisticated capturing method allows the capture of almost all particles in a uniform longitudinal distribution by turning on the the rf voltage very slowly [8.13], a procedure which is also called *adiabatic capture*.

Other matching problems occur when the injected beam is not continuous. A beam from a booster synchrotron or linear accelerator may be already bunched but may have a bunch length which is shorter than the rf wavelength or we may want to convert a bunched beam with a significant momentum spread into an unbunched beam with small momentum spread. Whatever the desired modification of the distribution of the beam in phase space may be, there are procedures to allow the change of particular distributions while keeping the overall emittance constant.

For example, to accept a bunched beam with a bunch length shorter than the rf wavelength in the same way as a continuous beam by matching only the momentum acceptance would cause *phase space filamentation* as

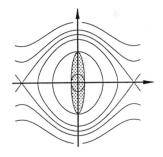

Fig. 8.17. Mismatch for a bunched beam

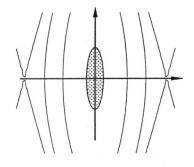

Fig. 8.18. Proper match for a bunched beam

shown in Fig. 8.17. In a proper matching procedure the rf voltage would be adjusted such that a phase space trajectory surrounds closely the injected beam (Fig. 8.18). In mathematical terms we would determine the bunch length $\widehat{\varphi}$ of the injected beam and would according to (8.70) adjust the rf voltage such that the corresponding momentum acceptance $\widehat{\delta} = (\Delta p/p_0)_{max}$ matches the momentum spread in the incoming beam.

Equation (8.70) represents a relation between the maximum momentum deviation and phase deviation for small amplitude phase space trajectories which allows us to calculate the bunch length as a function of external parameters. Methods have been discussed in transverse particle dynamics which allow the manipulation of conjugate beam parameters in phase space while keeping the beam emittance constant. Specifically, within the limits of constant phase space we were able to exchange beam size and transverse momentum or beam divergence by appropriate focusing arrangements to produce a wide parallel beam, for example, or a small beam focus.

In a similar way we are able in longitudinal phase space to manipulate within the limits of a constant beam emittance the bunch length and momentum spread. The focusing device in this case is the voltage in accelerating cavities.

Assume, for example, a particle bunch with a very small momentum spread but a long bunch length as shown in Fig. 8.19a. To transform such a bunch into a short bunch we would suddenly increase the rf voltage in a

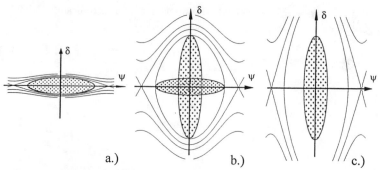

Fig. 8.19. Phase space rotation

time short compared to the synchrotron oscillation period. The whole bunch starts to rotate within the new bucket 8.19b exchanging bunch length for momentum spread. After a quarter synchrotron oscillation period, the bunch length has reached its shortest value and starts to increase again through further rotation of the bunch unless the rf voltage is suddenly increased a second time to stop the phase space rotation of the bunch (Fig. 8.19c). Before this second adjustment of the rf voltage the bunch boundary does not coincide with a phase space trajectory causing the whole bunch to rotate. The rf voltage therefore must be increased to such a value that all particles on the bunch boundary follow the same phase space trajectory.

This phase space manipulation can be conveniently expressed by repeated application of (8.41). The maximum momentum deviation $(\widehat{\Delta p/p_{\mathrm{o}}})_{\mathrm{o}}$ and the maximum phase deviation $\widehat{\varphi}_{\mathrm{o}}$ for the starting situation in Fig. 8.19a are related by

$$\left.\frac{\widehat{\Delta p}}{p_{\mathrm{o}}}\right|_{\mathrm{o}} = \frac{\Omega_{\mathrm{o}}}{h\,\omega_{\mathrm{rev}}\,|\eta_{\mathrm{c}}|}\,\widehat{\varphi}_{\mathrm{o}}\,, \tag{8.78}$$

where Ω_{o} is the starting synchrotron oscillation frequency for the rf voltage V_{o}. To start bunch rotation the rf voltage is increased to V_1 (Fig. 8.19b) and after a quarter synchrotron oscillation period at the frequency $\Omega_1 \propto \sqrt{V_1}$ the phase deviation $\widehat{\varphi}_{\mathrm{o}}$ has transformed into the momentum deviation

$$\left.\frac{\widehat{\Delta p}}{p_{\mathrm{o}}}\right|_{1} = \frac{\Omega_1}{h\,\omega_{\mathrm{rev}}\,|\eta_{\mathrm{c}}|}\,\widehat{\varphi}_{\mathrm{o}}\,. \tag{8.79}$$

At the same time the original momentum error $\widehat{\Delta p}/p_{\mathrm{o}}|_{\mathrm{o}}$ has become a phase error $\widehat{\varphi}_1$ given by

$$\left.\frac{\widehat{\Delta p}}{p_{\mathrm{o}}}\right|_{\mathrm{o}} = \frac{\Omega_1}{h\,\omega_{\mathrm{rev}}\,|\eta_{\mathrm{c}}|}\,\widehat{\varphi}_1\,. \tag{8.80}$$

Now we need to stop further phase space rotation of the whole bunch. This can be accomplished by increasing a second time the rf voltage during a time short compared to the synchrotron oscillation period in such a way that the new bunch length or $\widehat{\varphi}$ is on the same phase space trajectory as the new momentum spread $\Delta p/p_o|_1$ (Fig. 8.19c). The required rf voltage is then determined by

$$\left.\frac{\widehat{\Delta p}}{p_o}\right|_1 = \frac{\Omega_2}{h\,\omega_{\mathrm{rev}}\,|\eta_c|}\,\widehat{\varphi}_1 . \tag{8.81}$$

We take the ratio of (8.77,80) to get

$$\frac{\widehat{\varphi}_1\,\Omega_2}{\widehat{\varphi}_o\,\Omega_o} = \frac{\widehat{\Delta p/p_o}|_1}{\widehat{\Delta p/p_o}|_o}$$

and replace the ratio of the momentum spreads by the ratio of (8.78, 79). With $\Omega_i \propto \sqrt{V_i}$ and $\ell \propto \widehat{\varphi}$ we get finally for the reduction of the bunch length the scaling law

$$\frac{\ell_1}{\ell_o} = \left(\frac{V_o}{V_2}\right)^{1/4} . \tag{8.82}$$

In other words the bunch length can be reduced by increasing the rf voltage in a two step process and the bunch length reduction scales like the fourth power of the rf voltage. This phase space manipulation is symmetric in the sense that a beam with a large momentum spread and a short bunch length can be converted into a bunch with a smaller momentum spread at the expense of the bunch length by reducing the rf voltage in two steps.

The *bunch length manipulation* described here is correct and applicable only for nonradiating particles. For radiating particles like electrons the bunch manipulation is easier due to damping effects. Equation (8.41) still holds but the momentum spread is independently determined by synchrotron radiation and the bunch length therefore scales simply proportional to the square root of the rf voltage.

Problems

Problem 8.1. A 500 MHz rf system is supposed to be used in a Wideroe type linac to accelerate protons from a 1 MeV Van de Graaf accelerator. Determine the length of the first three drift tubes for an accelerating voltage at the gaps of 0.5 MeV while assuming that the length of the tubes shall not be less than three times the tube diameter of 2 cm. Describe the operating conditions from an rf frequency point of view.

Problem 8.2. A proton beam is injected in short bunches into a storage ring and the rf system in the storage ring is turned off. Derive an expression for the *debunching time*, or the time it takes for the bunched proton beam to spread out completely.

Problem 8.3. Calculate the synchrotron oscillation frequency for the Fermilab 8 GeV proton booster. The maximum momentum is $cp_{max} = 8.9$ GeV, the harmonic number $h = 84$ the rf voltage $V_{rf} = 200$ kV, transition energy $\gamma_{tr} = 5.4$ and rf frequency at maximum momentum $f_{rf} = 52.8$ MHz. Calculate and plot the rf and synchrotron oscillation frequency as a function of momentum from and injection momentum of 400 MeV to maximum momentum of 8.9 GeV while the synchronous phase is $\psi_s = 45°$. What is the momentum acceptance at injection and at maximum energy? How long does the acceleration last?

Problem 8.4. Specify a synchrotron of your choice made up of FODO cells for the acceleration of relativistic particles. Assume an rf system to provide an accelerating voltage equal to 10^{-4} of the maximum particle energy in the synchrotron. During acceleration the synchrotron oscillation tune shall remain less than $\nu_s \leq 0.02$. What are the numerical values for the rf frequency, harmonic number, rf voltage, synchronous phase angle and acceleration time in your synchrotron? In case of an electron synchrotron also include synchrotron radiation and determine the maximum achievable energy as limited by magnetic field limitation or rf voltage limits. In case of a proton synchrotron determine the change in the bunch length during acceleration.

Problem 8.5. The momentum acceptance in a synchrotron is reduced as the synchronous phase is increased. Derive a relationship between the maximum acceleration rate and momentum acceptance. How does this relationship differ for protons and radiating electrons?

Problem 8.6. Derive an expression for and plot the synchrotron frequency as a function of oscillation amplitude within the separatrices. What is the synchrotron frequency at the separatrices?

Problem 8.7. Consider a high-energy electron storage ring with a FODO lattice. Determine the equilibrium energy spread as derived in Chap.10 and specify rf parameters which will be sufficient to compensate for synchrotron radiation losses and provide an energy acceptance for all particles in a gaussian energy distribution up to $7\sigma_\varepsilon/E_o$. What is the synchrotron tune and the bunch length in your storage ring?

Problem 8.8. Sometimes it is desirable to produce short bunches, even only temporary in a storage ring either to produce short x-ray pulses or for quick ejection from a damping ring into a linear collider. By a sudden change of the rf voltage the bunch can be made to rotate in phase space. Determine analytical the shortest possible bunch length as a function of the rf voltage increase considering a finite energy spread. For how many turns would the short bunch remain within 50% of its shortest value?

Problem 8.9. Consider a pill box cavity with copper walls for a storage ring and choose a frequency between 300 MHz and 1000 MHz. Derive an expression for the wall losses due to the fundamental field only and derive an expression for the *shunt impedance* of the cavity defined by $R_{cy} = V_{rf}^2/P_{cy}$ where V_{rf} is the maximum rf voltage and P_{cy} the cavity wall losses. What are the rf losses if this cavity is used in the ring of problem 8.7? Assume that you can cool only about 100kW/m of cavity length. How many cavities would you need for your ring example?

Problem 8.10. In electron linear accelerators operating at 3 GHz accelerating fields of more than 50 MeV/m can be reached. Why can such high fields not be used in a storage ring? Discuss quantitatively, while scaling linac parameters to the frequency of your choice in the storage ring.

9 Synchrotron Radiation

Ever since J.C. Maxwell formulated his unifying electro magnetic theory in 1873, the phenomenon of electromagnetic radiation has fascinated the minds of theorists as well as experimentalists. The idea of displacement currents was as radical as it was important to describe electromagnetic waves. It was only fourteen years later when G. Hertz in 1887 succeeded to generate, emit and receive again electromagnetic waves, thus, proving experimentally the existence of electromagnetic waves and the validity of Maxwell's equations. The sources of electromagnetic radiation are oscillating electric charges and currents in a system of metallic wires.

Conceptually it was also only logic to expect electromagnetic waves to be emitted from free oscillating charges as well. Mathematically, however, significant problems appeared in the formulation of the radiation field from free charges. It became obvious that the radiation field at the observation point would depend on the dynamics of all radiating charges not at the time of observation but rather at the time the observed fields were emitted due to the finite speed of propagation of electromagnetic fields. It was Liénard [9.1] in 1898 and independently in 1900 Wiechert [9.2] who were first to formulate the concept of retarded potentials, now called the *Liénard-Wiechert potentials* for point charges like electrons. These retarded potentials relate the scalar and vector potential of electromagnetic fields at the observation point to the location of the emitting charges and currents at the time of emission. Using these potentials, Liénard was able to calculate the energy lost by electrons while circulating in a homogenous magnetic field.

Schott formulated and published in 1907 [9.3– 5] his classical theory of radiation from an electron circulating in a homogenous magnetic field. Although he was mainly interested in the spectral distribution of radiation and hoped to find an explanation for atomic radiation spectra. Verifying Liénard's conclusion on the energy loss he derived the angular and spectral distribution and the polarization of the radiation. Since this attempt to explain atomic spectra failed his paper was basically forgotten and many of his findings have been rediscovered forty years later.

It is interesting to note that these early pioneers were right in their theoretical description of electromagnetic radiation from free electrons even though the special theory of relativity was not known yet. Only because Maxwell's equations are invariant with respect to Lorentz transformations could the theory of electromagnetic waves be developed before relativity was known and accepted.

The theory of electromagnetic radiation from free electrons became fashionable again in the mid forties in parallel with the successful development of circular high-energy electron accelerators. At this time powerful betatrons have been put into operation [9.6] and it was Ivanenko and Pomeranchouk [9.7], who first in 1944 pointed out a possible limit to the betatron principle and maximum energy due to energy loss to electromagnetic radiation. This prediction was used by Blewett to calculate the radiation energy loss per turn in a newly constructed 100 MeV betatron at General Electric and to derive the change in the electron beam orbit due to this energy loss. In 1946 he measured [9.8] the shrinkage of the orbit due to radiation losses and the results agreed with predictions. On April 24, 1947 visible radiation was observed for the first time at the 70 MeV synchrotron built at General Electric [9.9– 11]. Since then this radiation is called *synchrotron radiation*.

The energy loss of particles to synchrotron radiation causes technical and economic limits for circular electron or positron accelerators. As the particle energy is driven higher and higher, more and more rf power must be supplied to the beam not only to accelerate particles but also to overcome energy losses due to synchrotron radiation. The limit is reached when the radiation power grows to high enough levels exceeding technical cooling capabilities or exceeding the funds available to pay for the high cost of electrical power. To somewhat ameliorate this limit, high-energy electron accelerators have been constructed with ever increasing circumferences to allow a more gentle bending of the particle beam. Since the synchrotron radiation power scales like the square of the particle energy the circumference must scale similar for a constant amount of rf power. Usually a compromise is reached by increasing the circumference less and adding more rf power in spaces along the ring lattice made available by the increased circumference. In general the maximum energy in large circular electron accelerators is limited by the available rf power while the maximum energy of proton or ion accelerators and low energy electron accelerators is more likely limited by the maximum achievable magnetic fields in bending magnets.

What is a nuisance for one group of researchers can be a gift for another group. Synchrotron radiation is emitted tangentially from the particle orbit and within a highly collimated angle of $\pm 1/\gamma$. The spectrum reaches from radio frequencies up to photon energies of many thousand electron volts, the radiation is polarized and the intensities greatly exceed most other available radiation sources specifically in the vacuum ultra violet to x-ray region.

With these properties synchrotron radiation was soon recognized to be a powerful research tool for material sciences, crystallography, surface physics, chemistry, biophysics, and medicine to name only a few areas of research. While in the past most of this research was done parasitically on accelerators built and optimized for high-energy physics the usefulness of synchrotron radiation for research has become important in its own right to justify the construction and operation of dedicated synchrotron radiation sources all over the world.

9.1 Physics of Synchrotron Radiation

Phenomenologically, synchrotron radiation is the consequence of a finite value for the velocity of light. Electric fields extend into space from charged particles in uniform motion. When charged particles become accelerated, however, parts of these fields cannot catch up with the particle anymore and give rise to synchrotron radiation. This happens more so as the particle velocity approaches the velocity of light.

The emission of light can be described by applying Maxwell's equations to moving charged particles. The mathematical derivation of the theory of radiation from Maxwell's equations is straightforward although quite elaborate and we will postpone this to Vol. II of this text. In this chapter we will follow a more intuitive discussion [9.12] which reveals more visually the physics of synchrotron radiation from basic physical principles.

9.1.1 Coulomb Regime

Electromagnetic radiation occurs whenever there are electric and magnetic fields with components orthogonal to each other such that the Poynting vector

$$\mathbf{S} = \frac{c}{4\pi} [\mathbf{E} \times \mathbf{B}] \tag{9.1}$$

is nonzero. This Poynting vector is defined as the electromagnetic energy flow per unit time through a unit surface element. Applying this to a stationary electrostatic charge, we note the Coulomb fields extending radially from the charge to infinity. Because the electric charge is stationary there is no magnetic field and therefore there is no radiation.

This situation is also true in the rest frame of a charge in uniform motion because in this frame the charge is at rest just like the example discussed before. In the laboratory system, however, the field components are different. Since the charge is moving, it constitutes an electric current which generates a magnetic field. From the Lorentz transformations (1.22) of electromagnetic fields between a fixed laboratory system L and a system L^* in uniform motion along the s-direction we have the relations

$$
\begin{aligned}
E_x &= \gamma E_x^*, & B_x &= +\gamma \beta E_y^*, \\
E_y &= \gamma E_y^*, & B_y &= -\gamma \beta E_x^*, \\
E_s &= E_s^*, & B_s &= 0,
\end{aligned}
\tag{9.2}
$$

and the components of the Poynting vector become

$$\frac{4\pi}{c} S_x = \gamma \beta E_s^* E_x^*,$$

$$\frac{4\pi}{c} S_y = \gamma \beta E_s^* E_y^*, \tag{9.3}$$

$$\frac{4\pi}{c} S_s = -\gamma^2 \beta E_r^{*2},$$

where $*$ indicates that the quantity be measured in the moving system L^*, $E_r^{*2} = E_x^{*2} + E_y^{*2}$ and $\beta = v_s/c$. The Poynting vector is nonzero and describes the flow of the field energy in the environment of a moving charged particle. The fields drop off rapidly with distance from the particle and the radiation is therefore confined close to the location of the particle. This part of the electromagnetic radiation is called the *Coulomb regime* and is responsible for the transport of electric energy along transmission lines. We will ignore this regime in our further discussion of synchrotron radiation because we are interested in radiation far away from radiating charges.

9.1.2 Radiation Regime

Charged particles in uniform motion produce radiation only in the Coulomb regime. We consider a coordinate system moving with a constant velocity equal to that of the particle before acceleration. Before acceleration the charge is at rest and the electric field lines extend radially to infinity. Now we consider the modification of this radial field distribution for accelerated particle motion. Acceleration of a charge causes it to move with respect to its "own" reference system and consequently generates a distortion of the purely radial electric fields of a uniformly moving charge. It is this distortion that gives rise to radiation. We assume a charge in uniform motion for times $t \leq 0$ apply an accelerating force at time $t = 0$ for a time ΔT and observe the charged particle and its fields in the moving frame of reference. Due to the acceleration the charge moves in its reference system during the time ΔT from point A to point B and as a consequence the field lines become distorted within a radius $c\Delta t$ from the original location A of the particle. The effects on the fields are shown schematically in Fig. 9.1 for an acceleration along the motion of the particle.

At time $t = 0$ all electric field lines extend radially from the charge at point A to infinity. During acceleration new radial field lines emerge from the charge now at locations between A and B. The new field lines must join the old field lines which, due to the finite velocity of light, are still unperturbed at a distance $c\Delta T$ and larger. As long as the acceleration lasts, a nonradial field component is created. Furthermore the moving charge creates an azimuthal magnetic field $B_\varphi^*(t)$ and the Poynting vector becomes nonzero causing the emission of radiation from an accelerated electrical charge.

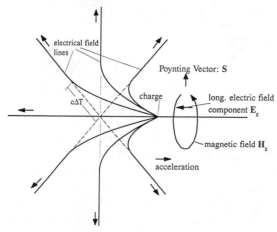

Fig. 9.1. Distortion of electrostatic fields by longitudinal particle acceleration and creation of synchrotron radiation

The electrical field perturbation is proportional to the electrical charge q and the acceleration a^*. We also note that the nonradial field component which causes the perturbation of the otherwise pure radial field distribution, varies like $\sin \Theta^*$, where Θ^* is the angle between the line of observation and the direction of particle motion. The field propagates radially outward with a field strength decaying linearly with distance R and we get finally for the electric field perturbation normal to the direction of observation

$$E_\perp^* \; = \; \frac{q\, a^*}{c^2\, R} \sin \Theta^*, \qquad (9.4)$$

where we have added a factor c^{-2} to be dimensionally correct. The radiation is emitted predominantly normal to the direction of acceleration as we would have expected from classical radiation theory. Since the associated magnetic field is proportional to the electric field, we find the Poynting vector to be proportional to the square of the electric field

$$\mathbf{S} \; = \; \frac{c}{4\pi}\, E^{*2}\, \mathbf{n}^*, \qquad (9.5)$$

where $\mathbf{n}^*$ is the unit vector in the direction of observation from the observer toward the radiation source. The result is consistent with our earlier finding of no radiation being emitted from a charge at rest as we reduce the acceleration to zero $a^* \to 0$.

Acceleration may not only occur in the longitudinal direction but also in the direction transverse to the velocity of the particle as shown in Fig. 9.2. The distortion of field lines in this case creates primarily transverse or radial field components with the maximum distortion in the forward

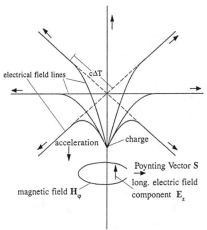

Fig. 9.2. Distortion of electrostatic fields by transverse particle acceleration and creation of synchrotron radiation

direction normal to acceleration. With Θ^* the angle between the direction of motion and observation we have a radiation field of the form

$$E_{\perp}^* \propto \frac{qa^*}{c^2 R} \cos \Theta^* \, . \tag{9.6}$$

This case of transverse acceleration describes the appearance of *synchrotron radiation* created by charged particles being deflected in magnetic fields. The radiation is emitted primarily in the forward direction tangential to the particle motion.

Obviously there would be no new fields created by acceleration if the velocity of propagation for electromagnetic fields were infinite $c \to \infty$. In this case the radial fields at all distances from the charge would instantly move in synchronism with the movement of the charge. Only the Coulomb regime would exist. By the same token no radiation is emitted for a uniform motion of the charge since again the field lines move with the charge and do not get distorted.

Although the acceleration and the creation of radiation fields is not periodic, we may Fourier decompose the radiation pulse and obtain a spectrum of plane waves

$$E^* = E_o^* e^{i\Phi^*} \, , \tag{9.7}$$

where the phase is defined by

$$\Phi^* = \omega^* \left[t - \frac{1}{c}(x^* n_x^* + y^* n_y^* + z^* n_z^*) \right] \, . \tag{9.8}$$

The phase of an electromagnetic wave is invariant and does not depend on the frame of reference. We have therefore the equality $\Phi^* = \Phi$ or

$$\omega \left[ct - (xn_x + yn_y + zn_z) \right] = \omega^* [ct^* - (x^* n_x^* + y^* n_y^* + z^* n_z^*)] \qquad (9.9)$$

between the phases as measured in both the laboratory L and the particle frame of reference L*. To derive the relationships between similar quantities in both systems we use the *Lorentz transformations* noting that the particle reference frame is the frame, where the particle or radiation source is at rest while the laboratory system is moving. The Lorentz transformations are therefore

$$
\begin{aligned}
x^* &= x, \\
y^* &= y, \\
s^* &= \gamma s - \gamma \beta ct, \\
ct^* &= \gamma ct - \gamma \beta s,
\end{aligned}
\qquad (9.10)
$$

where $\beta = v_s/c$ is the velocity of the moving system. Inserting these transformations into (9.9) and sorting coefficients we get for the oscillation frequency

$$\omega = \omega^* \gamma (1 + \beta n_z^*), \qquad (9.11)$$

which expresses the relativistic *Doppler effect*. Looking parallel to the direction of particle motion $n_z^* = 1$ the observed oscillation frequency is increased by the factor $(1 + \beta)\gamma \approx 2\gamma$ for highly relativistic particles. The Doppler effect is reduced if the radiation is viewed normal to the direction of motion when $n_z^* = 0$. For viewing angles in between these two extremes we set $n_z^* = \cos \Theta^*$.

9.1.3 Spatial Distribution of Synchrotron Radiation

Together with the transformation of frequencies we obtain from (9.9, 10) also the transformation for the direction of observation

$$n_x = \frac{n_x^*}{\gamma(1 + \beta n_z^*)} \qquad n_y = \frac{n_y^*}{\gamma(1 + \beta n_z^*)} \qquad n_z = \frac{\beta + n_z^*}{1 + \beta n_z^*}. \qquad (9.12)$$

This transformation defines the spatial distribution of radiation in the laboratory system. In the case of transverse acceleration the radiation in the particle rest frame is distributed like $\cos^2 \Theta^*$ about the direction of motion. This distribution becomes greatly collimated into the forward direction in the laboratory system. With $n_x^{*2} + n_y^{*2} = \sin^2 \Theta^* \approx \Theta^{*2}$ and $n_z^* = \cos \Theta^*$ we find

$$\Theta \approx \frac{\Theta^*}{\gamma(1 + \beta \cos \Theta^*)}. \qquad (9.13)$$

In other words, radiation from relativistic particles is collimated in the forward direction with an rms angle of about

$$\Theta_{\text{rms}} \approx \frac{1}{\gamma}. \tag{9.14}$$

This angle is very small for highly relativistic particles as for example an electron beam in a storage ring, where γ is of the order of 10^3 - 10^4.

9.1.4 Radiation Power

Integrating the Poynting vector (9.5) over a closed surface enclosing the radiating charge we get with (9.4) or (9.6) and $\mathbf{n}^* d\mathbf{A}^* = R^2 \sin \Theta^* d\Theta^* d\Phi^*$ the radiation power

$$P = \int \mathbf{S} d\mathbf{A}^* = \frac{2q^2}{3c} \dot{\beta}^{*2}, \tag{9.15}$$

where we have set for the acceleration $a^* = c\dot{\beta}^*$. This equation can be written in invariant form noting that the momentum of the radiation field is constant since radiation is emitted symmetrically into all directions and $d\mathbf{p} = 0$. With the energy $dU = Pdt$ we form the four vector $(dU/dt, cd\mathbf{p}/dt)$ which is invariant with respect to Lorentz transformations.

To express the four acceleration in (9.15) we use the more familiar momentum and get with $m\dot{\mathbf{v}}^* = d\mathbf{p}^*/d\tau$, where $\tau = t/\gamma$ is the time in the reference system $\mathcal{L}^*$ of the charged particle, m is the mass of the charged particle, and $\mathbf{p}$ its momentum. The radiation power in this representation is

$$P = \frac{2}{3} \frac{q^2}{m^2 c^3} \left(\frac{dp_\mu^*}{d\tilde{\tau}} \frac{dp^{*\mu}}{d\tilde{\tau}} \right). \tag{9.16}$$

Replacing in (9.16) the square of the four momentum by quantities in the laboratory frame of reference,

$$\frac{dp_\mu}{d\tilde{\tau}} \frac{dp^\mu}{d\tilde{\tau}} = \left(\frac{d\mathbf{p}}{d\tilde{\tau}} \right)^2 - \frac{1}{c^2} \left(\frac{dE}{d\tilde{\tau}} \right)^2$$

we get with $\mathbf{p} = \gamma m \mathbf{v}$ and $E = \gamma m c^2$ the radiation power

$$P = \frac{2}{3} \frac{q^2}{c} \gamma^2 \left[\left(\frac{d\gamma \boldsymbol{\beta}}{dt} \right)^2 - \left(\frac{d\gamma}{dt} \right)^2 \right].$$

After evaluation of the derivatives and some manipulation this becomes

$$P = \frac{2}{3} \frac{q^2}{c} \gamma^6 \left(\dot{\beta}^2 - [\boldsymbol{\beta} \times \dot{\boldsymbol{\beta}}]^2 \right), \tag{9.17}$$

where we have made use of the vector equation $\beta^2 \dot{\beta}^2 - (\boldsymbol{\beta}\dot{\boldsymbol{\beta}})^2 = [\boldsymbol{\beta} \times \dot{\boldsymbol{\beta}}]^2$. Equation (9.17) expresses the radiation power in a simple way and allows us to calculate other radiation characteristics based on beam parameters in the laboratory system. The radiation power is greatly determined by the geometric path of the particle trajectory through the quantities β and $\dot{\beta}$. Specifically if this path has strong oscillatory components we expect a corresponding synchrotron radiation power spectrum. This aspect will be discussed later and in Vol. II in great detail. Here we distinguish only between acceleration parallel $\dot{\beta}_\parallel$ or perpendicular $\dot{\beta}_\perp$ to the propagation β of the charge and set therefore

$$\dot{\beta} = \dot{\beta}_\parallel + \dot{\beta}_\perp .\tag{9.18}$$

Insertion into (9.17) shows the total radiation power to be composed of separate contributions from parallel and orthogonal acceleration. Separating both contributions we get the *synchrotron radiation power* for both parallel and transverse acceleration respectively

$$P_\parallel = \frac{2}{3}\frac{q^2}{c}\gamma^6\dot{\beta}_\parallel^2 ,\tag{9.19}$$

$$P_\perp = \frac{2}{3}\frac{q^2}{c}\gamma^4\dot{\beta}_\perp^2 .\tag{9.20}$$

Expressions have been derived that define the radiation power for parallel acceleration like in a linear accelerator or orthogonal acceleration found in circular accelerators or deflecting systems. We note a similarity for both contributions except for the energy dependence. At highly relativistic energies the same acceleration force leads to much less radiation if the acceleration is parallel to the motion of the particle compared to orthogonal acceleration. Parallel acceleration is related to the accelerating force by $\dot{\mathbf{v}}_\parallel = (1/\gamma^3\,(\mathrm{d}\mathbf{p}_\parallel/\mathrm{d}t)$ and after insertion into (9.19) the radiation power due to parallel acceleration becomes

$$P_\parallel = \frac{2}{3}\frac{q^2}{m^2c^3}\left(\frac{\mathrm{d}\mathbf{p}_\parallel}{\mathrm{d}t}\right)^2 .\tag{9.21}$$

The radiation power for acceleration along the propagation of the charged particle is therefore independent of the energy of the particle and depends only on the accelerating force or with $(\mathrm{d}\mathbf{p}_\parallel/\mathrm{d}t) = c\beta(\mathrm{d}E/\mathrm{d}x)$ on the energy increase per unit length of accelerator. Different from circular electron accelerators we encounter therefore no energy limit in a linear accelerator at very high energies.

In contrast very different radiation characteristics exist for transverse acceleration as it happens, for example, during the transverse deflection of a charged particle in a magnetic field. The transverse acceleration $\dot{\mathbf{v}}_\perp$ is expressed by the Lorentz force

$$\frac{\mathrm{d}\mathbf{p}_\perp}{\mathrm{d}t} = \gamma m\dot{\mathbf{v}}_\perp = \frac{e}{c}[\mathbf{v}\times\mathbf{B}]\tag{9.22}$$

and after insertion into (9.20) the radiation power from transversely accelerated particles becomes

$$P_\perp = \frac{2}{3} \frac{q^2}{m^2 c^3} \gamma^2 \left(\frac{d\mathbf{p}_\perp}{dt}\right)^2 . \tag{9.23}$$

From (9.21, 23) we find that the same accelerating force leads to a much higher radiation power by a factor γ^2 for transverse acceleration compared to longitudinal acceleration. For all practical purposes technical limitations prevent the occurrence of sufficient longitudinal acceleration to generate noticeable radiation. We express in (9.23) the momentum change by (9.22) and replace the deflecting magnetic field B by the bending radius ρ to get the instantaneous synchrotron radiation power

$$P_\gamma = \frac{2}{3} e^2 c \frac{\beta^4 \gamma^4}{\rho^2} , \tag{9.24}$$

or in more practical units,

$$P_\gamma(\text{GeV/sec}) = \frac{c\, C_\gamma}{2\pi} \frac{E^4}{\rho^2} . \tag{9.25}$$

Here we use the definition of Sand's radiation constant for electrons [9.13]

$$C_\gamma = \frac{4\pi}{3} \frac{r_c}{(mc^2)^3} = 8.8575 \times 10^{-5} \frac{\text{m}}{\text{GeV}^3} , \tag{9.26}$$

where r_c is the classical particle radius and mc^2 the rest energy of the particle.

From here on we will stop considering longitudinal acceleration unless specifically mentioned and eliminate, therefore, the index $\perp$ setting for the radiation power $P_\perp = P_\gamma$. We also restrict from now on the discussion to singly charged particles and set $q = e$ ignoring extremely high energies, where multiple charged ions may start to radiate.

The electromagnetic radiation of charged particles in transverse magnetic fields is proportional to the fourth power of the particle momentum $\beta\gamma$ and inversely proportional to the square of the bending radius ρ. The synchrotron radiation power increases very fast for high-energy particles and provides the most severe limitation to the maximum energy achievable in circular accelerators. We note, however, also a strong dependence on the kind of particles involved in the process of radiation. Because of the much heavier mass of protons compared to the lighter electrons we find appreciable synchrotron radiation only in electron accelerators. The radiation power of protons actually is smaller compared to that for electrons by the fourth power of the mass ratio or by the factor

$$\frac{P_e}{P_p} = 1836^4 = 1.36 \times 10^{13} . \tag{9.27}$$

In spite of this enormous difference measurable synchrotron radiation has been predicted by Coisson [9.14] and was indeed detected at the 400 GeV proton synchrotron, SPS, at CERN in Geneva [9.15, 16]. Substantial synchrotron radiation is expected in the *Superconducting Super Collider* [9.17], SSC, for proton energies above about 10 TeV.

Knowledge of the synchrotron radiation power allows us now to calculate the energy loss per turn of a particle in a circular accelerator by integrating the radiation power along the circumference of the circular accelerator

$$\Delta E = \oint P_\gamma dt = \frac{2}{3} e^2 \beta^3 \gamma^4 \oint \frac{ds}{\rho^2}. \tag{9.28}$$

In an isomagnetic lattice, where the bending radius is the same for all bending magnets $\rho = $ const, the integration around a circular accelerator can be performed and the energy loss per turn due to synchrotron radiation is

$$\Delta E = \oint P_\gamma dt = P_\gamma \frac{2\pi\rho}{\beta c} = \frac{4\pi}{3} e^2 \beta^3 \frac{\gamma^4}{\rho}. \tag{9.29}$$

The integration obviously is to be performed only along those parts of the circular accelerator, where synchrotron radiation occurs, or along bending magnets only. In more practical units, the energy loss of relativistic electrons per revolution in a circular accelerator with an isomagnetic lattice and a bending radius ρ is given by

$$\Delta E(GeV) = C_\gamma \frac{E^4}{\rho}. \tag{9.30}$$

From this energy loss per turn for each particle we calculate the total synchrotron radiation power for a beam of N_e particles. First we note that the total synchrotron radiation power for a single particle is its energy loss multiplied by the revolution frequency of the particle around the circular orbit. If C is the circumference of the orbit we have for the revolution frequency $f_{\rm rev} = \beta c / C$ and for the circulating particle current $I = e f_{\rm rev} N_e$. The total *synchrotron radiation power* is then

$$P_\gamma = \frac{\Delta E}{e} I \tag{9.31}$$

or in more practical units

$$P_\gamma(MW) = 8.86 \times 10^{-2} \frac{E^4(\text{GeV}^4)}{\rho(\text{m})} I(\text{A}). \tag{9.32}$$

The total synchrotron radiation power scales like the fourth power of the particle energy and is inversely proportional to the bending radius. The strong dependence of the radiation on the particle energy causes severe practical limitations on the maximum achievable energy in a circular accelerator.

9.1.5 Synchrotron Radiation Spectrum

Synchrotron radiation from relativistic charged particles is emitted over a wide spectrum of photon energies. The basic characteristics of this spectrum can be derived from simple principles as suggested by Jackson [9.18]. For an observer synchrotron light has the appearance similar to the light coming from a lighthouse. Although the light is emitted continuously an observer sees only a periodic flash of light as the aperture mechanism rotates in the lighthouse. Similarly, synchrotron light emitted from relativistic particles will appear to an observer as a single flash if it comes from a bending magnet in a transport line passed through by a particle only once or as a series of equidistant light flashes as bunches of particles orbit in a circular accelerator.

Since the duration of the light flashes is very short the observer notes a broad spectrum of frequencies as his eyes or instruments Fourier analyze the pulse of electromagnetic energy. The spectrum of synchrotron light from a circular accelerator is composed of a large number of harmonics with fundamental frequency equal to the revolution frequency of the particle in the circular accelerator. These harmonics reach a cutoff, where the period of the radiation becomes comparable to the duration of the light pulse. Even though the aperture of the observers eyes or instruments are assumed to be infinitely narrow we still note a finite duration of the light flash.

This is a consequence of the finite opening angle of the radiation as illustrated in Fig. 9.3. Synchrotron light emitted by a particle travelling along the orbit cannot reach the observer before it has reached the point P_0 when those photons emitted on one edge of the radiation cone at an angle $-1/\gamma$ aim directly toward the observer. Similarly, the last photons to reach the observer are emitted from point P_1 at an angle of $+1/\gamma$. Between point P_0 and point P_1 we have therefore a deflection angle of $2/\gamma$. The duration of the light flash for the observer is not the time it takes the particle to travel from point P_0 to point P_1 but must be corrected for the time of flight for the photon emitted at P_0. If particle and photon would travel toward the observer with exactly the same velocity the light pulse would be infinitely short. However, particles move slower following a slight detour and therefore the duration of the light pulse equals the time difference between the first photons from point P_0 arriving at the observer and the last photons being

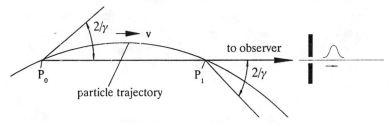

Fig. 9.3. Time structure of synchrotron radiation

emitted by the particles at point P_1. At the time $t = 0$ when the particle reaches point P_o the first photon can be observed at point P_1 at the time

$$t_\gamma = \frac{2\,\rho\sin\frac{1}{\gamma}}{c}. \tag{9.33}$$

The last photon to reach the observer is emitted when the particle arrives at point P_1 at the time

$$t_e = \frac{2\,\rho}{\beta c\,\gamma}. \tag{9.34}$$

The duration of the light pulse δt is therefore given by the difference of the times (9.33,34)

$$\delta t = t_e - t_\gamma = \frac{2\,\rho}{\beta c\,\gamma} - \frac{2\,\rho\sin\frac{1}{\gamma}}{c}. \tag{9.35}$$

The sine function can be expanded for small angles keeping linear and third order terms only and the duration of the light pulse at the location of the observer is after some manipulation

$$\delta t = \frac{4\,\rho}{3\,c\,\gamma^3}. \tag{9.36}$$

This compression by a factor of γ^2 is a consequence of the *Doppler effect* since radiation is emitted from a moving source. The finite pulse length limits the expansion into Fourier harmonics and amplitudes of the spectrum drop off significantly above a frequency of about

$$\omega_{\max} \approx \frac{\pi}{\delta t} \approx \frac{3\,\pi}{4}\,c\,\frac{\gamma^3}{\rho}. \tag{9.37}$$

This is a reasonably good representation for the width of the synchrotron radiation spectrum from highly relativistic particles. In a more rigorous derivation of the spectrum we will define a critical frequency

$$\omega_c = \frac{3}{2}\,c\,\frac{\gamma^3}{\rho}, \tag{9.38}$$

which is a convenient mathematical scaling parameter and indicates the photon frequency beyond which the intensity starts to drop significantly. Both frequencies apparently describe equally well the width of the radiation spectrum but for consistency the definition of the critical frequency is used exclusively for that purpose. In more practical units, the *critical photon frequency* is

$$\omega_c = C_c\frac{E^3}{\rho} \tag{9.39}$$

with

$$C_c = \frac{3c}{2(mc^2)^3} = 3.37 \times 10^{18} \frac{m}{sec\,GeV^3}\,, \tag{9.40}$$

where the numerical value has been calculated for electrons. The *critical photon energy* $\epsilon_c = \hbar\omega_c$ is then given by

$$\epsilon_c(keV) = 2.218\frac{E^3(GeV^3)}{\rho(m)} = 0.0665E^2(GeV^2)B(kG)\,. \tag{9.41}$$

The synchrotron radiation spectrum from relativistic particles in a circular accelerator is made up of harmonics of the particle revolution frequency ω_o and extends to values up to and beyond the critical frequency (9.38). Generally a real synchrotron radiation beam from say a storage ring will not display this harmonic structure. The distance between harmonics is extremely small compared to the extracted photon frequencies in the VUV and x-ray regime while the line width is finite due to the energy spread and beam emittance. For a single pass of particles through a bending magnet in a beam transport line we observe the same spectrum. Specifically, the maximum frequency is the same assuming similar parameters.

Synchrotron radiation is emitted in a particular spatial and spectral distribution, both of which will be derived in Vol. II, and we will use here only some of these results. A useful parameter to characterize the photon intensity is the *photon flux per unit solid angle* into a frequency bin $\Delta\omega/\omega$ and from a circulating beam current I defined by

$$\frac{d^2N_{ph}}{d\theta\,d\psi} = C_\Omega E^2 I\frac{\Delta\omega}{\omega}\frac{\omega^2}{\omega_c^2}K_{2/3}^2(\xi)F(\xi,\theta)\,, \tag{9.42}$$

where ψ is the angle in the deflecting plane and θ the angle normal to the deflecting plane,

$$C_\Omega = \frac{3\,\alpha}{4\pi^2\,e\,(mc^2)^2} = 1.3273 \times 10^{16}\frac{photons}{sec\,mrad^2\,GeV^2\,A}\,, \tag{9.43}$$

α the fine structure constant and

$$F(\xi,\theta) = \begin{cases} 1, & \text{for } \theta = 0\,, \\ (1+\gamma^2\theta^2)^2\left(1 + \frac{\gamma^2\theta^2}{1+\gamma^2\theta^2}\frac{K_{1/3}^2(\xi)}{K_{2/3}^2(\xi)}\right) & \text{otherwise.} \end{cases} \tag{9.44}$$

The functions $K_{1/3}(\xi)$ and $K_{2/3}(\xi)$ displayed in Fig. 9.4, are modified Bessel's functions with the argument

$$\xi = \tfrac{1}{2}\frac{\omega}{\omega_c}(1+\gamma^2\theta^2)^{3/2}\,. \tag{9.45}$$

Synchrotron radiation is highly polarized in the plane normal, (σ-mode), and parallel, (π-mode), to the deflecting magnetic field. The relative flux in

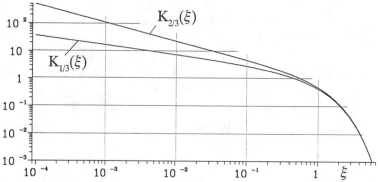

Fig. 9.4. Modified Bessel's functions $K_{1/3}(\xi)$ and $K_{2/3}(\xi)$.

both polarization directions is given by the two components in the second bracket of function $F(\xi,\theta)$ in (9.44). The first component is equal to unity and determines the photon flux for the polarization normal to the magnetic field or σ-mode, while the second term relates to the polarization parallel to the magnetic field which is also called the π-mode. Equation (9.42) expresses both the spectral and spatial photon flux for both the σ-mode radiation in the forward direction within an angle of about $\pm 1/\gamma$ and for the π-mode off axis.

For highly relativistic particles the synchrotron radiation is collimated very much in the forward direction and we may assume that all radiation in the nondeflecting plane is accepted by the experimental beam line. In this case we are interested in the photon flux integrated over all angles θ. This integration will be performed in Volume II with the result

$$\frac{dN_{ph}}{d\psi} = \frac{4\alpha}{9}\gamma\frac{\Delta\omega}{\omega}\frac{I}{e}S\left(\frac{\omega}{\omega_c}\right), \tag{9.46}$$

where ψ is the deflection angle in the bending magnet, α the fine structure constant and the function $S(x)$ is defined by

$$S\left(\frac{\omega}{\omega_c}\right) = \frac{9\sqrt{3}}{8\pi}\frac{\omega}{\omega_c}\int\limits_{\omega/\omega_c}^{\infty} K_{5/3}(\bar{x})d\bar{x}, \tag{9.47}$$

with $K_{5/3}(x)$ a modified Bessel's function. The function $S(\frac{\omega}{\omega_c})$ is shown in Fig. 9.5. In practical units the angle integrated photon flux is

$$\frac{dN_{ph}}{d\psi} = C_\psi EI\frac{\Delta\omega}{\omega}S\left(\frac{\omega}{\omega_c}\right) \tag{9.48}$$

with

$$C_\psi = \frac{4\alpha}{9\,e\,mc^2} = 3.967 \times 10^{19}\ \frac{\text{photons}}{\text{sec rad A GeV}}. \tag{9.49}$$

314

The spectral distribution depends only on the particle energy, the critical frequency ω_c and a purely mathematical function. This result has been derived originally by Ivanenko and Sokolov [9.19] and independently by Schwinger [9.20]. Specifically it should be noted that the spectral distribution, if normalized to the critical frequency, does not depend on the particle energy and can therefore be represented by a universal distribution shown in Fig. 9.5.

The energy dependence is contained in the cubic dependence of the critical frequency acting as a scaling factor for the actual spectral distribution. The synchrotron radiation spectrum in Fig. 9.5 is rather uniform up to the critical frequency beyond which the intensity falls off rapidly. This synchrotron radiation spectrum has been verified experimentally soon after such radiation sources became available [9.21, 22].

Equation (9.46) is not well suited for quick calculation of the radiation intensity at a particular frequency. We may, however, express (9.46) in much simpler form for very low and very large frequencies making use of limiting expressions of Bessel's functions for large and small arguments.

For small arguments $x = \omega/\omega_c \ll 1$ we may apply an asymptotic approximation [9.23] for the modified Bessel's function and get instead of (9.48)

$$\frac{\mathrm{d}N_{\mathrm{ph}}}{\mathrm{d}\psi} \approx C_\psi EI \frac{\Delta\omega}{\omega} 1.3333 \left(\frac{\omega}{\omega_c}\right)^{1/3}. \tag{9.50}$$

The approximation $S(x) \approx 1.3333 \, x^{1/3}$ is also shown in Fig. 9.5. Similarly we get for high photon frequencies $x = \omega/\omega_c \gg 1$

$$\frac{\mathrm{d}N_{\mathrm{ph}}}{\mathrm{d}\psi} \approx C_\psi EI \frac{\Delta\omega}{\omega} 0.77736 \frac{\sqrt{x}}{e^x}. \tag{9.51}$$

Both approximations are included in Fig. 9.5 and display actually a rather good representation of the real spectral radiation distribution. Specifically, we note the slow increase in the radiation intensity at low frequencies and the exponential drop off above the critical frequency.

9.1.6 Photon Beam Divergence

The expressions for the photon fluxes (9.42,46) provide the opportunity to calculate the spectral distribution of the photon beam divergence. Photons are emitted into a narrow angle and we may represent this narrow angular distribution by a gaussian distribution. The effective width of a gaussian distribution is $\sqrt{2\pi}\sigma_\theta$ and we set

$$\frac{\mathrm{d}N_{\mathrm{ph}}}{\mathrm{d}\psi} \approx \frac{\mathrm{d}^2\mathcal{N}}{\mathrm{d}\theta\mathrm{d}\psi} \sqrt{2\pi}\sigma_\theta.$$

With (9.42,48) the angular divergence of the forward lobe of the photon beam or for a beam polarized in the σ-mode is

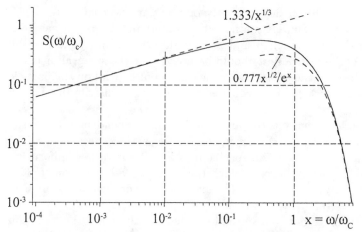

Fig. 9.5. Universal function of the synchrotron radiation spectrum $S(\omega/\omega_c)$

$$\sigma_\theta(\mathrm{mrad}) = \frac{C_\psi}{\sqrt{2\pi}C_\Omega}\frac{1}{E}\frac{S(x)}{x^2 K^2_{2/3}(x/2)} = \frac{f(x)}{E(\mathrm{GeV})}, \tag{9.52}$$

where $x = \omega/\omega_c$. For the forward direction $\theta \approx 0$ the function $f(x) = \sigma_\theta(\mathrm{mrad})E(\mathrm{GeV})$ is shown in Fig. 9.6 for easy numerical calculations.

For wavelengths $\omega \ll \omega_c$ (9.52) can be greatly simplified to become in more practical units

$$\sigma_\theta(\mathrm{mrad}) \approx \frac{0.54626}{E(\mathrm{GeV})}\left(\frac{\omega_c}{\omega}\right)^{1/3} = \frac{7.124}{[\rho(\mathrm{m})\epsilon_{\mathrm{ph}}(\mathrm{eV})]^{1/3}}, \tag{9.53}$$

where ρ is the bending radius and ϵ_{ph} the photon energy. The photon beam

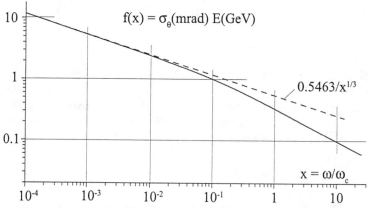

Fig. 9.6. Scaling function $f(x) = \sigma_\theta(mrad)E(GeV)$ for the photon beam divergence in (9.52)

316

divergence for low photon energies compared to the critical photon energy is independent of the particle energy and scales inversely proportional to the third root of the bending radius and photon energy.

9.2 Coherent Radiation

Synchrotron radiation is emitted into a broad spectrum with the lowest frequency equal to the revolution frequency and the highest frequency not far above the critical photon energy. Detailed observation of the whole radiation spectrum, however, may reveal significant differences to these theoretical spectra at the low frequency end. At low photon frequencies we may observe an enhancement of the synchrotron radiation beyond intensities predicted by the theory of synchrotron radiation as discussed so far. We note from the definition of the Poynting vector that the radiation power is a quadratic effect with respect to the electric charge. For photon wavelengths equal and longer than the bunch length, we expect therefore all particles within a bunch to radiate coherently and the intensity to be proportional to the square of the number N_e of particles rather than linearly proportional to N_e, as is the case for high frequencies. This quadratic effect can greatly enhance the radiation since the bunch population can be from 10^8 to 10^{11} electrons.

Generally such radiation is not emitted from a storage ring beam because radiation with wavelengths longer than the vacuum chamber dimensions are greatly damped and will not propagate along a metallic beam pipe [9.24]. This radiation shielding is fortunate for storage ring operation since it eliminates an otherwise significant energy loss mechanism. Actually, since this shielding affects all radiation of sufficient wavelength both the ordinary synchrotron radiation and the coherent radiation is suppressed. In Vol. II we will discuss the shielding effect in more detail and establish the transition scaling from shielding at low frequencies to propagation at higher frequencies.

New developments in storage ring physics, however, may make it possible to reduce the bunch length by as much as an order of magnitude below presently achieved short bunches of the order of 10 mm. Such bunches would then be much shorter than vacuum chamber dimensions and the emission of coherent radiation in some limited frequency range would be possible. Much shorter electron bunches of the order of 1 - 2 mm and associated coherent radiation can be produced in linear accelerators [9.25, 26], and specifically with bunch compression [9.27] a significant fraction of synchrotron radiation is emitted spontaneously as coherent radiation [9.28].

In this section we will discuss the physics of spontaneous coherent synchrotron radiation while distinguishing two kinds of coherence in synchrotron radiation, the *temporal coherence* and the *spatial coherence*. Temporal coherence occurs when all radiating electrons are located within a short bunch of the order of the wavelength of the radiation. In this case the radiation from all electrons is emitted with about the same phase. For spatial coherence the electrons may be contained in a long bunch but the transverse beam emittance must be smaller than the radiation wavelength. In either case there is a smooth transition from incoherent radiation to coherent radiation as determined by a form factor which depends on the bunch length or transverse emittance.

9.2.1 Temporal Coherent Synchrotron Radiation

To discuss the appearance of temporal coherent synchrotron radiation, we consider the radiation emitted from each particle within a bunch. The radiation field at a frequency ω from a single electron is

$$\mathcal{E}_k \propto e^{i(\omega t + \varphi_k)}, \tag{9.54}$$

where φ_k describes the position of the k-th electron with respect to the bunch center. With z_k the distance from the bunch center the phase is

$$\varphi_k = \frac{2\pi}{\lambda} z_k. \tag{9.55}$$

Here we assume that the cross section of the particle beam is small compared to the distance to the observer such that the path length differences from any point of the beam cross section to observer are small compared to the shortest wavelength involved.

The radiation power is proportional to the square of the radiation field and summing over all electrons we get

$$P(\omega) \propto \sum_{k,j}^{N_e} \mathcal{E}_k \mathcal{E}_j^* \propto \sum_{k,j}^{N_e} e^{i(\omega t + \varphi_k)} e^{-i(\omega t + \varphi_j)} = \sum_{k,j}^{N_e} e^{i(\varphi_k - \varphi_j)}$$

$$= N_e + \sum_{k \neq j}^{N_e} e^{i(\varphi_k - \varphi_j)}. \tag{9.56}$$

The first term N_e on the r.h.s. of (9.56) represents the ordinary incoherent synchrotron radiation with a power proportional to the number of radiating particles. The second term averages to zero for all but long wavelengths. The actual coherent radiation power spectrum depends on the particular particle distribution in the bunch. For a storage ring bunch it is safe to assume a gaussian particle distribution and we use therefore the density distribution

$$\Psi_g(z) = \frac{N_e}{\sqrt{2\pi}\sigma} \exp\left(-\frac{z^2}{2\sigma^2}\right), \tag{9.57}$$

where σ is the standard value of the gaussian bunch length. Instead of summing over all electrons we integrate over all phases and folding the density distribution (9.57) with the radiation power (9.56) we get with (9.55)

$$P(\omega) \propto N_{\rm e} + \frac{N_{\rm e}(N_{\rm e} - 1)}{2\pi\sigma^2} I_1 I_2 , \qquad (9.58)$$

where the integrals I_1 and I_2 are defined by

$$I_1 = \int_{-\infty}^{\infty} \exp\left(-\frac{z^2}{2\sigma^2} + {\rm i}\, 2\pi \frac{z}{\lambda}\right) {\rm d}z,$$

$$I_2 = \int_{-\infty}^{\infty} \exp\left(-\frac{w^2}{2\sigma^2} + {\rm i}\, 2\pi \frac{w}{\lambda}\right) {\rm d}w ,$$

and $z = \lambda\varphi_k/2\pi$ and $w = \lambda\varphi_j/2\pi$. The factor $(N - 1)$ reflects the fact that we integrate only over different particles. Both integrals are equal to the Fourier transform for a gaussian particle distribution. With

$$\int_{-\infty}^{\infty} \exp\left(-\frac{z^2}{2\sigma^2} + {\rm i}\, 2\pi \frac{z}{\lambda}\right) {\rm d}z = \sqrt{2\pi}\, \sigma \exp\left(-2\pi^2 \frac{\sigma^2}{\lambda^2}\right)$$

we get from (9.58) for the total radiation power at the frequency $\omega = 2\pi/\lambda$

$$P(\omega) = p(\omega)\left[N_{\rm e} + N_{\rm e}\,(N_{\rm e} - 1)\,g^2(\sigma)\right], \qquad (9.59)$$

where $p(\omega)$ is the radiation power from one electron and the Fourier transform,

$$g(\sigma) = \exp\left(-2\pi^2 \frac{\sigma^2}{\lambda^2}\right) \qquad (9.60)$$

becomes with the effective bunch length

$$\ell = \sqrt{2\pi}\sigma , \qquad (9.61)$$

finally

$$g(\ell) = \exp\left(-\pi\frac{\ell^2}{\lambda^2}\right) . \qquad (9.62)$$

The coherent radiation power falls off rapidly for wavelengths as short or even shorter than the effective bunch length ℓ.

In Fig. 9.7 the relative coherent radiation power is shown as a function of the effective bunch length in units of the radiation wavelength [9.29]. The fast drop off is evident and for an effective bunch length of about $\ell \approx 0.6\lambda$ the radiation power is reduced to about 10% of the maximum power for very short bunches, $\ell \to 0$.

Particle beams from a linear accelerator have often a more compressed particle distribution of a form between a gaussian and a rectangular distribution. If we take the extreme of a rectangular distribution

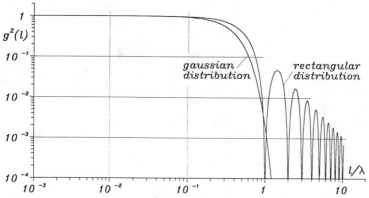

Fig. 9.7. Coherent synchrotron radiation power spectrum $g^2(\ell)$ as a function of the bunch length ℓ [9.29].

$$\Psi_r(z) = \begin{cases} 1 & \text{for } -\tfrac{1}{2}\ell < z < \tfrac{1}{2}\ell, \\ 0 & \text{otherwise}, \end{cases} \tag{9.63}$$

we expect to extend the radiation spectrum since the corners and sharp changes of the particle density require a broader spectrum in the Fourier transform. Following the procedure for the gaussian beam we get for a rectangular particle distribution the Fourier transform

$$g(\ell) = \frac{\sin x}{x}, \tag{9.64}$$

where $x = \pi(\ell/\lambda)$. Fig. 9.7 also shows the relative coherent radiation power for this distribution and we note a significant but scalloping extension to higher radiation frequencies.

Experiments have been performed with picosecond electron bunches from linear accelerators both at Tohoku University [9.25] and at Cornell University [9.26] which confirm the appearance of this coherent part of synchrotron radiation.

9.2.2 Spatially Coherent Synchrotron Radiation

Synchrotron radiation is emitted from a rather small area equal to the cross section of the electron beam. In the extreme and depending on the photon wave length the radiation may be spatially coherent because the beam cross section in phase space is smaller than the wave length. This possibility to create spatially coherent radiation is important for many experiments specifically for holography and we will discuss in more detail the conditions for the particle beam to emit such radiation.

Reducing the particle beam cross section in phase space by diminishing the particle beam emittance reduces also the source size of the photon beam. This process of reducing the beam emittance is, however, effective only to some point. Further reduction of the particle beam emittance would have no effect on the photon beam emittance because of *diffraction* effects. A point like photon source appears in an optical instrument as a disk with concentric illuminated rings. For synchrotron radiation sources it is of great interest to maximize the photon beam *brightness* which is the photon density in phase space. On the other hand designing a lattice for a very small beam emittance can causes beam stability problems. It is therefore prudent not to push the particle beam emittance to values much less than the diffraction limited photon beam emittance. In the following we will therefore define diffraction limited photon beam emittance as a guide for low emittance lattice design.

For highly collimated synchrotron radiation it is appropriate to assume Fraunhofer diffraction. Radiation from an extended light source appears diffracted in the image plane with a radiation pattern which is characteristic for the particular source size and radiation distribution as well as for the geometry of the apertures involved. For simplicity, we will use the case of a round aperture being the boundaries of the beam itself although in most cases the beam cross section is more elliptical. In spite of this simplification, however, we will obtain all basic physical properties of diffraction which are of interest to us.

We consider a circular light source with diameter $2a$. The radiation field at point P in the image plane is then determined by the *Fraunhofer diffraction integral* [9.30]

$$U(P) = C \int_0^a \int_0^{2\pi} e^{-ik\,\rho\,w\cos(\Theta-\psi)}\,\mathrm{d}\Theta\,\rho\,\mathrm{d}\rho\,. \tag{9.65}$$

Here k is the wave number of the radiation and w is the sine of the angle between the light ray and the optical axis as shown in Fig. 9.8.
With $\alpha = \Theta - \psi$ and the definition of the zeroth order Bessel's function

$$J_0(x) = \frac{1}{2\pi} \int_0^{2\pi} e^{-i\,x\cos\alpha}\,\mathrm{d}\alpha\,, \tag{9.66}$$

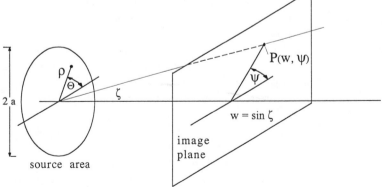

Fig. 9.8. Diffraction geometry

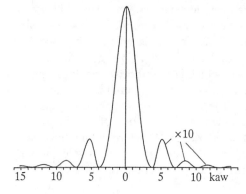

Fig. 9.9. Fraunhofer diffraction for a circular uniformly illuminated light source

and (9.65) can be expressed by the integral

$$U(P) = 2\pi C \int_0^a J_0(k\,\rho\,w)\rho\mathrm{d}\rho\,, \tag{9.67}$$

which can be solved analytically as well with the identity

$$\int_0^x y\,J_0(y)\,\mathrm{d}y = x\,J_1(x)\,.$$

The radiation intensity is proportional to the square of the radiation field and we get finally for the radiation intensity in the image plane at point P

$$I(P) = I_0\frac{2J_1^2(k\,a\,w)}{(k\,a\,w)^2}\,, \tag{9.68}$$

where $I(P) = |U(P)|^2$ and $I_0 = I(w = 0)$ is the radiation intensity at the image center. This result has been derived first by Airy [9.31]. The radiation intensity from a light source of small circular cross section is distributed in the image plane due to diffraction into a central circle and concentric rings illuminated as shown in Fig. 9.9.

Tacitly, we have assumed that the distribution of emission at the source is uniform which is generally not correct for a particle beam. A gaussian distribution is more realistic resembling the distribution of independently radiating particles. We must be careful in the choice of the scaling parameter. The relevant quantity for the Fraunhofer integral is not the actual particle beam size at the source point but rather the apparent beam size and distribution. By folding the particle density distribution with the argument of the Fraunhofer diffraction integral we get the radiation field from a round, gaussian particle beam,

$$U_g(P) = \mathrm{const} \int_0^\infty \mathrm{e}^{-\frac{\rho^2}{2\sigma_r^2}}\,J_0(k\,\rho\,\sigma_{r'})\rho\mathrm{d}\rho\,, \tag{9.69}$$

where σ_r is the apparent standard source radius and $w = \sigma_{r'}$ the radiation divergence. Introducing the variable $x = \rho/\sqrt{2}\sigma_r$ and replacing

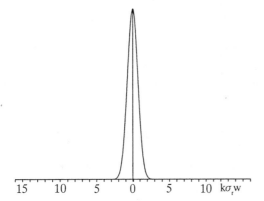

Fig. 9.10. Fraunhofer diffraction for a gaussian luminescence at the light source

$$k\rho\sigma_{r'} = 2x\sqrt{\frac{k^2\sigma_r^2\sigma_{r'}^2}{2}} = 2x\sqrt{z}$$

we get from (9.69)

$$U_{\mathrm{g}}(P) = \mathrm{const} \int_0^\infty \mathrm{e}^{-x^2} x J_0(2x\sqrt{z})\mathrm{d}x \tag{9.69}$$

and after integration

$$U_{\mathrm{g}}(P) = \mathrm{const}\,\exp[-\,{}^1\!/_2\,(k\,\sigma_r\,\sigma_{r'})^2]\,. \tag{9.71}$$

The diffraction pattern from a gaussian light source does not exhibit the ring structure of a uniform source (Fig. 9.10). The radiation field assumes rather the form of a gaussian distribution with a standard width of

$$k\sigma_r\sigma_{r'} = 1\,. \tag{9.72}$$

Similar to the particle beam characterization through its emittance we may do the same for the photon beam and doing so for the horizontal or vertical plane we have with $\sigma_{x,y} = \sigma_r/\sqrt{2}$ and $\sigma_{x',y'} = \sigma_{r'}/\sqrt{2}$ the photon beam emittance

$$\epsilon_{\mathrm{ph},x,y} = \tfrac{1}{2}\sigma_r\sigma_{r'} = \frac{\lambda}{4\pi}\,. \tag{9.73}$$

This is the *diffraction limited photon emittance* and reducing the electron beam emittance below this value would not lead to an additional reduction in the photon beam emittance. To produce a spatially coherent or diffraction limited radiation source the particle beam emittance must be less than the diffraction limited photon emittance

$$\epsilon_{x,y} \le \frac{\lambda}{4\pi}\,. \tag{9.74}$$

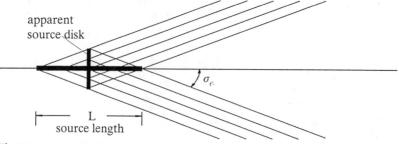

apparent
source disk

$\sigma_{r'}$

L
source length

Fig. 9.11. Apparent photon source size

Obviously, this condition is easier to achieve for long wavelengths. For a spatially coherent radiation source for visible light, for example, the electron beam emittance must be smaller than about $5 \times 10^{-8}\,\pi$ rad-m.

After having determined the diffraction limited photon emittance we may also determine the apparent photon beam size and divergence. The photon source extends over some finite length L along the particle path which could be either the path length required for a deflection angle of $2/\gamma$ or a much longer length in the case of an undulator radiation source to be discussed in the next section. With $\sigma_{r'}$ the diffraction limited beam divergence the photons seem to come from a disc with diameter (Fig. 9.11)

$$D = \sigma_{r'} L. \tag{9.75}$$

On the other hand we know from diffraction theory the correlation

$$D \sin \sigma_{r'} \approx D \sigma_{r'} = \lambda \tag{9.76}$$

and equating both equations gives the diffraction limited photon beam divergence

$$2\sigma_{r'} = \sqrt{\frac{\lambda}{L}}. \tag{9.77}$$

With this we get finally from (9.72) also the diffraction limited image size

$$\sigma_r = \frac{\sqrt{\lambda L}}{\pi}. \tag{9.78}$$

The apparent diffraction limited, radial photon beam size and divergence depend both on the photon wavelength of interest and the length of the source.

9.2.3 Spectral Brightness

The optical quality of a photon beam is characterized by the spectral brightness defined as the six-dimensional volume occupied by the photon beam in phase space

$$B = \frac{N_{\text{ph}}}{4\pi^2 \sigma_x \sigma_{x'} \sigma_y \sigma_{y'} (\mathrm{d}\omega/\omega)}, \tag{9.79}$$

where N_{ph} is the photon flux defined in (9.48). For bending magnet radiation there is a uniform angular distribution in the deflecting plane and we must therefore replace the gaussian divergence $\sigma_{x'}$ by the total angle $\Delta\psi$ accepted by the photon beam line or experiment.

The particle beam emittance must be minimized to achieve maximum spectral photon beam brightness. However, unlimited reduction of the particle beam emittance will, at some point, not further increase the brightness. Because of diffraction effects the photon beam emittance need not be reduced below the limit (9.73) discussed in the previous section. For a negligible particle beam emittance and deflection angle $\Delta\psi$ the maximum spectral brightness is therefore from (9.73, 79)

$$B_{\text{max}} = N_{\text{ph}} \frac{4}{\lambda^2} \frac{1}{\mathrm{d}\omega/\omega}. \tag{9.80}$$

For a realistic synchrotron light source the finite beam emittance of the particle beam must be taken into account as well which is often even the dominant emittance being larger than the diffraction limited photon beam emittance. We may add both contributions in quadrature and have for the total source parameters

$$\sigma_{\text{tot},x} = \sqrt{\sigma_{b,x}^2 + \tfrac{1}{2}\sigma_r^2}, \qquad \sigma_{\text{tot},x'} = \sqrt{\sigma_{b,x'}^2 + \tfrac{1}{2}\sigma_{r'}^2},$$

$$\sigma_{\text{tot},y} = \sqrt{\sigma_{b,y}^2 + \tfrac{1}{2}\sigma_r^2}, \qquad \sigma_{\text{tot},y'} = \sqrt{\sigma_{b,y'}^2 + \tfrac{1}{2}\sigma_{r'}^2}, \tag{9.81}$$

where σ_b are the respective particle beam parameters.

9.2.4 Matching

In case of a finite particle beam emittance the photon beam brightness is reduced. The amount of reduction, however, depends on the *matching to the photon beam*. As discussed earlier, the photon beam size and divergence are independent parameters and the particle beam size and divergence should be adjusted such that the resulting total beam parameter is minimized. The particle beam parameters are

$$\sigma_{b,x,y}^2 = \epsilon_{x,y}\beta_{x,y} \qquad \text{and} \qquad \sigma_{b,x',y'}^2 = \frac{\epsilon_{x,y}}{\beta_{x,y}}, \tag{9.82}$$

where $\beta_{x,y}$ is the betatron function at the photon source location. The product

$$\sigma_{\text{tot},x}\sigma_{\text{tot},x'} = \sqrt{\epsilon_x\beta_x + \tfrac{1}{2}\sigma_r^2}\sqrt{\frac{\epsilon_x}{\beta_x} + \tfrac{1}{2}\sigma_{r'}^2} \tag{9.83}$$

has a minimum for

$$\beta_{x,y} = \frac{\sigma_r}{\sigma_{r'}} = \frac{L}{2\pi}.$$

The values of the horizontal and vertical betatron functions should be adjusted according to (9.85) for optimum photon beam brightness. In case the particle beam emittances are much larger than the diffraction limited photon beam emittance, this minimum is very shallow and almost nonexistent in which case the importance of matching becomes practically irrelevant.

9.3 Insertion Devices

Deflection of a highly relativistic particle beam causes the emission of a broad spectrum of synchrotron radiation as shown in Fig. 9.5. The width of the spectrum is characterized by the critical photon energy and depends only on the particle energy and the radius of the bending magnet. Generally the radiation is produced in the bending magnets of a storage ring, where a high beam current is stored with a long beam lifetime of several hours. However, in order to adjust to special experimental needs other magnetic devices are being used as well to produce synchrotron radiation. We call such devices *insertion devices* since they do not contribute to the overall deflection of the particle beam in the circular accelerator. Their effect is localized and the overall deflection in an insertion device is zero. We will discuss some features of the different radiation sources and point the interested reader to Vol. II for more detailed discussion of the radiation characteristics from bending magnets and insertion devices.

9.3.1 Bending Magnet Radiation

The radiation from bending magnets is emitted tangentially from any point along the curved path and produces therefore a swath of radiation around the storage ring as shown in Fig. 9.12. In the vertical, nondeflecting plane, however, the radiation is very much collimated with a typical opening angle of $\pm 1/\gamma$.

Bending magnets, being a part of the geometry of the storage ring, however, cannot be varied to optimize for desired photon beam characteristics. While the lower photon spectrum is well covered even for rather low energy storage rings the x-ray region requires high beam energies and/or high magnetic fields.

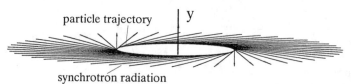

Fig. 9.12. Radiation swath from bending magnets in an electron storage ring

9.3.2 Wave Length Shifter

To meet the need for harder x-ray radiation in a low energy storage ring it is customary to install in a magnet free section of the ring a *wave length shifter*. Such a device consists of three ordinary dipole magnets with a high field central magnet and two lower field magnets with opposite field direction on either side to compensate the beam deflection by the central pole as shown in Fig. 9.13. The particle beam traversing this wave length shifter is deflected up and down or left and right in such a way that no net deflection remains.

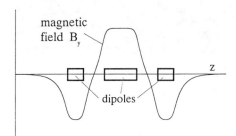

Fig. 9.13. Magnetic field distribution along the beam path for a wave length shifter

The condition for the longitudinal distribution of the field in case of a horizontally deflecting wave length shifter is

$$\int_{-\infty}^{+\infty} B_y(y = 0, z) \mathrm{d}z = 0.$$
(9.86)

A wavelength shifter with such field properties is neutral on the geometry of the particle beam path and therefore can be made in principle as strong as necessary or technically feasible. In most cases such wave length shifters are constructed as superconducting magnets with high fields to maximize the critical photon energy for the given particle beam energy. Some limitations apply for such devices as well as for any other insertion devices. The end fields of magnets can introduce particle focusing and nonlinear field components may introduce aberrations and instability. Effects due to such field errors have been discussed in Chap. 4 and will be discussed in more detail specifically for insertion devices in Vol. II. The radiation characteristics are similar to that of a bending magnet of equal field strength. Specifically we note the broad photon spectrum which extends up to photon energies as determined by the field of the insertion device.

9.3.3 Wiggler Magnet Radiation

The principle of a wave length shifter is extended in the case of a wiggler magnet. Such a magnet consists of a series of equal dipole magnets with alternating magnetic field direction. Again the end poles must be configured

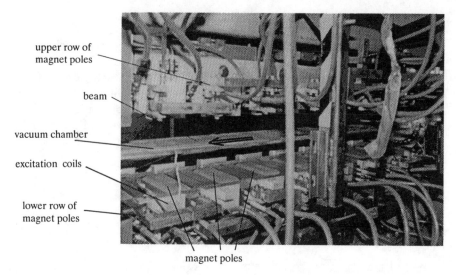

upper row of magnet poles

beam

vacuum chamber

excitation coils

lower row of magnet poles

magnet poles

Fig. 9.14. Electromagnetic wiggler magnet with a maximum field strength of 18 kG and eight poles [9.32]

to make the total device neutral to the geometry of the particle beam path such that $\int B_y(y = 0, z)\, \mathrm{d}z = 0$.

The advantage of using many magnet poles is to increase the photon flux. Each of N magnet poles produces a fan of radiation in the forward direction and the total photon flux is N times larger than that from a single pole. Wiggler magnets may be constructed as electromagnets with strong fields to function both as a wave length shifter and flux enhancer.

An example of a 18 kG electromagnetic wiggler magnet [9.32] is shown in Fig. 9.14. Here we note the flat vacuum chamber containing the particle beam and the upper and lower series of eight magnet poles in a retracted position. During operation both rows of wiggler poles are moved close to the top and bottom of the flat vacuum chamber to minimize gap and maximize the field. Strong fields can be obtained from electromagnets, but the space requirement for the excitation coils limits the number of poles within a given length. Electromagnetic wigglers are therefore primarily used, where a high critical photon energy is required.

To maximize photon flux wiggler magnets are constructed from permanent magnet materials. This construction technique results in rather short field periods allowing many wiggler poles for a given space in the storage ring lattice. In Fig. 9.15 a 54-pole wiggler magnet [9.33] is shown based on permanent magnets.

The longitudinal field distribution assumes the form of an alternating step function for magnet poles which are long compared to the gap aperture. Most wiggler magnets are however optimized to enhance the photon flux which requires many short magnet poles. In this case the magnetic field distribution is sinusoidal with a period length λ_p

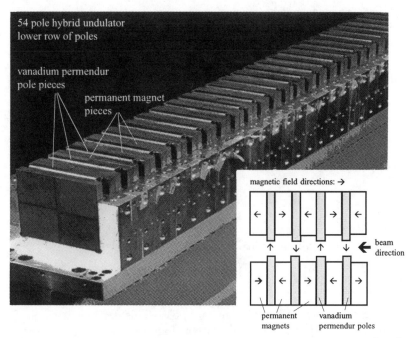

54 pole hybrid undulator
lower row of poles

vanadium permendur
pole pieces

permanent magnet
pieces

magnetic field directions: →

beam
direction

permanent
magnets

vanadium
permendur poles

Fig. 9.15. Row of 54 wiggler poles based on permanent magnet technology [9.33]

$$B_y(x, y = 0, z) = B_\mathrm{o} \sin 2\pi \frac{z}{\lambda_\mathrm{p}}.$$ (9.87)

The deflection angle per half pole is

$$\vartheta = \frac{B_\mathrm{o}}{B\rho} \int_0^{\lambda_\mathrm{p}/4} \sin 2\pi \frac{z}{\lambda_\mathrm{p}} \mathrm{d}z = \frac{B_\mathrm{o}}{B\rho} \frac{\lambda_\mathrm{p}}{2\pi},$$

where $B\rho$ is the beam rigidity. Multiplying this with the beam energy γ we define the wiggler *strength parameter* K

$$K = \gamma\vartheta = 0.934 \, B(\text{Tesla}) \, \lambda_\mathrm{p}(\text{cm}).$$ (9.88)

This wiggler strength parameter is generally much larger than unity. Conversely, a series of alternating magnet poles is called a wiggler magnet if the strength parameter $K \gg 1$.

The magnetic field strength can be varied in both electromagnetic wigglers as well as in permanent magnet wigglers. While this is obvious for electromagnets we note in the case of permanent magnets that the magnetic field strength in the gap depends on the distance between magnets poles or on the gap height. By varying mechanically the gap height of a permanent magnet wiggler the magnetic field strength can be varied as well.

The field strength also depends on the period length and on the design and magnet materials used. For a wiggler magnet constructed as a hybrid

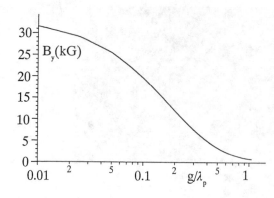

Fig. 9.16. On axis field strength in a Vanadium Permendur hybrid wiggler magnet as a function of gap aperture [9.34]

magnet with Vanadium Permendur poles the field strength along the mid-plane axis scales approximately like [9.34]

$$B_y(\text{kG}) \approx 33.3 \exp\left[-\frac{g}{\lambda_p}\left(5.47 - 1.8\,\frac{g}{\lambda_p}\right)\right] \quad \text{for} \quad g \lesssim \lambda_p, \qquad (9.89)$$

where g is the gap aperture between magnet poles. This dependency is also shown in Fig. 9.16 and we note immediately that the field strength drops off dramatically for magnet gaps of the order of a period length or greater.

On the other hand a significant field strength can be obtained for small gap apertures and it is therefore important to design for or find locations for the insertion device in a beam transport line, where the beam dimension normal to the deflection plane is very small.

9.3.4 Undulator Radiation

If the strength parameter is equal to about unity or less the wiggler magnet becomes an *undulator magnet* with rather particular radiation properties. Fundamentally an undulator magnet causes particles to be only very weakly deflected with an angle of less than $\pm 1/\gamma$ and consequently the transverse motion of particles is nonrelativistic. In this picture the electron motion viewed from the laboratory along the beam axis appears as a purely sinusoidal transverse oscillation similar to that in a linear radio antenna and the radiation emitted is therefore monochromatic with a period equal to the oscillation period.

In another equally valid view the static and periodic magnetic undulator field appears in the rest frame of the electron as a Lorentz contracted electromagnetic field or as monochromatic photons of wavelength $\lambda^* = \lambda_p/\gamma$. The emission of photons can therefore be described as Thomson scattering of photons by free electrons [9.35] and resulting in monochromatic radiation in the direction of the particle path.

Viewed from the laboratory system the radiation is Doppler shifted and applying (9.11) we get the wave length of the backscattered photons

$$\lambda_\gamma = \frac{\lambda_\mathrm{p}}{\gamma^2 \left(1 + \beta n_z^*\right)}. \tag{9.90}$$

Viewing the radiation parallel to the forward direction (9.12) becomes with $\Theta^* \ll 1$, $n_z = \cos \Theta^* \approx 1 - \tfrac{1}{2}\Theta^{*2}$ and $\beta \approx 1$

$$1 + \beta n_z^* = \frac{\beta + n_z^*}{n_z} \approx \frac{2 - \tfrac{1}{2}\Theta^{*2}}{n_z}. \tag{9.91}$$

Since $n_z \approx 1$ the wavelength of the emitted radiation is

$$\lambda_\gamma = \frac{\lambda_\mathrm{p}}{\gamma^2} \frac{1}{2 - \tfrac{1}{2}\Theta^{*2}} \approx \frac{\lambda_\mathrm{p}}{2\gamma^2}(1 + \tfrac{1}{4}\Theta^{*2}).$$

Transforming with (9.13) the angle Θ^* of the particle trajectory with respect to the direction of observation to the laboratory system, $\Theta^* = 2\gamma\Theta$ we distinguish two cases. One case, $\Theta = \mathrm{const}$, describes the particle motion in a *helical undulator*, where the magnetic field being normal to the undulator axis rotates about this axis. The more common case is that of a *flat undulator*, where the particle motion follows a sinusoidal path in which case we set $\langle \Theta^2 \rangle = \tfrac{1}{2}\vartheta^2$. Depending on the type of undulator we get with the undulator strength parameter $K = \gamma\vartheta$ for the wave length of radiation from a helical undulator

$$\lambda_\gamma = \frac{\lambda_\mathrm{p}}{2\gamma^2}(1 + K^2) \tag{9.92}$$

and for a flat undulator

$$\lambda_\gamma = \frac{\lambda_\mathrm{p}}{2\gamma^2}(1 + \tfrac{1}{2}K^2). \tag{9.93}$$

In more practical units (9.93) becomes

$$\lambda_\gamma(\text{Å}) = 13.056 \cdot \frac{\lambda_\mathrm{p}(\mathrm{cm})}{E^2(\mathrm{GeV}^2)}(1 + \tfrac{1}{2}K^2) \tag{9.94}$$

or in terms of photon energy

$$\epsilon_\gamma(\mathrm{eV}) = 950 \cdot \frac{E^2(\mathrm{GeV}^2)}{\lambda_\mathrm{p}(\mathrm{cm})(1 + \tfrac{1}{2}K^2)}. \tag{9.95}$$

The expressions for the photon wavelength and energy for undulator radiation derived so far are true for the first harmonic of radiation or *fundamental radiation* which is the only radiation for $K \ll 1$. As the undulator parameter increases, however, the oscillatory motion of the particle in the undulator deviates from a pure sinusoidal oscillation. For $K > 1$ the transverse motion becomes relativistic, causing a deformation of the sinusoidal motion and the creation of higher harmonics. These harmonics appear at integral multiples of the fundamental radiation energy. Only odd harmonics are emitted in the

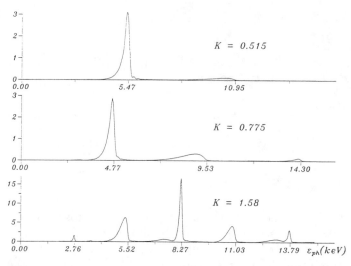

Fig. 9.17. Measured radiation spectrum from an undulator for different strength parameters K [9.36]. The intensity at low photon energies are reduced by absorption in a BE window

forward direction, $\theta \approx 0$, while the even harmonics are emitted into a small angle to the axis as will be discussed in more detail in Vol. II.

The electron motion through the undulator with N periods includes N oscillations and so does the radiation. Consequently, the line width of the radiation is

$$\frac{\Delta\lambda}{\lambda} = \frac{1}{N}. \tag{9.96}$$

Typical measured undulator spectra are shown in Fig. 9.17 for increasing undulator strength [9.36].

For $K < 1$ we note only one line in the radiation spectrum with more appearing as K is increased. As the undulator strength approaches and exceeds unity the transverse motion of the electrons becomes relativistic distorting the purely sinusoidal motion. Higher harmonics of the oscillatory motion give cause for the appearance of higher harmonics in the radiation spectrum. As the undulator strength is further increased more and more harmonics appear, each of them growing wider and finally merging into the well-known broad spectrum of a bending or wiggler magnet.

The concentration of all radiation into one or few lines is very desirable for many experiments utilizing monochromatic photon beams since radiation is produced only in the vicinity of the desired wavelength at high brightness. Radiation at other wavelengths creating undesired heating effects on optical elements and samples is greatly eliminated.

332

9.4 Back Scattered Photons

The principle of Thomson backscattering of the static undulator fields can be expanded to that of photon beams colliding head on with the particle beam. In the electron system of reference the electromagnetic field of this photon beam looks fundamentally no different than the electromagnetic field from the undulator magnet. We may therefore apply similar arguments to determine the wavelength of back scattered photons. The basic difference of both effects is that in the case of back scattered photons the photon beam moves with the velocity of light towards the electron beam and therefore the electron *sees* twice the Lorentz contracted photon frequency and expect therefore a beam of back scattered photons which is twice Doppler shifted. That extra factor of two does not apply for undulator radiation since the undulator field is static and the relative velocity with respect to the electron beam is c. If λ_p is the wavelength of the incident radiation, the wavelength of the backscattered photons is

$$\lambda_\gamma = \frac{\lambda_p}{4\gamma^2}\left(1 + \tfrac{1}{2}\gamma^2\vartheta^2\right), \tag{9.97}$$

where ϑ is the angle between the direction of observation and the particle beam axis. Scattering for example a high intensity laser beam from high-energy electrons produces a monochromatic beam of hard x-rays which is highly collimated within an angle of $\pm 1/\gamma$. If $\lambda_p = 10\mu m$ and the particle energy is 100 MeV the wave length of the backscattered x-rays would be 1.3 Å or the photon energy would be 9.5 keV.

9.4.1 Radiation Intensity

The intensity of the backscattered photons can be calculated in a simple way utilizing the Thomson scattering cross section [9.35]

$$\sigma_{Th} = \frac{8\pi}{3}r^2 = 6.65 \times 10^{-25}\,\text{cm}^2. \tag{9.98}$$

The total scattering event rate or the number of back scattered photons per unit time is then

$$N_{sc} = \sigma_{Th}\,\mathcal{L}, \tag{9.99}$$

where $\mathcal{L}$ is called the *luminosity*. The value of the luminosity is independent of the nature of the physical reaction and depends only on the intensities and geometrical dimensions of the colliding beams.

The number of backscattered photons or the luminosity can be determined by folding the particle density in one beam with the incident "particles" per unit time of the other beam. Obviously only those parts of the beam cross sections count which overlaps with the cross section of the other beam. For simplicity we assume a gaussian distribution in both beams and assume that both beam cross sections are the same. We further consider the particle beam as the target for the photon beam. With N_e electrons in each bunch of the particle beam within a cross section of $4\pi\sigma_x\sigma_y$ the particle density is $N_e/(4\pi\sigma_x\sigma_y)$.

We consider now for simplicity a photon beam with the same time structure as the electron beam. If this is not the case only that part of the photon beam which actually collides with the particle beam within the collision zone may be considered. For an effective photon flux $\dot{N}_{\mathrm{ph}}$ the luminosity is

$$\mathcal{L} = \frac{N_e\,\dot{N}_{\mathrm{ph}}}{4\pi\,\sigma_x\,\sigma_y}. \tag{9.100}$$

Although the Thomson cross section and therefore the photon yield is very small this technique can be used to produce photon beams with very specific characteristics. By analyzing the scattering distribution this procedure can also be used to determine the degree of polarization of an electron beam in a storage ring.

So far it was assumed that the incident and scattered photon energies are much smaller than the particle energy in which case it was appropriate to use the classical case of Thomson scattering. However, we note that equation (9.97) diverges in the sense that shorter and shorter wavelength radiation can be produced if only the particle energy is increased more and more. The energy of the backscattered photons, on the other hand, increases quadratically with the particle energy and therefore at some energy the photon energy becomes larger than the particle energy which is unphysical. In case of large photon energies with respect to the particle energy, the Compton corrections must be included [9.37–39].

The Compton cross section is given by [9.40]

$$\sigma_{\mathrm{C}} = \frac{3\sigma_{\mathrm{Th}}}{4x}\left[\left(1 - \frac{4}{x} - \frac{8}{x^2}\right)\ln(1+x) + \tfrac{1}{2} + \frac{8}{x^2} - \frac{1}{2(1+x)^2}\right], \tag{9.101}$$

where $x = \{4\gamma\hbar\omega_0/(mc^2)\}\cos^2(\alpha_0/2)$, $\gamma = E/(mc^2)$ the particle energy in units of the rest energy $\hbar\omega_0$ the incident photon energy and α_0 the angle between particle motion and photon direction with $\alpha_0 = 0$ for head on collision. The energy spectrum of the scattered photons is then [9.40]

$$\frac{d\sigma_C}{dy} = \frac{3\sigma_T h}{4x}\left\{1 - y + \frac{1}{1-y} - \frac{4y}{x(1+y)} + \frac{4y^2}{x^2(1-y)^2}\right\}, \tag{9.102}$$

where $y = \hbar\omega/E \leq y_{\max}$ is the scattered photon energy in units of the particle energy and $y_{\max} = x/(1+x)$ is the maximum energy of the scattered photons.

Problems

Problem 9.1. Consider an electron storage ring at an energy of 800 MeV, a circulating current of 1 amp and a bending radius of $\rho = 1.784$ m. Calculate the energy loss per turn, the critical energy and the total synchrotron radiation power. Plot the radiation spectrum and determine the frequency range available for experimentation. At what frequency in units of the critical frequency has the intensity dropped to 1 % of the maximum? What beam energy would be required to produce x-rays from this ring at a critical photon energy of 10 keV? Is that feasible or would the ring have to be larger? What would the new bending radius have to be?

Problem 9.2. The Superconducting Super Collider, SSC, is a proton accelerator up to energies of 20 TeV. The circumference is 68 km and the bending magnet fields are $B = 6.6$ Tesla. Calculate the energy loss per turn due to synchrotron radiation and the critical photon energy. What is the synchrotron radiation power for a circulating beam of 73 ma? Calculate the spatial radiation distribution for photons with $\epsilon_{\rm ph} = \epsilon_{\rm cr}$ in the vertical plane and determine the typical opening angle in units of $1/\gamma$. Calculate the longitudinal and transverse damping times.

Problem 9.3. Derive the equations of transformation for the frequency (9.11), and the direction of observations (9.12), from (9.9).

Problem 9.4. Specify main parameters for a synchrotron radiation source for digital subtractive angiography. In this medical procedure two hard x-ray beams are selected from monochromators, one just below and the other just above the K-edge of iodine. With each beam an x-ray picture of, for example, the human heart with peripheral arteries is taken, while the blood stream contains some iodine. Both pictures differ only where there is iodine because of the very different absorption coefficient for both x-ray beams and displaying the difference of both pictures shows the blood vessels alone. Select parameters for beam energy, wiggler magnet field, number of poles and beam current to produce an x-ray beam at 33keV of 2.0×10^{14} photons/sec/0.1%BW into an opening angle of 25 mrad, while keeping the 66keV and 99keV contamination to less than 1% of the 33keV radiation.

Problem 9.5. Assume a proton storage ring in space surrounding the earth at an average radius of 22,000 km. Further assume that 1 MW of rf power is available to compensate for synchrotron radiation. What is the maximum proton energy that can be reached with permanent magnets producing a maximum field of 5 kG? Is the energy limited by the maximum magnetic field or synchrotron radiation losses? Calculate the energy loss per turn, critical photon energy, and the total radiation power. Could you observe synchrotron radiation from this ring at the critical energy on earth? What are the answers in case of electrons?

Problem 9.6. Consider the storage ring of problem 9.5 and store a beam of muons at the maximum possible energy as limited by rf power or magnet field. What is the muon beam lifetime if we consider an energy acceptance of the storage ring of ±3 % ?

Problem 9.7. Consider an electron beam from an S-band linear accelerator operating at 3 GHz and producing bunches with $2 \cdot 10^8$ electrons per microbunch. The micro bunches have been compressed to an rms bunch length of 20 f-sec. Such bunches passing through a magnetic field can emit coherent radiation. Determine the optimum magnetic field to create a photon beam most concentrated into the forward direction. What is the magnetic field, the critical photon energy and the opening angle of the photon radiation? Calculate for a few photon energies the photon flux per micro bunch.

Problem 9.8. The photons created in problem 9.7 are now reflected back again to collide with a later electron bunch. We assume the electron and photon beam to have an rms beam diameter of 0.5 mm at the collision point. Calculate the backscattered photon flux if the electron beam consists of 5 μ-sec macro pulses at a repetition rate of 60 Hz.

10 Particle Beam Parameters

Particle beams are characterized by a set of quantifying parameters being either constants of motion or functions varying from point to point along a beam transport line. The parameters may be a single particle property like the betatron function which applies to all particles within a beam or quantities that are defined only for a collection of particles like the beam emittances or beam intensity. We will define and derive expressions for such beam parameters and use them to characterize particle beams and develop methods for manipulation of such parameters.

10.1 Definition of Beam Parameters

Particle beams and individual particles are characterized by a number of parameters which we use in beam dynamics. We will define such parameters first before we discuss the determination of their numerical value.

10.1.1 Beam Energy

Often we refer to the energy of a particle beam although we actually describe only the nominal energy of a single particle within this beam. Similarly, we speak of the beam momentum, beam kinetic energy or the velocity of the beam, when we mean to say that the beam is composed of particles with nominal values of these quantities. We found in earlier chapters that the most convenient quantity to characterize the "energy" of a particle is the momentum for transverse beam dynamics and the kinetic energy for acceleration. To unify the nomenclature it has become common to use the term *energy* for both quantities noting that the quantity of pure momentum should be multiplied with the velocity of light, cp, to become dimensionally correct. Thus, the particle momentum is expressed in the dimension of an energy without being numerically identical either to the total energy or the kinetic energy but approaching both for highly relativistic energies.

10.1.2 Time Structure

A true collective beam parameter is the time structure of the particle stream. We make the distinction between a continuous beam being a continuous flow of particles and a bunched beam. Whenever particles are accelerated by means of rf fields a *bunched beam* is generated, while continuous beams can in general be sustained only by dc accelerating fields or when no acceleration is required as is true for a proton beam in a storage ring. A *pulsed beam* consists of a finite number of bunches or a continuous stream of particles for a finite length of time. For example, a beam pulse from a linear accelerator is made up of a string of *micro bunches* generated by rf accelerating fields.

10.1.3 Beam Current

The beam intensity or beam current is expressed in terms of an electrical current using the common definition of the ratio of the electrical charge passing by a current monitor per unit time. For bunched beams the time span during which the charge is measured can be either shorter than the duration of the bunch or the beam pulse or may be long compared to both. Depending on which time scale we use, we define the *bunch current* or *peak current*, the *pulse current* or the *average current* respectively.

In Fig. 10.1 the general time structure of bunched beams is shown. The smallest unit is the *microbunch* which is separated to the next microbunch by the wavelength of the accelerating rf field or a multiple thereof. The microbunch current or *peak current* $\hat{I}$ is defined as the total microbunch charge q divided by the microbunch duration τ_μ,

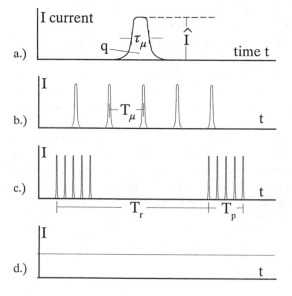

Fig. 10.1. Definitions for time structure and pulse currents.
a) peak current, $\hat{I} = q/\tau_\mu$, where τ_μ is the microbunch duration and q the charge per microbunch
b) pulse current
$I_p = \hat{I}\tau_\mu/T_\mu = q/T_\mu$, where T_μ is the microbunch period.
c) average current
$\langle I \rangle = I_p T_p/T_r$ with T_p the pulse duration and $1/T_r$ the *pulse repetition rate*.
d) continuous beam current

$$\hat{I} = \frac{q}{\tau_\mu}.\tag{10.1}$$

The micro pulse duration must be specially defined to take a nonuniform charge distribution of the particular accelerator into account. A series of microbunches form a *beam pulse* which is generally determined by the duration of the rf pulse. In a conventional S-band electron linear accelerator the rf pulse duration is of the order of a few micro seconds while a superconducting linac can produce a continuous stream of microbunches thus eliminating the pulse structure of the beam. An electrostatic accelerator may produce pulsed beams if the accelerating voltage is applied only for short time intervals. The *pulse current* I_p is defined as the average current during the duration of the pulse. If the duration of the micro bunch is τ_μ and the time between successive microbunches T_μ the pulse current is

$$I_p = \hat{I}\frac{\tau_\mu}{T_\mu} = \frac{q}{T_\mu}.\tag{10.2}$$

The average beam current, finally, is the beam current averaged over a complete cycle of the particular accelerator.

$$\langle I \rangle = I_p\frac{T_p}{T_r} = \frac{q}{T_r}\frac{T_p}{T_\mu} = \frac{n_\mu q}{T_r},\tag{10.3}$$

where n_μ is the number of microbunches per pulse and q the charge in a microbunch. In a beam transport line this is the total charge passing by per unit time, where the unit time is at least as long as the distance between beam pulses. In a circular accelerator it is the total circulating charge divided by the revolution time. For the experimenter using particles from a cycling synchrotron accelerator the average current is the total charge delivered to the experiment during a time long compared to the cycling time divided by that time.

The "beam on – beam off" time is measured by the *duty factor* defined as the fraction of actual beam time to total time at the experimental station. Depending on the application, it is desirable to have a high duty factor where the particles come more uniformly distributed in time compared to a low duty factor where the same number of particles come in short bursts.

10.1.4 Beam Dimensions

Of great importance for the design of particle accelerators is the knowledge of beam size parameters like transverse dimensions, bunch length and energy spread as well as the particle intensity distribution in six-dimensional phase space. In this respect electron beams behave in general different from beams of heavier particles like protons which is a consequence of the synchrotron radiation and the effects of quantized emission of photons on the dynamic parameters of the electrons. Where such radiation effects are negligible beams of any kind of particles evolve the same way along a beam line. Specifically we have seen that in such cases the beam emittances are a constant of motion and the beam sizes are therefore modulated only by the variation of the betatron and dispersion functions as determined by the focusing structure. The particle distribution stays constant while rotating in phase space. This is true for the transverse as well as for the longitudinal and energy parameters.

A linear variation of beam emittances with energy is introduced when particles are accelerated or decelerated. We call this variation *adiabatic damping*, where the beam emittances scale inversely proportional with the particle momentum and the transverse beam sizes, divergences, bunch length and energy spread scale inversely to the square root of the particle momentum. This adiabatic damping actually is not a true damping process where the area in phase space is reduced. It rather reflects the particular definition of beam emittances with respect to the canonical dimensions of phase space. In transverse beam dynamics, for example, we are concerned with geometric parameters and a phase space element would be expressed by the product $\Delta u \Delta u'$. *Liouville's theorem*, however, requires the use of canonical variables, momentum and position, and the same phase space element is $\Delta u \Delta p_u$, where $\Delta p_u = p_0 u'$ and u is any of the three degrees of freedom. Acceleration increases the particle momentum p_0 and as a consequence the geometric emittances $\Delta u \Delta u'$ must be reduced to keep the product $\Delta u \Delta p_u$ constant. This reduction of the geometric emittance by acceleration is called adiabatic damping and occurs in all three degrees of freedom.

More consistent with Liouville's theorem of constant phase space density is the *normalized emittance* defined by

$$\epsilon_n = \beta \gamma \epsilon, \tag{10.4}$$

where γ is the particle energy in units of the rest energy and $\beta = v/c$. This normalized emittance obviously has the appropriate definition to stay constant under the theorem of Liouville.

It is often difficult and not practical to define a beam emittance for the whole beam. Whenever the beam is fuzzy at the edges it may not make sense to include all particles into the definition of the beam emittance and provide expensive aperture for the fuzzy part of the beam. In such cases one might define the beam emittance as 95% of the total beam intensity or whatever seems appropriate. Relativistic electron beams in circular accelerators are particularly fuzzy due to the quantized emission of synchrotron radiation. As a consequence, the particle distribution assumes a gaussian distribution. In Vol. II we will discuss the evolution of the beam emittance due to statistical effects in great detail and derive the particle distribution from the Fokker-Planck equation. In this case we define the beam emittance for that part of the beam which is contained within one standard unit of the gaussian distribution.

The beam emittance for particle beams is primarily defined by the characteristic source parameters and source energy. Given perfect matching between different accelerators and beam lines during subsequent acceleration, this source emittance is reduced inversely proportional to the particle momentum by adiabatic damping and stays constant in terms of normalized emittance. This describes accurately the ideal situation for proton and ion beams, for nonrelativistic electrons and electrons in linear accelerators as long as statistical effects are absent. A variation of the emittance occurs in the presence of statistical effects in the form of collisions with other particles or emission of synchrotron radiation and we will concentrate here in more detail on the evolution of beam emittances in highly relativistic electron beams.

Statistical processes cause a spreading of particles in phase space or a continuous increase of beam emittance. In cases where this diffusion is due to the particle density, the emittance increase may stop for all practical purposes because the scattering occurrence drops to lower and lower values as the beam emittance increases. Such a case appears in *intrabeam scattering* [10.1– 3], where particles within the same bunch collide and exchange energy. Specifically when particles exchange longitudinal momentum into transverse momentum and gain back the lost longitudinal momentum from the accelerating cavities. The beam "heats" up transversely which becomes evident in the increased beam emittance and beam sizes.

Statistical perturbations due to synchrotron radiation, however, lead to truly equilibrium states where the continuous excitation due to quantized emission of photons is compensated by damping. Discussing first the effect of damping will prepare us to combine the results with statistical perturbations leading to an equilibrium state of the beam dimensions.

10.2 Damping

Emission of synchrotron radiation causes the appearance of a reaction force on the emitting particle which must be taken into account to accurately describe particle dynamics. In doing so we note from the theory of synchrotron radiation that the energy lost into synchrotron radiation is lost through the emission of many photons and on average we may assume that the energy loss is continuous. Specifically we assume that single photon emissions occur fast compared to the oscillation period of the particle such that we may treat the effect of the recoil force as an impulse.

In general we must consider the motion of a particle in all three degrees of freedom or in six-dimensional phase space. The appearance of damping stems from the emission of synchrotron radiation in general, but the physics of damping in the longitudinal degree of freedom is different from that in the transverse degrees of freedom. The rate of energy loss into synchrotron radiation depends on the particle energy itself and is high at high energies and low at low energies. As a consequence, a particle with a higher than ideal energy will loose more energy to synchrotron radiation than the ideal particle and a particle with lower energy will loose less energy. The combined result is that the energy difference between such three particles has been reduced, an effect that shows up as damping of the beam energy spread. With the damping of the energy spread we observe also a damping of its conjugate variable, the phase.

In the transverse plane we note that the emission of a photon leads to a loss of longitudinal as well as transverse momentum since the particle performs betatron oscillations. The total lost momentum is, however, replaced in the cavity only in the longitudinal direction. Consequently, the combined effect of emission of a photon and the replacement of the lost energy in accelerating cavities leads to a net loss of transverse momentum or transverse damping.

Although damping mechanisms are different for transverse and longitudinal degrees of freedom, the total amount of damping is limited and determined by the amount of synchrotron radiation. This correlation of damping decrements in all degrees of freedom was derived first by Robinson [10.4] for general accelerating fields as long as they are not so strong that they would appreciably affect the particle orbit.

10.2.1 Robinson Criterion

Following Robinson's idea we will derive what is now known as *Robinson's damping criterion* by observing the change of a six dimensional vector in phase space due to synchrotron radiation and acceleration. The components of this vector are the four transverse coordinates (x, x', y, y'), the energy deviation ΔE, and the longitudinal phase deviation from the synchronous phase $\varphi = \psi - \psi_s$. Consistent with *smooth approximation* a continuous distribution of synchrotron radiation along the orbit is assumed as well as continuous acceleration to compensate energy losses. During the short time dt the six-dimensional vector

$$
\mathbf{u} = \begin{bmatrix} x \\ x' \\ y \\ y' \\ \varphi \\ \Delta E \end{bmatrix}
\tag{10.5}
$$

will change by an amount proportional to dt. We may expand the transformations into a Taylor series keeping only linear terms and express the change of the phase space vector in form of a matrix transformation

$$
\Delta \mathbf{u} = \mathbf{u}_1 - \mathbf{u}_o = dt \, \boldsymbol{M} \, \mathbf{u}_o .
\tag{10.6}
$$

From the eigenvalue equation for this transformation matrix,

$$\mathcal{M}\mathbf{u}_j = \lambda_j \mathbf{u}_j \,,$$

where $\mathbf{u}_j$ are the eigenvectors, λ_i the eigenvalues being the roots of the characteristic equation $\det(\mathcal{M} - \lambda\mathcal{I})$ and $\mathcal{I}$ the unity matrix. From (10.6) we get

$$\mathbf{u}_1 = (1 + \mathcal{M}\,\mathrm{d}t)\mathbf{u}_o = (1 + \lambda_j\,\mathrm{d}t)\mathbf{u}_o \approx \mathbf{u}_o e^{\lambda_j\,\mathrm{d}t}\,. \tag{10.7}$$

Since the eigenvectors must be real the eigenvalues come in conjugate complex pairs

$$\lambda_j = \alpha_\kappa \pm \mathrm{i}\beta_\kappa \,,$$

where $\kappa = 1, 2, 3$ and

$$\sum_{j=1}^{j=6} \lambda_j = 2 \sum_{\kappa=1}^{\kappa=3} \alpha_\kappa \,. \tag{10.8}$$

The quantities α_κ cause damping or excitation of the eigenvectors depending on whether they are negative or positive, while the β_κ contribute only a frequency shift of the oscillations.

Utilizing the transformation matrix $\mathcal{M}$ we derive expressions for the eigenvalues by evaluating the expression $\frac{\mathrm{d}}{\mathrm{d}\tau}\det(\tau\mathcal{M} - \lambda\mathcal{I})|_{\tau=0}$ in two different ways. With $\mathcal{M} = \lambda_j\mathcal{I}$ we get

$$\frac{\mathrm{d}}{\mathrm{d}\tau}\det\left[(\tau\lambda_i - \lambda)\mathcal{I}\right] = \frac{\mathrm{d}}{\mathrm{d}\tau}\prod_{j=1}^{j=6}(\tau\lambda_j - \lambda)|_{\tau=0} = -\lambda^5 \sum_{j=1}^{j=6}\lambda_j \,. \tag{10.9}$$

On the other hand we may execute the differentiation on the determinate directly and get

$$\frac{\mathrm{d}}{\mathrm{d}\tau}\det(\tau\mathcal{M} - \lambda\mathcal{I})|_{\tau=0}$$

$$= \begin{vmatrix} m_{11} & m_{12} & m_{13} & \cdots \\ \tau m_{21} & \tau m_{22} - \lambda & \tau m_{23} & \cdots \\ \tau m_{31} & \tau m_{32} & \tau m_{33} - \lambda & \cdots \\ \cdots & \cdots & \cdots & \cdots \end{vmatrix}_{\tau=0} +$$

$$+ \begin{vmatrix} \tau m_{11} - \lambda & \tau m_{11} & \tau m_{13} & \cdots \\ m_{21} & m_{22} & m_{23} & \cdots \\ \tau m_{31} & \tau m_{32} & \tau m_{33} - \lambda & \cdots \\ \cdots & \cdots & \cdots & \cdots \end{vmatrix}_{\tau=0} + \cdots \tag{10.10}$$

$$= -\lambda^5 m_{11} \ldots - \lambda^5 m_{66} = -\lambda^5 \sum_{j-1}^{j=6} m_{jj} \,.$$

Comparing (10.9, 10) we note the relation

$$\sum_{j-1}^{j=6} \lambda_j = \sum_{j-1}^{j=6} m_{jj} = 2 \sum_{\kappa=1}^{\kappa=3} \alpha_\kappa \qquad (10.11)$$

between eigenvalues, matrix elements, and damping decrements. To further identify the damping we must determine the transformation. The elements m_{11}, m_{33}, and m_{55} are all zero because the particle positions (x, y, φ) are not changed by the emission of a photon or by acceleration during the time dt:

$$m_{11} = 0 \qquad m_{33} = 0 \qquad m_{55} = 0. \qquad (10.12)$$

The slopes, however, will change. Since synchrotron radiation is emitted in the forward direction we have no direct change of the particle trajectory due to the emission process. We ignore at this point the effects of a finite radiation opening angle, $\theta = \pm 1/\gamma$, and show in connection with the derivation of the vertical beam emittance that this effect is negligible while determining damping. Acceleration will change the particle direction because the longitudinal momentum is increased while the transverse momentum stays constant, see Fig. 10.2.

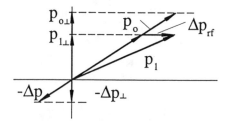

Fig. 10.2. Reduction of the slope of trajectories by acceleration. For simplicity we assume here that the energy loss $-\Delta p_\gamma$ due to the emission of a photon is immediately replaced by acceleration

As shown in Fig. 10.2 a particle with a total momentum p_o and a transverse momentum $p_{o\perp}$ due to betatron oscillation emits a photon of energy ε_γ. This process leads to a loss of momentum of $-\Delta p = \varepsilon_\gamma/\beta$ and a loss of transverse momentum of $-\Delta p_\perp$. Acceleration will again compensate for this energy loss. During acceleration the momentum is increased by $\Delta p_{rf} = +(P_{rf}/c\beta)\,dt$, where P_{rf} is the rf power to the beam. The transverse momentum during this acceleration is not changed and we have therefore $(p_o - \Delta p)u_o' = [p_o - \Delta p + (P_{rf}/c\beta)\,dt]u_1'$, where u_o' and u_1' is the slope of the particle trajectory before and after acceleration, respectively. With $u' = \dot{u}/\beta c$ and $cp_o = \beta E_o$ we have to first order in Δp and $P_{rf}\,dt$

344

$$\dot{u}_1 = \frac{E_o}{E_o + P_{rf}\,dt}\,\dot{u}_o \approx \left(1 - \frac{P_{rf}\,dt}{E_o}\right)\dot{u}_o. \tag{10.13}$$

From (10.7) we get with (10.13) and using average values for the synchrotron radiation power around the ring with $u = x$ or $u = y$

$$m_{22} = -\frac{\langle P_{\gamma o}\rangle}{E_o} \qquad \text{and} \qquad m_{44} = -\frac{\langle P_{\gamma o}\rangle}{E_o}, \tag{10.14}$$

where we note that the rf power is equal to the nominal synchrotron radiation power $\langle P_{\gamma o}\rangle = U_o/T_o$. The energy variation of the particle is the combination of the energy loss $-P_\gamma\,dt$ and energy gain $P_{rf}\,dt$. With

$$P_\gamma(E) = P_\gamma(E_o) + \left.\frac{\partial P_\gamma}{\partial E}\right|_o \Delta E_o \quad \text{and} \quad P_{rf}(\psi) = P_{rf}(\psi_s) + \left.\frac{\partial P_{rf}}{\partial \psi}\right|_{\psi_s}\varphi,$$

where $\varphi = \psi - \psi_s$ we get

$$\begin{aligned}
\Delta E_1 &= \Delta E_o - P_\gamma(E)\,dt + P_{rf}(\psi)\,dt \\
&= \Delta E_o - \left.\frac{\partial P_\gamma}{\partial E}\right|_o \Delta E_o\,dt + \left.\frac{\partial P_{rf}}{\partial \psi}\right|_{\psi_s}\varphi\,dt
\end{aligned} \tag{10.15}$$

because $P_\gamma(E_o) = P_{rf}(\psi_s)$. Equation (10.15) exhibits two more elements of the transformation matrix

$$m_{65} = \left.\frac{\partial P_{rf}}{\partial \psi}\right|_{\psi_s} \qquad \text{and} \qquad m_{66} = -\left.\frac{\partial P_\gamma}{\partial E}\right|_o. \tag{10.16}$$

We have now all elements necessary to determine the damping decrements. From (10.12,14,16) we get for the sum of the damping decrements

$$\sum_{\kappa=1}^{\kappa=3}\alpha_\kappa = \tfrac{1}{2}\sum_{j=1}^{j=6}m_{jj} = -\frac{\langle P_{\gamma o}\rangle}{E_o} - \tfrac{1}{2}\left.\frac{\partial P_\gamma}{\partial E}\right|_o, \tag{10.17}$$

which depends only on the synchrotron radiation power and the particle energy. This result was first derived by Robinson [10.4] and is known as *Robinson's damping criterion* .

We may separate the damping decrements. For a plane circular accelerator without vertical bending magnets and coupling, the vertical damping decrement $\alpha_v = \alpha_2$ can be extracted. Since the vertical motion is not coupled to either the horizontal or the synchrotron oscillations we get from (10.14, 17)

$$\alpha_v = -\tfrac{1}{2}\frac{\langle P_{\gamma o}\rangle}{E_o}. \tag{10.18}$$

The damping decrement for synchrotron oscillations has been derived earlier in Chap. 8 to be from (8.29) with $U/T_o = \langle P_\gamma\rangle$

$$\alpha_s = -\frac{1}{2} \frac{\mathrm{d}\langle P_\gamma \rangle}{\mathrm{d}E}\bigg|_{\mathrm{o}}.$$ (10.19)

The horizontal damping decrement finally can be derived from Robinson's damping criterion (10.17) and the two other decrements (10.18, 19) to be

$$\alpha_\mathrm{h} = -\frac{1}{2} \frac{\langle P_\gamma \rangle}{E_\mathrm{o}} - \frac{1}{2} \frac{\partial P_\gamma}{\partial E}\bigg|_{\mathrm{o}} + \frac{1}{2} \frac{\mathrm{d}\langle P_\gamma \rangle}{\mathrm{d}E}\bigg|_{\mathrm{o}}.$$ (10.20)

We may further evaluate the total and partial differential of the synchrotron radiation power P_γ with energy E. From $P_\gamma \propto E^2 B^2$ we get for the partial differential

$$\frac{\partial P_\gamma}{\partial E}\bigg|_{\mathrm{o}} = 2 \frac{\langle P_{\gamma\mathrm{o}} \rangle}{E_\mathrm{o}}.$$ (10.21)

The total differential of the synchrotron radiation power depends not only on the particle energy directly but also on the variation of the magnetic field with energy as seen by the particle. A change in the particle energy causes a shift in the particle orbit where the dispersion function is nonzero and this shift may move the particle to a location with different field strength. To include all energy dependent contributions we inspect the definition

$$\langle P_\gamma \rangle = \frac{1}{T_\mathrm{o}} \oint P_\gamma \mathrm{d}\tau$$

and note that

$$c\,\mathrm{d}\tau = \mathrm{d}\sigma = \left(1 + \frac{\eta}{\rho} \frac{\Delta E}{E_\mathrm{o}}\right) \mathrm{d}s$$

and therefore

$$\langle P_\gamma \rangle = \frac{1}{cT_\mathrm{o}} \oint P_\gamma \left(1 + \frac{\eta}{\rho} \frac{\Delta E}{E_\mathrm{o}}\right) \mathrm{d}s.$$ (10.22)

Differentiating (10.22) with respect to the energy we get

$$\frac{\mathrm{d}\langle P_\gamma \rangle}{\mathrm{d}E}\bigg|_{\mathrm{o}} = \frac{1}{cT_\mathrm{o}} \oint \left[\frac{\mathrm{d}P_\gamma}{\mathrm{d}E}\bigg|_{\mathrm{o}} + P_{\gamma\mathrm{o}} \frac{\eta}{\rho E_\mathrm{o}}\right] \mathrm{d}s,$$ (10.23)

where

$$\frac{\mathrm{d}P_\gamma}{\mathrm{d}E}\bigg|_{\mathrm{o}} = 2\frac{P_{\gamma\mathrm{o}}}{E_\mathrm{o}} + 2\frac{P_{\gamma\mathrm{o}}}{B_\mathrm{o}} \frac{\mathrm{d}B}{\mathrm{d}x} \frac{\mathrm{d}x}{\mathrm{d}E} = 2\frac{P_{\gamma\mathrm{o}}}{E_\mathrm{o}} + 2\frac{P_{\gamma\mathrm{o}}}{E_\mathrm{o}} \rho\, k\, \eta.$$

Collecting all components we get finally for the synchrotron oscillation damping decrement (10.19)

$$\alpha_s = -\frac{1}{2} \frac{\mathrm{d}P_\gamma}{\mathrm{d}E}\bigg|_{\mathrm{o}} = -\frac{1}{2} \frac{\langle P_{\gamma\mathrm{o}} \rangle}{E_\mathrm{o}} (2 + \vartheta),$$ (10.24)

where we used $\langle P_{\gamma o} \rangle \propto \oint \mathrm{d}s/\rho^2$ and $P_{\gamma o} \propto 1/\rho^2$ for

$$\vartheta = \frac{\oint \frac{\eta}{\rho^3}\left(1 + 2\,\rho^2\,k\right)\,\mathrm{d}s}{\oint \mathrm{d}s/\rho^2}. \tag{10.25}$$

Similarly we get from (10.20) for the horizontal damping decrement

$$\alpha_{\mathrm{h}} = -\frac{1}{2}\frac{\langle P_{\gamma} \rangle}{E_{\mathrm{o}}}\left(1 - \vartheta\right). \tag{10.26}$$

Both the synchrotron and betatron oscillation damping can be modified by a particular choice of lattice. From (10.25) we note the contribution η/ρ^3 which is caused by sector magnets. Particles with higher energies follow a longer path in a sector magnet and therefore radiate more. Consequently synchrotron damping is increased with ϑ. This term vanishes for rectangular magnets and must be modified appropriately for wedge magnets. For a rectangular magnet

$$\vartheta_{\mathrm{rect}} = \frac{\oint 2\,(\eta\,k/\rho)\,\mathrm{d}s}{\oint \mathrm{d}s/\rho^2} \tag{10.27}$$

and for wedge magnets

$$\vartheta_{\mathrm{wedge}} = \frac{\sum_i \left[\theta_{\mathrm{o}}\eta_{\mathrm{o}}/\rho^2 + \int_i 2\,(\eta\,k/\rho)\,\mathrm{d}s + \theta_{\mathrm{e}}\eta_{\mathrm{e}}/\rho^2\right]_i}{\oint \mathrm{d}s/\rho^2}. \tag{10.28}$$

Here we add all contributions from all magnets i in the ring. The edge angles at the entrance θ_{o} and exit θ_{e} are defined to be positive going from a rectangular magnet toward a sector magnet.

The second term in the nominator of (10.25) becomes significant for *combined function* magnets and vanishes for *separated function* magnets. Specifically a strong focusing gradient, $k > 0$, combined with beam deflection, $\rho \neq 0$, can contribute significantly to ϑ. For $\vartheta = 1$ all damping in the horizontal plane is lost and turns to antidamping or excitation of betatron oscillations for $\vartheta > 1$. This occurs, for example, in older combined function synchrotrons. At low energies, however, the beam in such lattices is still stable due to strong *adiabatic damping* and only at higher energies when synchrotron radiation becomes significant will horizontal antidamping take over and dictate an upper limit to the feasibility of such accelerators.

Conversely, vertical focusing, $k < 0$, can be implemented into bending magnets such that the horizontal damping is actually increased since $\vartheta < 0$. However, there is a limit for the stability of synchrotron oscillations for $\vartheta = 2$.

In summary the damping decrements for betatron and synchrotron oscillations can be expressed by

$$\alpha_h = -\frac{1}{2}\frac{\langle P_\gamma \rangle}{E_o}(1-\vartheta) = -\frac{1}{2}\frac{\langle P_\gamma \rangle}{E_o}J_h,$$

$$\alpha_s = -\frac{1}{2}\frac{\langle P_\gamma \rangle}{E_o}(2+\vartheta) = -\frac{1}{2}\frac{\langle P_\gamma \rangle}{E_o}J_s, \tag{10.29}$$

$$\alpha_v = -\frac{1}{2}\frac{\langle P_\gamma \rangle}{E_o} \qquad\quad = -\frac{1}{2}\frac{\langle P_\gamma \rangle}{E_o}J_v,$$

where the factors J_i are the *damping partition numbers*,

$$J_h = 1-\vartheta,$$
$$J_v = 1, \tag{10.30}$$
$$J_s = 2+\vartheta,$$

and *Robinson's damping criterion* can be expressed by

$$\sum_i J_i = 4. \tag{10.31}$$

Damping can be obtained in circular electron accelerators in all degrees of freedom. In transverse motion particles oscillate in the potential created by quadrupole focusing and any finite amplitude is damped by synchrotron radiation damping. Similarly, longitudinal synchrotron oscillations are contained by a potential well created by the rf fields and the momentum compaction and finite deviations of particles in energy and phase are damped by synchrotron radiation damping. We note that the synchrotron oscillation damping is twice as strong as transverse damping. In a particular choice of lattice, damping rates can be shifted between different degrees of freedom and special care must be exercised when combined function magnets or strong sector magnets are introduced into a ring lattice.

10.3 Particle Distribution in Phase Space

The particle distribution in phase space is rarely uniform. To determine the required aperture in a particle transport system avoiding excessive losses we must, however, know the particle distribution. Proton and ion beams involve particle distributions which due to Liouville's theorem do not change along a beam transport system, except for the variation of the betatron and dispersion function. The particle distribution can therefore be determined for example by measurements of beam transmission through a slit for varying openings. If this is done at two points about 90° apart in betatron phase space, angular as well as spatial distribution can be determined.

This procedure can be applied also to electrons in a transport system. The distribution changes, however, significantly when electrons are injected into a circular accelerator. We will discuss the physics behind this violation of Liouville's theorem and determine the resulting electron distribution in phase space.

10.3.1 Equilibrium Phase Space

Relativistic electron and positron beams passing through bending magnets emit synchrotron radiation, a process that leads to quantum excitation and damping. As a result the original beam emittance at the source is completely replaced by an equilibrium emittance that is unrelated to the original source characteristics. Postponing a rigorous treatment of statistical effects to Vol. II we concentrate here on a more visual discussion of the reaction of synchrotron radiation on particle and beam parameters.

Energy Spread: Emission of photons causes primarily a change of particle energy leading to an energy spread within the beam. To evaluate the effect of quantized emission of photons on the beam energy spread, we observe particles undergoing synchrotron oscillations. A particle may have at time t_o an energy deviation

$$\Delta E = A_o \, e^{i\Omega(t-t_o)} . \tag{10.32}$$

Emission of a photon with energy ε at time t_1 causes a perturbation to (10.32) and the particle continues to undergo synchrotron oscillations with a new amplitude

$$A = A_o \, e^{i\Omega(t-t_o)} - \varepsilon \, e^{i\Omega(t-t_1)} = A_1 \, e^{i\Omega(t-t_1)} . \tag{10.33}$$

The change in oscillation amplitude due to the emission of one photon of energy ε can be derived from (10.33) by multiplying the second equality by its imaginary conjugate and we get

$$A_1^2 = A_o^2 + \varepsilon^2 - 2\varepsilon \, A_o \cos[\Omega(t_1 - t_o)] . \tag{10.34}$$

Because the times at which photon emission occurs is random we have for the average increase in oscillation amplitude due to the emission of a photon of energy ε

$$\langle \Delta A^2 \rangle = \langle A_1^2 - A_o^2 \rangle = \varepsilon^2 . \tag{10.35}$$

The rate of change in amplitude per unit time due to quantum excitation while averaging around the ring is

$$\left\langle \left. \frac{dA^2}{dt} \right|_q \right\rangle_s = \int_0^\infty \varepsilon^2 \dot{n}(\varepsilon) \, d\varepsilon = \left\langle \dot{N}_{\rm ph} \langle \varepsilon^2 \rangle \right\rangle_s , \tag{10.36}$$

where $n(\varepsilon)$ is the number of photons of energy ε emitted per unit time and energy bin $d\varepsilon$. Along the beam line this can be expressed on average by the total photon flux $\dot{N}_{\rm ph}$ multiplied by the average value of the square of the photon energy. Damping causes a reduction in the synchrotron oscillation amplitude and we get with $A = A_0 e^{-\alpha_s t}$

$$\left\langle \frac{{\rm d}A^2}{{\rm d}t}\Big|_{\rm d} \right\rangle_s = -2\alpha_s \langle A^2 \rangle . \tag{10.37}$$

Both quantum excitation and damping lead to an equilibrium state

$$\left\langle \dot{N}_{\rm ph}\langle \varepsilon^2 \rangle \right\rangle_s - 2\alpha_s \langle A^2 \rangle = 0 \tag{10.38}$$

or

$$\langle A^2 \rangle = \tfrac{1}{2}\tau_s \left\langle \dot{N}_{\rm ph}\langle \varepsilon^2 \rangle \right\rangle_s , \tag{10.39}$$

where $\tau_s = 1/\alpha_s$ is the synchrotron oscillation damping time. Due to the *central limit theorem* of statistics the energy distribution due to statistical emission of photons assumes a gaussian distribution with the standard root mean square energy spread $\sigma_\varepsilon^2 = \tfrac{1}{2}\langle A^2 \rangle$. The photon spectrum has been derived in Chap. 9 and the integral in (10.36) can be evaluated to give [10.5]

$$\dot{N}_{\rm ph}\langle \varepsilon^2 \rangle = \frac{55}{24\sqrt{3}}(P_{\gamma 0}\varepsilon_{\rm c}). \tag{10.40}$$

Replacing the synchrotron radiation power P_γ by its expression in (9.24) and the critical photon energy $\varepsilon_{\rm c} = \hbar\omega_{\rm c}$ by (9.38) we get

$$\dot{N}_{\rm ph}\langle \varepsilon^2 \rangle = \frac{55}{32\pi\sqrt{3}}\left[cC_\gamma \hbar c (mc^2)^4\right]\gamma^7 \frac{1}{\rho^3} \tag{10.41}$$

and the *equilibrium energy spread* becomes finally

$$\frac{\sigma_\varepsilon^2}{E^2} = \frac{\tau_s}{4E^2}\left\langle \dot{N}_{\rm ph}\langle \varepsilon^2 \rangle \right\rangle_s = C_{\rm q}\frac{\gamma^2}{J_s}\frac{\langle 1/\rho^3 \rangle_s}{\langle 1/\rho^2 \rangle_s} , \tag{10.40}$$

where

$$C_{\rm q} = \frac{55}{32\sqrt{3}}\frac{\hbar c}{mc^2} = 3.84 \times 10^{-13}\,{\rm m} \tag{10.43}$$

for electrons and positrons. The equilibrium energy spread in an electron storage ring depends only on the beam energy and the bending radius.

Bunch Length: The conjugate coordinate to the energy deviation is the phase and a spread of particle energy appears also as a spread in phase or as a longitudinal particle distribution and an equilibrium *bunch length*. From (8.70) we get

$$\sigma_\ell = \frac{c|\eta_{\rm c}|}{\Omega}\frac{\sigma_\varepsilon}{E_0} , \tag{10.44}$$

and replacing the synchrotron oscillation frequency by its expression (8.36) we get finally for the equilibrium bunch length in a circular electron accelerator

$$\sigma_\ell = \frac{\sqrt{2\pi}\,c}{\omega_{\mathrm{rev}}} \sqrt{\frac{\eta_c E_{\mathrm{o}}}{he\widehat{V}\cos\psi_{\mathrm{s}}}} \frac{\sigma_\varepsilon}{E_{\mathrm{o}}}. \tag{10.45}$$

We note that the equilibrium electron bunch length can be varied by varying the rf voltage and scales like $\sigma_\ell \propto 1/\sqrt{\widehat{V}}$ which is a much stronger dependence than the scaling obtained for nonradiating particles in Sect. 8.2.5. A very small bunch length can be obtained by adjusting the momentum compaction to a small value including zero. As the momentum compaction approaches zero, however, second order terms must be considered which will be discussed in detail in Vol. II. An electron storage ring where the momentum compaction is adjusted to be zero or close to zero is called an *isochronous ring* [10.6] or a *quasi isochronous ring* [10.7]. Such rings do not yet exist at this time but are intensely studied and problems are being solved in view of great benefits for research in high energy physics, synchrotron radiation sources and free electron lasers to produce short particle or light pulses.

Transverse Beam Emittance: The sudden change of particles energy due to the quantized emission of photons also causes a change in the characteristics of transverse particle motion. Neither position nor the direction of the particle trajectory is changed during the forward emission of photons along the direction of the particle propagation, ignoring for now the small transverse perturbation due to the finite opening angle of the radiation by $\pm 1/\gamma$. From beam dynamics, however, we know that different reference trajectories exist for particles with different energies. Two particles with energies cp_1 and cp_2 follow two different reference trajectories separated, at the position s along the beam transport line by a distance

$$\Delta x = \eta(s)\frac{cp_1 - cp_2}{cp_{\mathrm{o}}}, \tag{10.46}$$

where $\eta(s)$ is the dispersion function and cp_{o} the reference energy. Although particles in general do not exactly follow these reference trajectories, they do perform betatron oscillations about these trajectories. The sudden change of the particle energy during the emission of a photon consequently leads to a sudden change in the reference path and thereby to a sudden change in the betatron oscillation amplitude.

Postponing a rigorous discussion of the evolution of phase space due to statistical perturbations to Vol. II we follow here a more intuitive path to determine the equilibrium transverse beam emittance. Similar to the discussion leading to the equilibrium energy spread we will observe the perturbations of photon emission to the transverse motion. In the case of longitudinal quantum excitation it was sufficient to consider the effect of photon emission on the particle energy alone since the particle phase is not changed by this process. In the transverse planes, however, this is generally not true and we must consider the variation of both conjugate coordinates due to the emission of a photon.

As a particle emits a photon it will not change its actual position and direction. However, the position of a particle with respect to the ideal reference orbit is the combination of its betatron oscillation amplitude and a chromatic contribution due to a finite energy deviation and dispersion. Variation of the particle position $u = u_\beta + \eta (\Delta E/E_o)$, and direction $u' = u'_\beta + \eta' (\Delta E/E_o)$ due to the emission of a photon of energy ε is described by

$$
\begin{aligned}
\delta u &= 0 = \delta u_\beta + \eta \frac{\varepsilon}{E_o} &\quad \text{or} \quad& \delta u_\beta = -\eta \frac{\varepsilon}{E_o}, \\
\delta u' &= 0 = \delta u'_\beta + \eta' \frac{\varepsilon}{E_o} &\quad \text{or} \quad& \delta u'_\beta = -\eta' \frac{\varepsilon}{E_o}.
\end{aligned}
\tag{10.47}
$$

We note the sudden changes in the betatron amplitudes and slopes because the sudden energy loss leads to a simultaneous change in the equilibrium orbit. This perturbation will modify the phase ellipse the particle moved on. The variation of the phase ellipse $\gamma u^2 + 2\alpha u u' + \beta u'^2 = a^2$ is expressed by

$$
\gamma \delta(u_\beta^2) + 2\alpha \delta(u_\beta u'_\beta) + \beta \delta({u'_\beta}^2) = \delta(a^2)
$$

and inserting the relations (10.47) we get terms of the form $\delta(u_\beta^2) = (u_{\beta o} + \delta u_\beta)^2 - u_{\beta o}^2$ etc. Emission of photons can occur at any betatron phase and we therefore average over all phases. As a consequence, all terms depending linearly on the betatron amplitude and its derivatives or variations thereof vanish. The variation of the phase ellipse or oscillation amplitude a averaged over all betatron phases and due to the emission of photons with energy ε becomes

$$
\langle \delta a^2 \rangle = \frac{\varepsilon^2}{E_o^2} \mathcal{H}(s),
\tag{10.48}
$$

where

$$
\mathcal{H}(s) = \beta \eta'^2 + 2\alpha \eta \eta' + \gamma \eta^2.
\tag{10.49}
$$

We average again over all photon energies, multiply by the total number of photons emitted per unit time and integrate over the whole ring to get the variation of the oscillation amplitude per turn

$$
\Delta \langle a^2 \rangle = \frac{1}{c E_o^2} \oint \dot{N}_{\mathrm{ph}} \langle \varepsilon^2 \rangle \mathcal{H}(s) \, ds.
\tag{10.50}
$$

The rate of change of the oscillation amplitude is then from (10.50)

$$\frac{\mathrm{d}\langle a^2\rangle}{\mathrm{d}t}\bigg|_q = \frac{1}{E_o^2}\left\langle \dot{N}_{\mathrm{ph}}\langle\varepsilon^2\rangle\mathcal{H}(s)\right\rangle_s,\tag{10.51}$$

where the index s indicates averaging around the ring. This quantum excitation of the oscillation amplitude is compensated by damping for which we have similar to (10.37)

$$\left\langle\frac{\mathrm{d}a^2}{\mathrm{d}t}\right\rangle\bigg|_d = -2\alpha_{\mathrm{h}}\langle a^2\rangle.\tag{10.52}$$

Equilibrium is reached when both quantum excitation and damping are of equal strength which occurs for

$$\frac{\sigma_u^2}{\beta_u} = \frac{\tau_u}{4E^2}\left\langle \dot{N}_{\mathrm{ph}}\langle\varepsilon^2\rangle\mathcal{H}\right\rangle_s.\tag{10.53}$$

Here we have used the definition of the standard width of a gaussian particle distribution $\sigma_u^2 = \langle u^2(s)\rangle = \frac{1}{2}a^2\beta_u$ with the betatron function β_u and $u = x$ or y. With (10.29,41) and (9.24) we get

$$\frac{\sigma_u^2}{\beta_u} = C_q\frac{\gamma^2}{J_u}\frac{\langle 1/\rho^3\mathcal{H}\rangle}{\langle 1/\rho^2\rangle},\tag{10.54}$$

which we define as the *equilibrium beam emittance*,

$$\frac{\sigma_u^2}{\beta_u} = \epsilon_u,\tag{10.55}$$

of a relativistic electron in a circular accelerator.

In (10.50) we decided to integrate the quantum excitation over a complete turn of a circular accelerator. This should not be taken as a restriction but rather as an example. If we integrate along an open beam transport line we would get the increase of the beam emittance along this beam line. This becomes especially important for very high energy linear colliders where beams are transported along the linear accelerator and some beam transport system in the final focus section just ahead of the collision point. Any dipole field along the beam path contributes to an increase of the beam emittance, whether it be real dipole magnets, dipole field errors, or small correction magnets for beam steering. Since there is no damping, the emittance growth is therefore in both planes

$$\Delta\epsilon_u = \frac{1}{2\,c\,E_o^2}\int \dot{N}_{\mathrm{ph}}\langle\varepsilon^2\rangle\mathcal{H}(s)\,\mathrm{d}s.\tag{10.56}$$

The function $\mathcal{H}$ is now evaluated with contributions to the dispersion functions both from real bending magnets and from dipole field errors and associated dipole correctors. Since such errors occur in both planes there is an emittance increase in both planes as well. With (10.41) the increase in beam emittance is finally

$$\Delta\epsilon_u = \frac{55\,C_\gamma\,\hbar c\,(mc^2)^2}{64\,\pi\,\sqrt{3}}\gamma^5 \int \frac{\mathcal{H}\,\mathrm{d}s}{\rho^3}, \tag{10.57}$$

where the integration is taken along the whole beam line. The perturbation of the beam emittance in a beam transport line increases with the fifth power of the particle energy. At very high energies we expect therefore a significant effect of dipole errors on the beam emittance even if the basic beam transport line is straight.

So far we have not yet distinguished between the horizontal and vertical plane since the evolution of the phase space does not depend on the particular degree of freedom. The equilibrium beam emittance, however, depends on machine parameters and circular accelerators are not constructed symmetrically. Specifically, accelerators are mostly constructed in a plane and therefore there is no deflection in the plane normal to the ring plane. Assuming bending only occurs in the horizontal plane, we may use (10.54) directly as the result for the horizontal beam emittance $u = x$.

In the vertical plane the bending radius $\rho_v \to \infty$ and the vertical beam emittance reduces to zero by virtue of the still active damping. Whenever we have ideal conditions like this it is prudent to consider effects that we may have neglected leading to less than ideal results. In this case we have neglected the fact that synchrotron radiation photons are emitted not strictly in the forward direction but rather into a small angle $\pm 1/\gamma$. Photons emitted at a slight angle exert a recoil on the particle normal to the direction of the trajectory. A photon emitted at an angle θ with respect to the direction of the trajectory and an azimuth ϕ causes a variation of the vertical slope by

$$\delta y' = -\theta\,\cos\phi\,\frac{\varepsilon}{E_\mathrm{o}}$$

while the position is not changed $\delta y = 0$. This leads to a finite beam emittance which can be derived analogous to the general derivation above

$$\frac{\sigma_y^2}{\beta_y} = \frac{\tau_y}{4E^2}\left\langle \dot{N}_\mathrm{ph}\langle \varepsilon^2\theta^2\cos^2\phi\rangle\,\beta_y\right\rangle_s. \tag{10.58}$$

We set

$$\langle \varepsilon^2\theta^2\cos^2\phi\rangle \approx \langle\varepsilon^2\rangle\langle\theta^2\cos^2\phi\rangle \approx \langle\varepsilon^2\rangle\frac{1}{2\gamma^2}$$

and get finally for the fundamental lower limit of the vertical beam emittance

$$\frac{\sigma_y^2}{\beta_y} = \epsilon_y = \frac{C_\mathrm{q}\bar{\beta}_y}{2J_y}\frac{\langle 1/\rho^3\rangle}{\langle 1/\rho^2\rangle}. \tag{10.59}$$

354

Very roughly $\epsilon_y/\epsilon_x = 1/\gamma^2 \ll 1$ and it is therefore justified to neglect this term in the calculation of the horizontal beam emittance. This fundamental lower limit of the equilibrium beam emittance is of the order of 10^{-13} m, assuming the betatron function and the bending radius to be of similar magnitude, and therefore indeed very small compared to actual achieved beam emittances in real accelerators. In reality we observe a larger beam emittance in the vertical plane due to coupling or due to vertical steering errors which create a small vertical dispersion and, consequently, a small vertical beam emittance. As a practical rule the vertical beam emittance is of the order of one percent of the horizontal beam emittance due to field and alignment tolerances of the accelerator magnets. For very small horizontal beam emittances, however, this percentage may increase because the vertical beam emittance due to vertical dipole errors becomes significant.

Sometimes it is necessary to include vertical bending magnets in an otherwise horizontal ring. In this case the vertical dispersion function is finite and so is $\mathcal{H}_y(s)$. The vertical emittance is determined by evaluating (10.54) while using the vertical dispersion function. Note, however, that all bending magnets must be included in the calculation of equilibrium beam emittances because it is immaterial whether the energy loss was caused in a horizontally or vertically bending magnets. The same is true for the damping term in the denominator. Differences in the horizontal and vertical beam emittance come from the different betatron and η-functions.

10.3.2 Transverse Beam Parameters

Beam parameters like width, height, length, divergence, and energy spread are not all fixed independent quantities, but rather depend on emittances and lattice and rf parameters. These multiple dependencies allow the adjustment of beam parameters, within limits, to be optimum for the intended application. In this section we will discuss such dependencies.

A particle beam at any point of a beam transport line may be represented by a few phase ellipses for different particle momenta as shown in Fig. 10.3. The phase ellipses for different momenta are shifted proportional to the dispersion function at that point and its derivative. Generally, the form and orientation of the ellipses are slightly different too due to chromatic aberrations in the focusing properties of the beam line. For the definition of beam parameters we need therefore the knowledge of the lattice functions including chromatic aberrations and the beam emittance and momentum spread.

Beam Sizes: The particle beam width or beam height is determined by the beam emittance, the value of the betatron function, the value of the dispersion function, and the energy spread. The betatron and dispersion functions vary along a beam transport line and depend on the distribution of the beam focusing elements. The beam sizes are therefore also functions

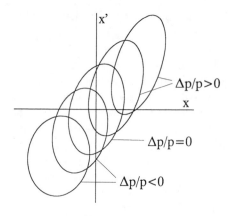

Fig. 10.3. Distribution of beam ellipses for a beam with finite emittance and momentum spread (schematic). The variation in the shape of the phase ellipses for different energies reflects the effect of chromatic aberrations

of the location along the beam line. From the focusing lattice these functions can be derived and the beam sizes be calculated.

The beam size of a particle beam is generally not well defined since the boundaries of a beam tends to become fuzzy. We may be interested in the beam size that defines all of a particle beam. In this case we look for that phase ellipse that encloses all particles and obtain the beam size in the form of the *beam envelope*. The beam half-width or half-height of this beam envelope is defined by

$$u_\beta(s) \;=\; \sqrt{\epsilon_u \beta_u(s)} \tag{10.60}$$

with $u = (x, y)$. If there is also a finite momentum spread within the beam particles the overall beam size or beam envelope is increased by the dispersion

$$u_\eta(s) \;=\; \eta_u(s)\,\frac{\Delta cp}{cp_\mathrm{o}} \tag{10.61}$$

and the total beam size is

$$u_\mathrm{tot}(s) \;=\; u_\beta(s) + u_\eta(s) \;=\; \sqrt{\epsilon_u \beta_u(s)} + \eta_u(s)\,\frac{\Delta cp}{cp_\mathrm{o}}. \tag{10.62}$$

This definition of the beam size assumes a uniform particle distribution within the beam and is used mostly to determine the acceptance or the *beam stay clear*, BSC, of a beam transport system. The acceptance of a beam transport system is defined as the maximum emittance a beam may have and still pass through the vacuum chambers of a beam line. In Fig. 10.3 this would be the area of that ellipse that encloses the whole beam including off momentum particles. In practice, however, we would choose a larger acceptance to allow for errors in the beam path.

Since the lattice functions vary along a beam line the required aperture, to let a beam with the maximum allowable emittance pass, is not the same everywhere along the system. To characterize the aperture variation con-

Table 10.1. Fraction of total beam intensity

	1-dimension	2-dimension	3-dimension
$1\,\sigma$:	68.26%	46.59%	31.81%
$2\,\sigma$:	95.44%	91.09%	86.93%
$\sqrt{6}\sigma$:	98.56%	97.14%	95.74%

sistent with the acceptance, a beam stay clear area, BSC, is defined as the required material free aperture of the beam line.

For a more precise description of the actual beam size the particle distribution must be considered. Most particle beams have a gaussian or near gaussian density distribution in all six dimensions of phase space and therefore the contributions to the beam parameters from different sources add in quadrature. The beam parameters for gaussian particle distributions are defined as the standard values of the gaussian distribution

$$\sigma_x, \sigma_{x'}, \sigma_y, \sigma_{y'}, \sigma_\delta, \sigma_\ell \,, \tag{10.63}$$

where most designations have been defined and used in previous chapters and where $\sigma_\delta = \sigma_\epsilon/cp_0$ and σ_ℓ the bunch length. Quoting beam sizes in units of σ can be misleading specifically, in connection with beam intensities. For example, a beam with a horizontal and vertical size of one sigma has a cross section of $2\sigma_x 2\sigma_y$ and includes only 46.59% of the beam. Therefore, beam intensities are often given for two sigma's or as in the case of proton and ion beam for $\sqrt{6}\,\sigma$'s. In Table 10.1 the fraction of the total beam intensity is compiled for a few generally used units of beam size measurement and for beam size, cross section, and volume.

The beam size for gaussian beams is thereby

$$\sigma_{\mathrm{u,tot}} \;=\; \sqrt{\epsilon_u \beta_u(s) + \eta^2(s)\sigma_\delta^2}\,. \tag{10.64}$$

Four parameters are required to determine the beam size in each plane although in most cases the vertical dispersion vanishes.

Beam Divergence: The angular distribution of particles within a beam depends on the rotation of the phase ellipse and we define analogous to the beam size an *angular beam envelope* by

$$\sigma_{\mathrm{u'},\mathrm{tot}} \;=\; \sqrt{\epsilon_u \gamma_u(s) + \eta'^2(s)\,\sigma_\delta^2}\,. \tag{10.65}$$

Again there is a contribution from the betatron motion, from a finite momentum spread and associated chromatic aberration. The horizontal and vertical beam divergencies are again determined by four parameters in each plane.

10.4 Variation of the Equilibrium Beam Emittance

In circular electron accelerators the beam emittance is determined by the emission of synchrotron radiation and the resulting emittance is not always equal to the desired value. In such situations methods to alter the equilibrium emittance are desired and we will discuss in the next sections such methods which may be used to either increase or decrease the beam emittance.

10.4.1 Beam Emittance and Wiggler Magnets

The beam emittance in an electron storage ring can be greatly modified by the use of wiggler magnets both to increase [10.8] or to decrease the beam emittance. A decrease in beam emittance has been noted by Tazzari [10.9] while studying the effect of a number of wiggler magnets in a low emittance storage ring design. Manipulation of the beam emittance in electron storage rings has become of great interest specifically, to obtain extremely small beam emittances, and we will therefore derive systematic scaling laws for the effect of wiggler magnets on the beam emittance as well as on the beam energy spread [10.10, 11].

The particle beam emittance in a storage ring is the result of two competing effects, the quantum excitation caused by the quantized emission of photons and the damping effect. Both effects lead to an equilibrium beam emittance observed in electron storage rings.

Independent of the value of the equilibrium beam emittance in a particular storage ring, it can be further reduced by increasing the damping without also increasing the quantum excitation. More damping can be established by causing additional synchrotron radiation through the installation of deflecting dipole magnets like strong wigglers magnets. In order to avoid quantum excitation of the beam emittance, however, the placement of wiggler magnets has to be chosen carefully. As discussed earlier, an increase of the beam emittance through quantum excitation is caused only when synchrotron radiation is emitted at a place in the storage ring where the dispersion function is finite. The emission of a photon causes a sudden energy loss and thereby also a sudden change of the particle's equilibrium trajectory which generally causes a corresponding increase in the betatron oscillation amplitude about the new equilibrium orbit.

Emittance reducing wiggler magnets therefore must be placed in areas around the storage ring where the dispersion vanishes to minimize quantum excitation. To calculate the modified equilibrium beam emittance we start from (10.56) and get with (10.41) an expression for the quantum excitation of the emittance which can be expanded to include wiggler magnets

$$\frac{d\epsilon}{dt}\bigg|_{q_o} = c\, C_q E^5 \left\langle \frac{\mathcal{H}}{\rho^3} \right\rangle_o , \tag{10.66}$$

where

$$C_Q = \frac{55}{24\sqrt{3}} \frac{r_e \hbar c}{(mc^2)^6} = 2.06 \times 10^{-11} \frac{\text{m}^2}{\text{GeV}^5}. \tag{10.67}$$

The quantity $\mathcal{H}$ is evaluated for the plane for which the emittance is to be determined, E is the particle energy, and ρ_o the bending radius of the regular ring magnets. The average $\langle \rangle$ is to be taken for the whole ring and the index $_o$ indicates that the average $\langle \mathcal{H}/\rho^3 \rangle_o$ be taken only for the ring proper without wiggler magnets.

Since the contributions of different magnets, specifically, of regular storage ring magnets and wiggler magnets are independent of each other, we may use the results of the basic ring lattice and add to the regular quantum excitation and damping the appropriate additions due to the wiggler magnets,

$$\left.\frac{d\epsilon}{dt}\right|_{qw} = c\, C_Q E^5 \left[\left\langle \frac{\mathcal{H}}{\rho^3} \right\rangle_o + \left\langle \frac{\mathcal{H}}{\rho^3} \right\rangle_w \right]. \tag{10.68}$$

Both, ring magnets and wiggler magnets produce synchrotron radiation and contribute to damping of the transverse particle oscillations. Again, we may consider both contributions separately and adding the averages we get the combined rate of emittance damping

$$\left.\frac{d\epsilon}{dt}\right|_{dw} = -2\,\epsilon_w\, C_d\, J_x\, E^3 \left[\left\langle \frac{1}{\rho^2} \right\rangle_o + \left\langle \frac{1}{\rho^2} \right\rangle_w \right], \tag{10.69}$$

where ϵ_w is the beam emittance with wiggler magnets,

$$C_d = \frac{c}{3} \frac{r_e}{(mc^2)^3} = 2.11 \times 10^3 \frac{\text{m}^2}{\text{GeV}^3 \text{sec}}, \tag{10.70}$$

and J_u the damping partition number with $u = x, y$. The equilibrium beam emittance is reached when the quantum excitation rate and the damping rates are of equal magnitude. We add therefore (10.68) and (10.69) and solve for the emittance

$$\epsilon_w = C_q \frac{\gamma^2}{J_u} \frac{\left\langle \frac{\mathcal{H}}{\rho^3} \right\rangle_o + \left\langle \frac{\mathcal{H}}{\rho^3} \right\rangle_w}{\left\langle \frac{1}{\rho^2} \right\rangle_o + \left\langle \frac{1}{\rho^2} \right\rangle_w}, \tag{10.71}$$

where C_q is defined in (10.43). With ϵ_o being the unperturbed beam emittance, $(\rho_w \to \infty)$, the relative emittance change due to the presence of wiggler magnets is

$$\frac{\epsilon_w}{\epsilon_o} = \frac{1 + \langle \mathcal{H}/\rho^3 \rangle_w / \langle \mathcal{H}/\rho^3 \rangle_o}{1 + \langle 1/\rho^2 \rangle_w / \langle 1/\rho^2 \rangle_o}. \tag{10.72}$$

Making use of the definition of average parameter values we get with the circumference of the storage ring $C = 2\pi R$

$$\left\langle \frac{\mathcal{H}}{\rho^3} \right\rangle_\text{o} = \frac{1}{C} \oint \frac{\mathcal{H}}{\rho_\text{o}^3} \, ds, \qquad \left\langle \frac{\mathcal{H}}{\rho^3} \right\rangle_\text{w} = \frac{1}{C} \oint \frac{\mathcal{H}}{\rho_\text{w}^3} \, ds,$$

$$\left\langle \frac{1}{\rho^2} \right\rangle_\text{o} = \frac{1}{C} \oint \frac{1}{\rho_\text{o}^2} \, ds, \quad \text{and} \quad \left\langle \frac{1}{\rho^2} \right\rangle_\text{w} = \frac{1}{C} \oint \frac{1}{\rho_\text{w}^2} \, ds. \tag{10.73}$$

Evaluation of these integrals for the particular storage ring and wiggler magnet employed gives from (10.72) the relative change in the equilibrium beam emittance. We note that the quantum excitation term scales like the cube while the damping scales only quadratically with the wiggler curvature. This feature leads to the effect that the beam emittance is always reduced for small wiggler fields and increases only when the third power terms become significant.

Concurrent with a change in the beam emittance a change in the momentum spread due to the wiggler radiation can be derived similarly,

$$\frac{\sigma_{\epsilon w}^2}{\sigma_{\epsilon o}^2} = \frac{1 + \langle 1/\rho^3 \rangle_\text{w} / \langle 1/\rho^3 \rangle_\text{o}}{1 + \langle 1/\rho^2 \rangle_\text{w} / \langle 1/\rho^2 \rangle_\text{o}}. \tag{10.74}$$

Closer inspection of (10.72, 74) reveals basic rules and conditions for the manipulations of beam emittance and energy spread. If the ring dispersion function is finite in the wiggler section, we have $\langle \mathcal{H}_\text{w} \rangle \neq 0$ which can lead to strong quantum excitation depending on the magnitude of the wiggler magnet bending radius ρ_w. This situation is desired if the beam emittance must be increased [10.8]. If wiggler magnets are placed into a storage ring lattice were the ring dispersion function vanishes, only the small dispersion function due to the wiggler magnets must be considered for the calculation of $\langle \mathcal{H}_\text{w} \rangle$ and therefore only little quantum excitation occurs. In this case the beam emittance can be reduced since the wiggler radiation contributes more strongly to damping and we call such magnets *damping wigglers* [10.10– 12].

Whenever wiggler magnets are used which are stronger than the ordinary ring magnets $\rho_\text{w} < \rho_\text{o}$ the momentum spread in the beam is increased. This is true for virtually all cases of interest.

Conceptual methods to reduce the beam emittance in a storage ring have been derived which are based on increased synchrotron radiation damping while avoiding quantum excitation effects. Optimum lattice parameters necessary to achieve this will be derived in the next section.

10.4.2 Damping Wigglers

The general effects of wiggler magnet radiation on the beam emittance has been described and we found that the beam emittance can be reduced if the wiggler is placed where $\eta = 0$ to eliminate quantum excitation $\langle \mathcal{H}_\text{w} \rangle = 0$. This latter assumption, however, is not quite correct. Even though we have chosen a place, where the storage ring dispersion function vanishes, we find the quantum excitation factor $\mathcal{H}_\text{w}$ to be not exactly zero once the wiggler

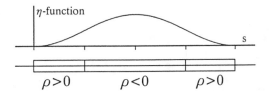

$\rho>0$ $\rho<0$ $\rho>0$

Fig. 10.4. Dispersion function in one period of a wiggler magnet

magnets are turned on because they create their own dispersion function. To calculate this dispersion function we assume a sinusoidal wiggler field [10.12]

$$B(z) = B_{\mathrm{w}} \cos k_{\mathrm{p}} z, \tag{10.75}$$

where $k_{\mathrm{p}} = 2\pi/\lambda_{\mathrm{p}}$ and λ_{p} the wiggler period length as shown in Fig.10.4. The differential equation for the dispersion function is then

$$\eta'' = \frac{1}{\rho} = \frac{1}{\rho_{\mathrm{w}}} \cos k_{\mathrm{p}} z \tag{10.76}$$

which can be solved by

$$\begin{aligned}
\eta(s) &= \frac{1}{k_{\mathrm{p}}^2 \rho_{\mathrm{w}}}(1 - \cos k_{\mathrm{p}} z), \\
\eta'(s) &= \frac{1}{k_{\mathrm{p}} \rho_{\mathrm{w}}} \sin k_{\mathrm{p}} z,
\end{aligned} \tag{10.77}$$

where we have assumed that the wiggler magnet is placed in a dispersion free location $\eta_{\mathrm{o}} = \eta_{\mathrm{o}}' = 0$. With this solution the first two equations (10.73) can be evaluated. To simplify the formalism we ignore the z-dependence of the lattice functions within the wiggler magnet setting $\alpha_x = 0$ and $\beta_x = $ const. Evaluating the integrals (10.73), we note that the absolute value of the bending radius ρ must be used along the integration path because the synchrotron radiation does not depend on the sign of the deflection. With this in mind we evaluate the integrals $\int_{\mathrm{o}}^{\pi}(\eta^2/|\rho^3|)\,\mathrm{d}z$ and $\int_{\mathrm{o}}^{\pi}(\eta'^2/|\rho^3|)\,\mathrm{d}z$. For each half period of the wiggler magnet the contribution to the integral is

$$\Delta \int\limits_{\mathrm{o}}^{\lambda_{\mathrm{p}}/2} \frac{\mathcal{H}}{\rho^3}\,\mathrm{d}z = \frac{36}{15}\frac{1}{\beta_x}\frac{1}{k_{\mathrm{p}}^5 \rho_{\mathrm{w}}^5} + \frac{4}{15}\frac{\beta_x}{k_{\mathrm{p}}^3 \rho_{\mathrm{w}}^5} \approx \frac{4}{15}\frac{\beta_x}{k_{\mathrm{p}}^3 \rho_{\mathrm{w}}^5}, \tag{10.78}$$

where the approximation $\lambda_{\mathrm{p}} \ll \beta_x$ was used. For the whole wiggler magnet with N_{w} periods the total quantum excitation integral is with the deflection angle per wiggler pole $\Theta_{\mathrm{w}} = 1/(\rho_{\mathrm{w}} k_{\mathrm{p}})$

$$\int_w \frac{\mathcal{H}}{\rho^3}\, dz \approx N_w \frac{8}{15} \frac{\beta_x}{\rho_w^2} \Theta_w^3 . \tag{10.79}$$

Similarly, the damping integral for the total wiggler magnet is

$$\int_w \frac{1}{\rho^2}\, dz = \pi N_w \,, \frac{\Theta_w}{\rho_w^2} . \tag{10.80}$$

Inserting expressions (10.73, 79, 80) into equation (10.72) we get for the emittance ratio

$$\frac{\epsilon_{xw}}{\epsilon_{xo}} = \frac{1 + \frac{8}{30\pi} N_w \frac{\beta_x}{\langle \mathcal{H}_o \rangle_\rho} \frac{\rho_o^2}{\rho_w^2} \Theta_w^3}{1 + \frac{1}{2} N_w \frac{\rho_o}{\rho_w} \Theta_w} , \tag{10.81}$$

where $\langle \mathcal{H}_o \rangle_\rho$ is the average value of $\mathcal{H}$ in the ring bending magnets excluding the wiggler magnets. We note from (10.81)that the beam emittance indeed can be reduced by wiggler magnets if Θ_w is kept small. For easier numerical calculation we replace $\langle \mathcal{H}_o \rangle_\rho$ by the unperturbed beam emittance which from (10.71) in the limit $\rho_w \to \infty$ is

$$\langle \mathcal{H}_o \rangle_\rho = \frac{J_x \rho_o \epsilon_{xo}}{C_q \gamma^2} \tag{10.82}$$

and get instead of (10.81)

$$\frac{\epsilon_{xw}}{\epsilon_{xo}} = \frac{1 + \frac{8\,C_q}{30\pi J_x} N_w \frac{\beta_x}{\epsilon_{xo}\,\rho_w} \gamma^2 \frac{\rho_o}{\rho_w} \Theta_w^3}{1 + \frac{1}{2} N_w \frac{\rho_o}{\rho_w} \Theta_w} . \tag{10.83}$$

The beam emittance is reduced by wiggler magnets whenever the condition

$$\frac{8}{15\pi} \frac{C_q}{J_x} \frac{\beta_x}{\epsilon_o\,\rho_w} \gamma^2 \Theta_w^2 \leq 1 \tag{10.84}$$

is fulfilled.

For large numbers of wiggler poles $N_w \to \infty$ the beam emittance reaches asymptotically a lower limit given by

$$\frac{\epsilon_{xw}}{\epsilon_{xo}} \to \frac{4}{30\pi} \frac{C_q}{J_x} \frac{\beta_x}{\epsilon_o\,\rho_w} \gamma^2 \Theta_w^2 . \tag{10.85}$$

In this limit the ultimate beam emittance is independent of the unperturbed beam emittance.

For many wiggler poles the increase in momentum spread also reaches an asymptotic limit which is given from (10.74)

$$\frac{\sigma_{\epsilon w}^2}{\sigma_{\epsilon o}^2} \to \frac{\rho_o}{\rho_w} = \frac{B_w}{B_o} , \tag{10.86}$$

where B_o is the magnetic field strength in the ring magnets. Beam stability and acceptance problems may occur if the beam momentum spread is allowed to increase too much and therefore inclusion of damping wigglers must be planned with some caution.

10.5 Variation of the Damping Distribution

Robinson's criterion provides an expression for the overall damping in six-dimensional phase space without specifying the distribution of damping in the three degrees of freedom. In accelerators we make an effort to decouple the particle motion in the three degrees of freedom as much as possible and as a result we try to optimize the beam parameters in each plane separately from the other planes for our application. Part of this optimization is the adjustment of damping and as a consequence of beam emittances to desired values. Robinson's criterion allows us to modify the damping in one plane at the expense of damping in another plane. This shifting of damping is done by varying *damping partition* numbers defined in (10.81).

From the definition of the ϑ parameter is is clear that damping partition numbers can be modified depending on whether the accelerator lattice is a combined function or a separated function lattice. Furthermore we may adjust virtually any distribution between partition numbers by choosing a combination of gradient and separated function magnets.

10.5.1 Damping Partition and rf Frequency

Actually such "gradients" can be introduced even in a separated function lattice. If the rf frequency is varied the beam will follow a path that meets the synchronicity condition. Increasing the rf frequency, for example, leads to a shorter wavelength and therefore the total path length in the ring need to be shorter. As a consequence of the principle of phase stability the beam energy is reduced and the beam follows a lower energy equilibrium orbit with the same harmonic number as the reference orbit for the reference energy. Decreasing the rf frequency leads just to the opposite effect. The off momentum orbits pass systematically off center through quadrupoles which therefore function like combined function gradient magnets.

To quantify this effect we use only the second term in the expression (10.79) for ϑ. The first term, coming from sector magnets, will stay unaffected. Displacement of the orbit in the quadrupoles will cause a bending with a bending radius

$$\frac{1}{\rho_q} = k\,\delta x\,. \tag{10.87}$$

An rf frequency shift causes a momentum change of

$$\frac{\Delta p}{p_0} = -\frac{1}{\alpha_c}\frac{\Delta f_{\mathrm{rf}}}{f_{\mathrm{rf}}}, \tag{10.88}$$

which in turn causes a shift in the equilibrium orbit of

$$\delta x = \eta\frac{\Delta p}{p_0} = -\frac{\eta}{\alpha_c}\frac{\Delta f_{\mathrm{rf}}}{f_{\mathrm{rf}}} \tag{10.89}$$

and the bending radius of the shifted orbit in quadrupoles is

$$\frac{1}{\rho_q} = k\delta x = k\eta\frac{\Delta p}{p_0} = -k\frac{\eta}{\alpha_c}\frac{\Delta f_{\mathrm{rf}}}{f_{\mathrm{rf}}}. \tag{10.90}$$

Inserted into the second term of (10.79), where ρ_a is the actual bending radius of the ring bending magnets we get

$$\Delta\vartheta = -\frac{1}{\alpha_c}\frac{\oint 2k^2\eta^2\,\mathrm{d}s}{\oint \mathrm{d}s/\rho_a^2}\frac{\Delta f_{\mathrm{rf}}}{f_{\mathrm{rf}}}. \tag{10.91}$$

We see that all quantities in (10.91) are fixed properties of the lattice and changing the rf frequency leads just to the effect we expected. Specifically, we note that all quadrupoles contribute additive irrespective of their polarity. We may apply this to a simple isomagnetic FODO lattice where all bending magnets and quadrupoles have the same absolute strength respectively with $\oint \mathrm{d}s/\rho^2 = 2\pi/\rho_a$. Integration of the nominator in (10.91) leads to

$$\oint 2k^2\eta^2\,\mathrm{d}s = 2k^2(\eta_{\max}^2 + \eta_{\min}^2)l_q 2n_c,$$

where l_q is half the quadrupole length in a FODO lattice, $\eta_{\max}$ and $\eta_{\min}$ the values of the η-function in the focusing QF and defocusing QD quadrupoles, respectively, and n_c the number of FODO cells in the ring. With all this we get finally for the variation of the ϑ parameter

$$\Delta\vartheta = -n_c\frac{2\rho_a}{\pi\alpha_c l_q}\frac{\eta_{\max}^2 + \eta_{\min}^2}{f^2}\frac{\Delta f_{\mathrm{rf}}}{f_{\mathrm{rf}}}. \tag{10.92}$$

Here we have used the focal length $f^{-1} = kl_q$. We replace in (10.92) the η functions by the expressions (6.67) derived for a FODO lattice, recall the relation $f = \kappa L$ and get finally [10.13]

$$\Delta\vartheta = -\frac{\rho_a}{\rho}\frac{1}{\alpha_c}\frac{L}{l_q}(4\kappa^2 + 1)\frac{\Delta f_{\mathrm{rf}}}{f_{\mathrm{rf}}}, \tag{10.93}$$

where ρ is again the average bending radius in the FODO cell as defined in Chap. 6. The variation of the ϑ parameter in a FODO lattice is the more sensitive to rf frequency variations the longer the cell compared to the quadrupole length and the weaker the focusing. For other lattices the expressions may not be as simple as for the FODO lattice but can always be computed numerically by integrations and evaluation of (10.91).

By varying the rf frequency and thereby the horizontal and longitudinal damping partition number we have found a way to either increase or decrease the horizontal beam emittance. The adjustments, however, are limited. To decrease the horizontal beam emittance we would increase the horizontal partition number and at the same time the longitudinal partition number would be reduced. The limit is reached when the longitudinal motion becomes unstable or in practical cases when the partition number becomes less than about 0.5. Other more practical limits may occur before stability limits are reached if the momentum change becomes too large to fit in the vacuum chamber aperture.

10.5.2 Robinson Wiggler

The horizontal betatron motion in a combined function synchrotron FODO lattice is not damped because $\vartheta > 1$. Beam stability in a synchrotron therefore exists only during acceleration when the antidamping is over compensated by adiabatic damping, and the maximum energy achievable in a combined function synchrotron is determined when the quantum excitation becomes too large to be compensated by adiabatic damping. In an attempt at the Cambridge Electron Accelerator, CEA, to convert the synchrotron into a storage ring the problem of horizontal beam instability was solved by the proposal [10.14] to insert a damping wiggler consisting of a series of poles with alternating fields and gradients designed such that the horizontal partition number becomes positive and $2 < \vartheta < 1$.

Such magnets can be used in a general way to vary the damping partition numbers without having to vary the rf frequency and thereby moving the beam away from the center of the beam line aperture.

10.5.3 Damping Partition and Synchrotron Oscillation

The damping partition number and, therefore, damping depends on the relative momentum deviation of the whole beam or of particles within a beam from the reference energy. During synchrotron oscillations significant momentum deviations can occur, specifically, in the tails of a gaussian distribution. Such momentum deviations, although only temporary, can lead to reduced damping or outright antidamping [10.15]. To quantify this effect we write (10.91) in the form

$$\Delta \vartheta = \frac{\oint 2\,k^2\,\eta^2\,\mathrm{d}s}{\oint \mathrm{d}s\,/\,\rho_a^2}\,\frac{\Delta p}{p_\mathrm{o}} = C_\mathrm{o}\frac{\Delta p}{p_\mathrm{o}}. \tag{10.94}$$

The momentum deviation is not a constant but rather oscillates with the synchrotron oscillation frequency,

$$\frac{\Delta p}{p_{\mathrm{o}}} = \frac{\Delta p}{p_{\mathrm{o}}}\bigg|_{\max} \sin \Omega t = \delta_{\max} \sin \Omega t \,, \tag{10.95}$$

where Ω is the synchrotron oscillation frequency. The damping partition number oscillates as well and the damping decrement is

$$\frac{1}{\tau} = \frac{1}{\tau_{x\mathrm{o}}}(1 - C_{\mathrm{o}}\delta_{\max} \sin \Omega t)\,. \tag{10.96}$$

If the perturbation is too large we have antidamping during part of the synchrotron oscillation period. As a consequence the beam is "breathing" in its horizontal and longitudinal dimensions while undergoing synchrotron oscillations. To quantify this we calculate similar to (10.53) the total rate of change of the betatron oscillation amplitude a^2, as defined by the phase space ellipse $\gamma u^2 + 2\alpha u u' + \beta u'^2 = a^2$, composed of quantum excitation and modified damping

$$\frac{\mathrm{d}\langle a^2 \rangle}{\mathrm{d}t} = \frac{\langle \dot{N}_{\mathrm{ph}}\langle \epsilon_{\gamma}^2 \rangle \mathcal{H}\rangle}{E_{\mathrm{o}}^2} - \frac{2\langle a^2 \rangle}{\tau}\,. \tag{10.97}$$

The amplitude a^2 has the dimension of an emittance and can be expressed in terms of a betatron amplitude by $a^2 = u_{\max}^2/\beta_u$. Replacing the varying damping time by $\tau^{-1} = \tau_{\mathrm{o}}^{-1}(1 - \delta_{\max}C_{\mathrm{o}} \sin \Omega t)$ (10.97) becomes

$$\frac{\mathrm{d}\langle a^2 \rangle}{\langle a^2 \rangle} = \frac{2}{\tau_{\mathrm{o}}}\delta_{\max}C_{\mathrm{o}} \sin \Omega t \, \mathrm{d}t \,,$$

which can be readily integrated to give

$$\langle a^2 \rangle = \langle a_{\mathrm{o}}^2 \rangle \exp\left[\frac{2\,\delta_{\max}C_{\mathrm{o}}}{\Omega\,\tau_{\mathrm{o}}}(1 - \cos \Omega t)\right]\,. \tag{10.98}$$

The effect is the largest for particles with large energy oscillations and may be sufficient to require an adjustment of the vacuum chamber aperture. On the other hand the effect on the core of the beam is generally very small since $\delta_{\max}$ is small.

10.5.4 Can We Eliminate the Beam Energy Spread?

To conclude the discussions on beam manipulation we conceive a way to eliminate the energy spread in a particle beam. From beam dynamics we know that the beam particles can be sorted according to their energy by introducing a dispersion function. The distance of a particle from the reference axis is the proportional to its energy and given by

$$x_{\delta} = D\delta\,, \tag{10.99}$$

where D is the value of the dispersion at the location under consideration and $\delta = \Delta E / E_o$ the energy error. For simplicity we make no difference between energy and momentum during this discussion. We consider now a cavity excited at a higher mode such that the accelerating field is zero at the axis, but varies linearly with the distance from the axis. If now the accelerating field, or after integration through the cavity, the accelerating voltage off axis is

$$V_{rf}(x_\delta) = -\frac{x_\delta}{D} E_o \,, \tag{10.100}$$

we have just compensated the energy spread in the beam. The particle beam has become monochromatic, at least to the accuracy assumed here. In reality the dispersion of the beam is not perfect due to the finite beam emittance.

We will discuss cavity modes in Vol. II and find that the desired mode exists indeed and the lowest order of such modes is the TM_{110}-mode. So far we have made no mistake and yet Liouville's theorem seems to be violated because this scheme does not change the bunch length and the longitudinal emittance has been indeed reduced by application of macroscopic fields.

The problem is that we are by now used to consider transverse and longitudinal phase space separate. While this separation is desirable to manage the mathematics of beam dynamics, we must not forget, that ultimately beam dynamics occurs in sixdimensional phase space. Since Liouville's theorem must be true, its apparent violation warns us to observe changes in other phase space dimensions. In the case of beam monochromatization we notice that the transverse beam emittance has been increased. The transverse variation of the longitudinal electric field causes by virtue of Maxwell's equations the appearance of transverse magnetic fields which deflect the particles transversely thus increasing the transverse phase space at the expense of the longitudinal phase space.

This is a general feature of electromagnetic fields which has been investigated by Panofsky and Wenzel [10.16] and their result is known as the *Panofsky-Wenzel theorem*. The Lorentz force due to the fields causes a change in the particle momentum which in the transverse direction is given by

$$c\mathbf{p}_\perp = \frac{e}{\beta} \int_o^d \left[\mathbf{E}_\perp + \frac{1}{c} (\mathbf{v} \times \mathbf{B})_\perp \right] dz \,. \tag{10.101}$$

Expressing the fields by the vector potential $\mathbf{E}_\perp = -\partial \mathbf{A}_\perp / \partial t$ and $\mathbf{B}_\perp = (\nabla \times \mathbf{A})_\perp$ the change in the transverse momentum can be expressed by

$$\frac{\partial}{\partial t} c\mathbf{p}_\perp = -e\nabla_\perp \int_o^d \mathbf{E}_z \, dz \,, \tag{10.102}$$

where the integration is taken over the length d of the cavity. This is the Panofsky-Wenzel theorem which states that transverse acceleration occurs whenever there is a transverse variation of the accelerating field. In conclusion we find that indeed we may monochromatize a particle beam with the use of for example a TM_{110}-mode, but only at the expense of transverse beam emittance.

Problems

Problem 10.1. Show that the horizontal damping partition number is negative in a fully combined function FODO lattice as employed in older synchrotron accelerators. Why, if there is horizontal antidamping in such synchrotrons, is it possible to retain beam stability during acceleration? What happens if we accelerate a beam and keep it orbiting in the synchrotron at some higher energy?

Problem 10.2. Future colliding beam facilities for high-energy physics experimentation are based on two linear accelerators aimed at each other and producing beams of very high energy for collision. In this arrangement synchrotron radiation is avoided compared to a storage ring. We assume that such beams can be directed to different detectors. Design an S-shaped beam transport system based on a FODO lattice which would allow the beams to be directed into a detector which is displaced by the distance D normal to the linac axis. The beams have an energy of $E_o = 1000$ GeV and a beam emittance of $\epsilon = 1.0 \cdot 10^{-12}$ m which should not be diluted in this beam transport system by more than 10%. Determine quadrupole and bending magnet parameters.

Problem 10.3. Strong focusing is required along a very high energy linear accelerator. Misalignments and path correction introduce dipole fields which are the source of synchrotron radiation and quantum excitation. Assume a normalized emittance of $\gamma \epsilon = 10^{-6}$ m and an initial beam energy of 10 GeV at the entrance to the linac. The high-energy linac has a circular aperture of 3 mm diameter. Design a FODO cell with sufficient focusing to contain this beam within a radius of 0.5 mm leaving the rest for path distortions. The distance between quadrupoles increases linearly with energy. Determine with statistical methods the number and strength of the quadrupoles for an acceleration of 100 MeV/m. Determine the alignment tolerances for these quadrupoles to keep the emittance increase due to quantum excitation in the dipole field from misaligned quadrupoles and due to correctors to 10%.

Problem 10.4. Consider the FODO lattice along the linear accelerator in problem 10.3 and determine the increase in beam energy spread due to synchrotron radiation from the finite beam size in quadrupoles.

Problem 10.5. Consider an electron beam in a 6 GeV storage ring with a bending radius of $\rho = 20$ m in the bending magnets. Calculate the rms energy spread σ_ϵ/E_o and the damping time τ_s. What is the probability for a particle to emit a photon with an energy of σ_ϵ and $2\sigma_\epsilon$. How likely is it that this particle emits another such photon within a damping time? In evaluating the particle distribution, do we need to consider multiple photon emissions?

Problem 10.6. Consider one of the storage rings in Table 6.1 and calculate the equilibrium beam emittance and energy spread. To manipulate the beam emittance we vary the rf frequency. Determine the maximum variation possible with this method.

Problem 10.7. A large hadron collider LHC is considered to be constructed in the LEP tunnel of 28 km circumference at CERN in Geneva. The maximum proton energy is 15 TeV. Determine the magnetic bending field required if 80% of the circumference can be used for bending magnets. Calculate the synchrotron radiation power for a circulating proton current of 200 mA, damping times, equilibrium beam emittance and energy spread.

Problem 10.8. Consider the problem of monochromatization discussed in Sect. 10.5 and prove directly without using Liouville's theorem, that the emittance in four-dimensional phase space (x, x', z, δ) remains constant indeed.

Problem 10.9. An electron beam circulating in a 1.5 GeV storage ring emits synchrotron radiation. The rms emission angle of photons is $1/\gamma$ about the forward direction of the particle trajectory. Determine the photon phase space distribution at the source point and at a distance of 10 m away while ignoring the finite particle beam emittance. Now assume a gaussian particle distribution with a horizontal beam emittance of $\epsilon_x = 0.15 \cdot 10^{-6}$ rad-m. Fold both the photon and particle distributions and determine the photon phase space distribution 10 m away from the source point if the electron beam size is $\sigma_x = 1.225$ mm, the electron beam divergence $\sigma_{x'} = 0.1225$ mrad and the source point is a symmetry point of the storage ring. Assume the dispersion function to vanish at the source point. For what photon wavelengths would the vertical electron beam size appear diffraction limited if the emittance coupling is 10% ?.

Problem 10.10. Prove the validity of Eq. 10.102

11 Beam Life Time

Particles travelling along a beam transport line or orbiting in a circular accelerator can be lost due to a variety of causes. We ignore the trivial cases of beam loss due to technical malfunctioning of beam line components or losses caused by either complete physical obstruction of the beam line or a mismatch of vacuum chamber aperture and beam dimensions. For a well designed beam transport line or circular accelerator we distinguish two main classes for beam loss which are losses due to scattering and losses due to instabilities. While particle losses due to scattering with other particles is a single particle effect leading to a gradual loss of beam intensity, instabilities can lead to catastrophic loss of part or all of the beam. In this chapter we will concentrate on single particle losses due to interactions with residual gas atoms.

The effect of particle scattering on the beam characteristics is different for beam transport lines compared to circular accelerators especially compared to storage rings. Since a beam passes through transport lines only once, we are not concerned about beam life time but rather with the effect of particle scattering on the transverse beam size. For storage rings, in contrast, we consider both the effect of scattering on the beam emittance as well as the overall effect on the beam lifetime. Since long lifetimes of the order of many hours are desired in storage rings even small effects can accumulate to significantly reduce beam performance. In proton rings continuous scattering with residual gas atoms or with other protons of the same beam can change the beam parameters considerably for lack of damping. Even for electron beams, where we expect the effects of scattering to vanish within a few damping times, we may observe an increase in beam emittance. This is specifically true due to intra beam scattering for dense low emittance beams at low energies when damping is weak.

Collisions of particles with components of residual gas atoms, losses due to a finite acceptance limited by the physical or dynamic aperture, collisions with other particles of the same beam, or with synchrotron radiation photons can lead to absorption of the scattered particles or cause large deflections leading to instable trajectories and eventual particle loss. The continuous loss of single particles leads to a finite beam lifetime and may in severe cases require significant hardware modifications or a different mode of operation to restore a reasonable beam lifetime.

Each of these loss mechanisms has a particular parameter characterizing and determining the severity of the losses. Scattering effects with residual gas atoms are clearly dominated by the vacuum pressure while scattering effects with other particles in the same beam depend on the particle density. Some absorption of particles at the vacuum chamber walls will always occur due to the gaussian distribution of particles in space. Even for nonradiating proton beams which are initially confined to a small cross section, we observe the development of a halo of particles outside the beam proper due to intra beam scattering. The expansion of this halo is obviously limited by the vacuum chamber aperture. In circular accelerators this aperture limitation may not only be effected by solid vacuum chambers but also by "soft walls" due to stability limits imposed by the dynamic aperture.

Longitudinal phase or energy oscillations are limited either by the available rf parameters determining the momentum acceptance or by the transverse acceptance at locations, where the dispersion function is nonzero whichever is more restrictive. A momentum deviation or spread translates at such locations into a widening of the beam and particles loss occurs if the momentum error is too large to fit within the stable aperture. Transverse oscillation amplitudes are limited by the stable transverse acceptance as limited by the vacuum chamber wall or by aberrations due to nonlinear fields.

11.1 Beam Lifetime and Vacuum

Particle beams are generally confined within evacuated chambers to avoid excessive scattering on residual gas atoms. Considering *multiple Coulomb scattering* alone the rms radial scattering angle of particles with momentum p and velocity β passing through a scattering material of thickness L can be described by [11.1]

$$\vartheta_{\mathrm{rms}} = Z \frac{20_{\mathrm{MeV}}}{\beta c p} \sqrt{\frac{L}{L_{\mathrm{r}}}}, \tag{11.1}$$

where Z is the charge multiplicity of the particle and L_{r} the radiation length of the scattering material. The scattering angle is the angle for which the intensity has fallen to a fraction $1/e$ of the peak intensity. We may integrate (11.1) and get the beam radius of a pencil beam after passing through a scatterer of thickness L

$$r = Z \frac{40_{\mathrm{MeV}}}{3 \beta c p} \frac{\sqrt{L^3}}{\sqrt{L_{\mathrm{r}}}}. \tag{11.2}$$

The beam emittance generated by scattering effects is then in both the horizontal and vertical plane just the product of the projections of the distance r of the particles from the reference path and the radial scattering angles ϑ

onto the respective plane. From (11.1, 2) the beam emittance growth due to Coulomb scattering in a scatterer of length L is then

$$\epsilon_{x,y}(\mathrm{rad\,m}) = Z^2 \frac{2}{3} \frac{[14(\mathrm{MeV})]^2}{(\beta cp)^2} \frac{L^2(\mathrm{m})}{L_r(\mathrm{m})}. \tag{11.3}$$

For atmospheric air the radiation length is $L_r = 300.5$ m and a pencil electron beam with a momentum of say $cp = 1000$ MeV passing through 20 m of atmospheric air would grow through scattering to a beam diameter of 6.9 cm or to a beam emittance of about 177 mrad mm in each plane. This is much too big an increase in beam size to be practical in a 20 m beam transport line let alone in a circular accelerator or storage ring, where particles are expected to circulate at nearly the speed of light for many turns like in a synchrotron or for many hours in a storage ring.

To avoid beam blow up due to scattering we obviously need to provide an evacuated environment to the beam with a residual gas pressure which must be the lower the longer the beam is supposed to survive scattering effects. This does not mean that beam transport in atmospheric pressure must be avoided at all cost. Sometimes it is very useful to let a beam pass though air to provide free access for special beam monitoring devices specifically at the end of a beam transport line before the beam is injected into a circular accelerator. Obviously, this can be done only if the scattering effects through very thin metallic windows and the short length of atmospheric air will not spoil the beam emittance too much.

11.1.1 Elastic Scattering

As particles travel along an evacuated pipe they occasionally collide with atoms of the residual gas. These collisions can be either on nuclei or electrons of the residual gas atoms. The physical nature of the collision depends on the mass of the colliding partners. Particles heavier than electrons suffer mostly an energy loss in collisions with the atomic shell electrons while they lose little or no energy during collisions with massive nuclei but are merely deflected from their path by elastic scattering. The lighter electrons in contrast suffer both deflection as well as energy losses during collisions.

In this section we concentrate on the elastic scattering process, where the energy of the fast particle is not changed. For the purpose of calculating particle beam lifetimes due to elastic or *Coulomb scattering* we ignore screening effects by shell electrons and mathematical divergence problems at very small scattering angles. The scattering process therefore is described by the classical *Rutherford scattering* with the differential cross section per atom in cgs units

$$\frac{\mathrm{d}\sigma}{\mathrm{d}\Omega} = \left(\frac{zZe^2}{2\beta cp}\right)^2 \frac{1}{\sin^4\frac{\theta}{2}}, \tag{11.4}$$

where z is the charge multiplicity of the incident particle eZ the charge of the heavy scattering nucleus, θ the scattering angle with respect to the incident path, Ω the solid angle with $d\Omega = \sin\theta\,d\theta\,d\varphi$, and φ the polar angle.

To determine the particle beam lifetime or the particle loss rate we will calculate the rate of events for scattering angles larger than a maximum value of $\widehat{\theta}$ which is limited by the acceptance of the beam transport line. Any particle being deflected by an angle larger than this maximum scattering angle will be lost. We integrate the scattering cross section over all angles greater than $\widehat{\theta}$ up to the maximum scattering angle π. With n scattering centers or atoms per unit volume and N beam particles the loss rate is

$$-\frac{dN}{dt} = c\beta n N \int_{\widehat{\theta}}^{\pi} \frac{d\sigma}{d\Omega}\,d\Omega. \tag{11.5}$$

Under normal conditions at $0°C$ and a gas pressure of 760 mm mercury the number of scattering centers in a homogeneous gas is equal to twice the *Avogadro number* $\mathcal{A}$ and becomes for an arbitrary gas pressure P

$$n = 2\mathcal{A}\frac{P(\text{Torr})}{760} = 2\cdot 2.68675 \times 10^{19}\,\frac{P(\text{Torr})}{760}. \tag{11.6}$$

The factor 2 comes from the fact that homogeneous gases are composed of two atomic molecules, where each atom acts as a separate scattering center. This assumption would not be true for single atomic noble gases which we do not consider here, but will be included in a later generalization. The integral on the r.h.s. of (11.5) becomes with (11.4)

$$\int_{\widehat{\theta}}^{\pi} \frac{\sin\theta\,d\theta}{\sin^4(\theta/2)} = \frac{2}{\tan^2(\widehat{\theta}/2)}. \tag{11.7}$$

Dividing (11.5) by N we find an exponential decay of beam intensity with time

$$N = N_o\,e^{-t/\tau}, \tag{11.8}$$

where the decay time constant or *beam lifetime* is

$$\tau^{-1} = c\beta\,2\,\mathcal{A}\,\frac{P(\text{Torr})}{760}\left(\frac{zZe^2}{2\,\beta\,cp}\right)^2\frac{4\pi}{\tan^2(\widehat{\theta}/2)}. \tag{11.9}$$

The maximum acceptable scattering angle $\widehat{\theta}$ is limited by the acceptance ϵ_A of the beam transport line. A particle being scattered by an angle θ at a location, where the betatron function has the value β_θ reaches a maximum betatron oscillation amplitude of $A = \sqrt{\beta_A\,\beta_\theta}\,\theta$ elsewhere along the beam transport line where the betatron function is β_A. The minimum value of A^2/β_A along the beam transport line is equal to the *ring acceptance*

$$\epsilon_A = \left.\frac{A^2}{\beta_A}\right|_{\text{min}} . \tag{11.10}$$

This ring acceptance may be limited by the *physical aperture* of the vacuum chamber or by the *dynamic aperture*. For simplicity we ignore here the variation of the betatron function and take an average value $\langle\beta\rangle$ at the location of the scattering event and get finally for the maximum allowable scattering angle

$$\widehat{\theta}^2 = \frac{\epsilon_A}{\langle\beta\rangle} . \tag{11.11}$$

This angle is generally rather small and we may set $\tan(\widehat{\theta}/2) \approx (\widehat{\theta}/2)$. Utilizing these definitions and approximations we obtain for the lifetime of a beam made up of singly charged particles $z = 1$ due to elastic Coulomb scattering expressed in more practical units

$$\tau_{\text{cs}}(\text{hours}) = 10.25 \frac{(cp)^2(\text{GeV}^2)\,\epsilon_A(\text{mm mrad})}{\langle\beta(\text{m})\rangle\,P(\text{nTorr})} , \tag{11.12}$$

where we have assumed that the residual gas composition is equivalent to nitrogen gas N_2 with $Z^2 \approx 49$. The *Coulomb scattering lifetime* is proportional to the ring acceptance or proportional to the square of the aperture A where A^2/β is a minimum.

The particle loss due to Coulomb scattering is most severe at very low energies and increases with the acceptance of the beam transport line. Further, the beam lifetime depends on the focusing in the transport line through the average value of the betatron function. If instead of averaging the betatron function we integrate the contributions to the beam lifetime along the transport line we find that the effect of the scattering event depends on the betatron function at the location of the collision and the probability that such a collision occurs at this location depends on the gas pressure there. Therefore, it is prudent to not only minimize the magnitude of the betatron functions alone but rather minimize the product βP along the transport line. Specifically, where large values of the betatron function cannot be avoided, extra pumping capacity should be provided to reach locally a low vacuum pressure for long Coulomb scattering lifetime.

We have made several simplifications and approximations by assuming a homogeneous gas and assuming that the maximum scattering angle be the same in all directions. In practical situations, however, the acceptance need not be the same in the vertical and horizontal plane. First we will derive the beam lifetime for nonisotropic aperture limits. We assume that the apertures in the horizontal and vertical plane allow maximum scattering angles of $\widehat{\theta}_x$ and $\widehat{\theta}_y$. Particles are then lost if the scattering angle θ into a polar angle φ exceeds the limits

$$\theta > \frac{\widehat{\theta}_x}{\cos\varphi} \quad \text{and} \quad \theta > \frac{\widehat{\theta}_y}{\sin\varphi}. \tag{11.13}$$

The horizontal aperture will be relevant for all particles scattered into a polar angle between zero and $\arctan(\widehat{\theta}_y/\widehat{\theta}_x)$ aperture will be relevant while particles scattered into a polar angle of $\arctan(\widehat{\theta}_y/\widehat{\theta}_x)$ and $\pi/2$ will be absorbed by the vertical aperture whenever the scattering angle exceeds the limit. We calculate the losses in only one quadrant of the polar variable and multiply the result by 4 since the scattering and absorption process is symmetric about the polar axis. The integral (11.7) becomes in this case

$$\int_{\widehat{\theta}}^{\pi} \frac{\sin\theta \, d\theta \, d\varphi}{\sin^4(\theta/2)} = 4 \int_0^{\arctan(\widehat{\theta}_y/\widehat{\theta}_x)} d\varphi \int_{\widehat{\theta}_x/\cos\varphi}^{\pi} \frac{\sin\theta \, d\theta}{\sin^4(\theta/2)}$$

$$+ 4 \int_{\arctan(\widehat{\theta}_y/\widehat{\theta}_x)}^{\pi/2} d\varphi \int_{\widehat{\theta}_y/\sin\varphi}^{\pi} \frac{\sin\theta \, d\theta}{\sin^4(\theta/2)}. \tag{11.14}$$

The solutions of the integrals are similar to that in (11.7) and we get

$$\int_{\widehat{\theta}}^{\pi} \frac{\sin\theta \, d\theta \, d\varphi}{\sin^4\theta/2} = \frac{8}{\widehat{\theta}_y^2} \left[\pi + (R^2+1)\sin(2\arctan R)\right.$$

$$\left. + 2(R^2-1)\arctan R\right], \tag{11.15}$$

where $R = \widehat{\theta}_y/\widehat{\theta}_x$.

Using (11.15) instead of (11.7) in (11.9) gives a more accurate expression for the beam lifetime due to Coulomb scattering. We note that for $R = 1$ we do not get exactly the lifetime (11.9) but find a lifetime that is larger by a factor of $1 + \pi/2$. This is because we used a rectangular aperture in (11.15) compared to a circular aperture in (11.7). The beam lifetime (11.12) becomes now for a rectangular acceptance

$$\tau_{cs}(\text{hours}) = 10.25 \frac{2\pi}{F(R)} \frac{(cp)^2(\text{GeV}^2)\,\epsilon_A(\text{mm mrad})}{\langle\beta(\text{m})\rangle\,P(\text{nTorr})}. \tag{11.16}$$

The function $F(R)$

$$F(R) = [\pi + (R^2+1)\sin(2\arctan R) + 2(R^2-1)\arctan R] \tag{11.17}$$

is shown in Fig. 11.1 and for some special cases the factor $2\pi/F(R)$ assumes the values:

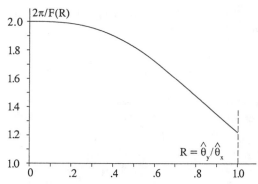

Fig. 11.1. Function F(R) to determine the acceptance for Coulomb scattering

Shape of Aperture	round	square	rectangular
Ratio $R = \widehat{\theta}_y/\widehat{\theta}_x$	1.00	1.00	$0 \rightarrow 1$
$2\pi/F(R)$	1.00	1.22	$2.00 \rightarrow 1.22$

Tacitly we have assumed that the vertical acceptance is smaller than the horizontal acceptance which in most cases is true. In cases, where $\widehat{\theta}_y > \widehat{\theta}_x$, we may use the same equations with x and y exchanged.

Particles performing large amplitude betatron oscillations form as a consequence of Coulomb scattering a halo around the beam proper. In case of an electron storage ring the particle intensity in this halo reaches an equilibrium between the constant supply of scattered electrons and synchrotron radiation damping.

The deviation of the particle density distribution from a gaussian distribution due to scattering can be observed and measured. In Fig. 11.2 beam lifetime measurements are shown for an electron beam in a storage ring as a function of a variable ring acceptance as established by a movable scraper. The abscissa is the actual position of the scraper during the beam lifetime measurement, while the variable for the ordinate is the aperture for which a pure gaussian particle distribution would give the same beam lifetime.

If the particle distribution had been purely gaussian the measured points would lie along a straight line. In reality, however, we observe an overpopulation of particles in the tails of the distribution for amplitudes larger than about 6 σ forcing the scraper to be located farther away from the beam center to get a beam lifetime equal to that of a pure gaussian distribution. This overpopulation or halo at large amplitudes is due to elastic Coulomb scattering on the residual gas atoms.

Since the acceptance of the storage ring is proportional to the square of the aperture at the scraper, we expect the beam lifetime due to Coulomb scattering to vary proportional to the square of the scraper position. This is

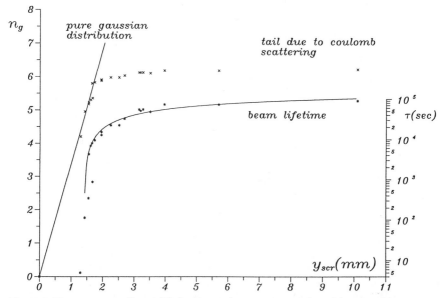

Fig. 11.2. Measurement of beam lifetime in an electron storage ring with a movable scraper. The curve on the left shows the Coulomb scattering halo for amplitudes larger than 6 σ indicating a strong deviation from a gaussian particle distribution. The curve on the right shows the beam lifetime as a function of scraper position.

shown in Fig. 11.3 for good vacuum and poor vacuum conditions. In the case of poor vacuum we find a saturation of the beam lifetime at large scraper openings which indicates that the scraper is no longer the limiting aperture in the ring. This measurement therefore allows an accurate determination of the *physical ring acceptance* or the *dynamic aperture* whichever is smaller.

So far we have assumed the residual gas to consist of homogeneous two atom molecules. This is not an accurate description of the real composition of the residual gas although on average the residual gas composition is equivalent to a nitrogen gas. Where the effects of a more complex gas composition becomes important, we apply (11.9) to each different molecule and atom of the residual gas. In (11.9) we replace the relevant factor $P Z^2$ by a summation over all gas components. If P_i is the partial pressure of the molecules i and Z_j the atomic number of the atom j in the molecule i we replace in (11.9)

$$P Z^2 \rightarrow \sum_{i,j} P_i Z_j^2 \tag{11.18}$$

and sum over all atoms i in the molecule j.

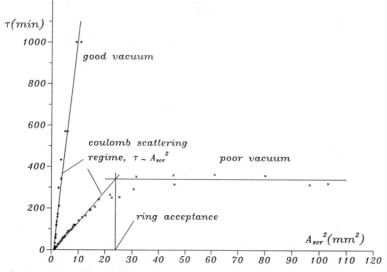

Fig. 11.3. Beam lifetime in an electron storage ring as a function of the acceptance. The transition of the curve on the right from a linear dependence of beam lifetime on the acceptance to a constant life time occurs when the acceptance due to the scraper position is equal to the ring acceptance.

11.1.2 Inelastic Scattering

Charged particles passing through matter become deflected by strong electrical fields from the atomic nuclei. This deflection constitutes an acceleration and the charged particles lose energy through emission of radiation which is called *bremsstrahlung*. If this energy loss is too large such that the particle energy error becomes larger than the storage ring energy acceptance the particle gets lost. We are therefore interested to calculate the probability for such large energy losses to estimate the beam lifetime.

The probability to suffer a relative energy loss $\delta = \mathrm{d}E/E_o$ due to such an *inelastic scattering* process has been derived by Bethe and Heitler [11.2, 3]. For extreme relativistic particles and full screening this probability per unit thickness of matter is [11.4]

$$\mathrm{d}P = 2\overline{\Phi}\, N\, \frac{\mathrm{d}\delta}{\delta}\,(1-\delta)\left[\left(\frac{2-2\delta+\delta^2}{1-\delta} - \frac{2}{3}\right)2\ln\frac{183}{Z^{1/3}} + \frac{2}{9}\right], \quad (11.19)$$

where N is the number of atoms per unit volume and the factor $\overline{\phi}$ is with the fine structure constant $\alpha = 1/137$

$$\overline{\Phi} = r_e^2\, Z^2\, \alpha\,, \tag{11.20}$$

where r_e is the classical electron radius. We integrate this probability over all energy losses larger than the energy acceptance of the storage ring $\delta \geq \delta_{\text{acc}}$ and get after some manipulation and setting $\delta_{\text{acc}} \ll 1$

$$P = 2\overline{\Phi} \int\limits_{\delta_{\text{acc}}}^{1-\delta_{\text{acc}}} \frac{d\delta}{\delta} (1-\delta) \left[\left(\frac{2-2\delta+\delta^2}{1-\delta} - \frac{2}{3} \right) 2 \ln \frac{183}{Z^{1/3}} + \frac{2}{9} \right]$$

$$\approx \frac{4}{3} - \ln \delta_{\text{acc}} \left(4\overline{\Phi} N \ln \frac{183}{Z^{1/3}} + \frac{3}{9} N \overline{\Phi} \right). \tag{11.21}$$

The *radiation length* L_r is defined as the distance over which the particle energy has dropped to $1/e$ due to inelastic scattering. For highly relativistic particles this length is given by [11.4]

$$\frac{1}{L_r} = \overline{\Phi} N \left(4 \ln \frac{183}{Z^{1/3}} + \frac{2}{9} \right). \tag{11.22}$$

Combining (11.20, 21), we find the simple solution that the probability for a particle to suffer a relative energy loss of more than δ_{acc} per radiation length is

$$P_{\text{rad}} = -\frac{4}{3} \ln \delta_{\text{acc}}. \tag{11.19}$$

To calculate the beam lifetime or beam decay rate due to bremsstrahlung we note that the probability for a particle loss per unit time is equal to the beam decay rate or equal to the inverse of the beam lifetime. The bremsstrahlung lifetime is therefore

$$\tau_{\text{bs}}^{-1} = -\frac{1}{N_o} \frac{dN}{dt} P \frac{c}{L_r} = -\frac{4}{3} \frac{c}{L_r} \ln \delta_{\text{acc}}. \tag{11.24}$$

The radiation length for gases are usually expressed for a standard temperature of 20°C and a pressure of 760 Torr. Under vacuum conditions the radiation length of the residual gas is therefore increased by the factor $760/P_{\text{Torr}}$. We recognize again the complex composition of the residual gas and define an effective radiation length by

$$\frac{1}{L_{r,\text{eff}}} = \sum_i \frac{1}{L_{r,i}}, \tag{11.25}$$

where $L_{r,i}$ is the radiation length for gas molecules i. From (11.23, 24) the beam lifetime due to bremsstrahlung for a composite residual gas is

$$\tau_{\text{bs}}^{-1} = -\frac{4}{3} c \sum_i \frac{\widetilde{P}_i}{L_{r,i}} \ln \delta_{\text{acc}}, \tag{11.26}$$

where $\widetilde{P}_i$ is the residual partial gas pressure for the gas molecules i. Although the residual gas of ultra high vacuum systems rarely includes a significant amount of nitrogen gas, the average value for $\langle Z^2 \rangle$ of the residual gas components is approximately 50 or equivalent to nitrogen gas. For all practical purposes we may therefore assume the residual gas to be nitrogen with a radiation length under normal conditions of $L_{r,N_2} = 290\,\mathrm{m}$ and scaling to the actual vacuum pressure P_{vac} we get for the beam lifetime

$$\tau_{bs}^{-1}\,(\mathrm{hours}^{-1}) = 0.00653\,P_{vac}(\mathrm{nTorr})\,\ln\frac{1}{\delta_{acc}}. \tag{11.27}$$

Basically the bremsstrahlung lifetime depends only on the vacuum pressure and the energy acceptance and the product of beam lifetime and vacuum pressure is a function of the energy acceptance $\delta_{acc} = \Delta\gamma/\gamma$,

$$\tau_{bs}(\mathrm{hour}) \cdot P(\mathrm{nTorr}) = \frac{419.29}{\ln(\gamma/\Delta\gamma)}. \tag{11.28}$$

In tabular form we get:

$\delta_{acc} = \Delta\gamma/\gamma$,	0.005	0.010	0.015	0.020	0.025
$\tau(\mathrm{hr})\,P(\mathrm{nTorr})$	79.15	91.05	99.84	107.2	113.7 .

The beam lifetime due to bremsstrahlung is only a weak logarithmic function of the energy acceptance and for quick estimates is about 100 hours divided by the average pressure in nTorr.

There are many more forms of interaction possible between energetic particles and residual gas atoms. Chemical, atomic, and nuclear reactions leading to the formation of new molecules like ozone, ionization of atoms or radioactive products contribute further to energy loss of the beam particles and eventual loss from the beam. These effects, however, are very small compared to the Coulomb scattering or bremsstrahlung losses and may therefore be neglected in the estimation of beam lifetime.

11.2 Ultra High Vacuum System

Accelerated particles interact strongly with residual gas atoms and molecules by elastic and inelastic collisions. To minimize particle loss due to such collisions we provide an evacuated beam pipe along the desired beam path. For open beam transport systems *high vacuum* of $10^{-5} - 10^{-7}$ Torr is sufficient. This is even sufficient for pulsed circular accelerators like synchrotrons, where the particles remain only for a short time. In storage rings, however, particles are expected to circulate for hours and therefore *ultra high vacuum* conditions must be created.

11.2.1 Thermal Gas Desorption

To reach very low gas pressures in the region of $10^{-6} - 10^{-7}$ Torr for high vacuum or even lower pressures in the regime of ultra high vacuum, *UHV*, we must consider the continuous desorption of gas molecules from the walls due to *thermal desorption*. Gas molecules adsorbed on the chamber surface are in thermal equilibrium with the environment and the thermal energy of the molecules assumes a statistically determined Boltzmann distribution. This distribution includes a finite probability for molecules to gain a large enough amount of energy to overcome the adsorption energy and be released from the wall.

The total gas flow from the wall due to this thermal gas desorption depends mostly on the preparation of the material. While for carefully cleaned surfaces the thermal desorption coefficient may be of the order of $10^{-12} - 10^{-13}$ Torr lt/sec/cm^2 a bakeout to $140 - 300°C$ can reduce this coefficient by another order of magnitude.

11.2.2 Synchrotron Radiation Induced Desorption

In high-energy electron or positron accelerators a significant amount of energy is emitted in the form of synchrotron radiation. This radiation is absorbed by vacuum chamber walls and causes not only a heating effect of the chamber walls but also the desorption of gas molecules adsorbed on the surface.

The physical process of photon induced gas desorption evolves in two steps [11.5]. First a photon hitting the chamber walls causes an electron emission with the probability $\eta_e(\varepsilon)$, where ε is the photon energy. Secondly, the emission as well as the later absorption of that photo electron can desorb neutral atoms from the chamber surface with the probability η_d. To calculate the total desorption in a storage ring we start from the differential synchrotron radiation photon flux (9.48) which we integrate over the ring circumference and write now in the form

$$\frac{dN(\varepsilon)}{dt} = 2\pi C_\psi \, E \, I_b \, \frac{\Delta\omega}{\omega} S(x) , \tag{11.29}$$

where $\varepsilon = \hbar\omega$ is the photon energy, I_b the beam current, E the beam energy and $S(x)$ a mathematical function defined by (9.47).

The photoelectron current $\dot{N}_e$ results from the folding of (11.29) with the photoelectron emission coefficient $\eta_e(\omega)$ for the material used to construct the vacuum chamber and the integration over all photon energies,

$$\dot{N}_e = 2\pi C_\psi \, E \, I_b \int_o^\infty \frac{\eta_e(\omega)}{\omega} \, S\left(\frac{\omega}{\omega_c}\right) \, d\omega . \tag{11.30}$$

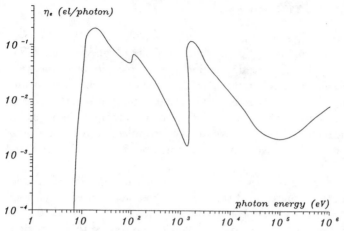

Fig. 11.4. Photon electron coefficient η_e for aluminum [11.6]

The photoelectron emission coefficient depends on the choice of the material for the vacuum chamber. Fig. 11.4 displays the photoelectron coefficient for aluminum as a function of the photon energy with a simplifying approximation suitable for numerical estimation of gas desoprtion [11.6].

We note there are virtually no photoelectrons for photon energies of less than 10 eV. At 1460 eV the K-edge of aluminum causes a sharp increase of the coefficient followed by a monotonous decrease for higher photon energies.

The photoelectron coefficient depends not only on the material of the photon absorber but also on the incident angle. The probability to release an electron from the surface is increased for shallow incidence of the photon. The *enhancement factor* $F(\Theta)$ represents the increase in the photoelectron-emission coefficient $\eta_e(\epsilon)$ due to a non normal incidence of a photon on the surface, where Θ is the angle between the photon trajectory and the normal to the absorbing material surface. For angles close to normal incidence the enhancement factor scales like the inverse of the sine of the angle and

$$F(\Theta) \;=\; \frac{1}{\sin \Theta}\,. \tag{11.31}$$

For larger angles, however, the enhancement factor falls off from the inverse sine dependence as has been determined by measurements [11.7] and reaches a maximum value of about seven for small angles.

The gas production is determined by the desorption rate Q, defined as the total number of neutral atoms released along the circumference from the chamber surface,

$$Q \;=\; 2\,\frac{22.4 \times 760}{6 \times 10^{23}}\,\dot{N}_e\,I_b\,\eta_d\,, \tag{11.32}$$

where Q is expressed in Torr lt/sec and η_d is the desorption coefficient. The factor 2 is due to the fact that a photo electron can desorb an atom while leaving as well as while arriving at a surface. With (11.32) we get the average vacuum pressure $\langle P \rangle$ from

$$\langle P \rangle = \frac{Q}{S}, \tag{11.33}$$

where S is the total installed pumping speed in the storage ring. For a reasonably accurate estimate of the photon flux we may use the small argument approximation (9.50) for photon energies $\varepsilon \leq \varepsilon_c$. Photons of higher energies generally do not contribute significantly to the desorption since there are only few. To obtain the photon flux we therefore need to integrate only from 10 eV to $\epsilon = \epsilon_c$ the differential photon flux (9.50) folded with the photoelectron-emission coefficient $\eta_e(\varepsilon)$.

The desorption coefficient η_d is largely determined by the treatment of the vacuum chamber like baking, beam cleaning, argon discharge cleaning, etc. For example in the aluminum chamber of the storage ring SPEAR [11.9] the desorption coefficient at 1.5 GeV was initially about $\eta_d \approx 5 \times 10^{-3}$ then 5×10^{-4} after one month of operation, 10^{-4} after two months of operation and reached about 3×10^{-6} after about one year of operation. These numbers are not to be viewed too generally, since the cleaning process depends strongly on the particular preparation of the surfaces. However, following well established cleaning procedures and handling of ultra high vacuum components these numbers can be of general guidance consistent with observations on other storage rings.

Laboratory measurements [11.6] show the following relationship between photoelectron current $I_{phe} = e\,N_e$, desorption coefficient η_d and total integrated beam time of a vacuum system

$$\eta_d = 7 \times 10^{-5} \frac{I_{phe}(A)}{t(hr)^{0.63}}. \tag{11.34}$$

New vacuum chambers release much gas when the first synchrotron radiation strikes the surface, but cleans as the *radiation cleaning* continues.

Problems

Problem 11.1. Consider a circular electron storage ring of your choice and specify beam energy, current, ring circumference and average vacuum chamber dimensions. Calculate the total thermal gas desorption and the total required pumping capacity in the ring. Now add synchrotron radiation and estimate the increase of pumping speed needed after say 100 ampere-hours of beam operation. Plot the average gas pressure as a function of integrated beam time.

12 Collective Phenomena

Transverse and longitudinal beam dynamics as discussed in earlier chapters is governed by purely single particle effects where the results do not depend on the presence of the other particles in the beam. We specifically excluded space charge effects to allow the detailed discussion of *single particle dynamics*. In many cases this assumption may not be appropriate and *collective effects* must be taken into account. Similar to the treatment of errors, we try to include collective effects as perturbations although a number of such effects cause serious beam instability which can hardly be considered a perturbation. However, the strength of collective beam instability depends on the beam intensity and we can always reduce the beam current such that the instability indeed is only a perturbation. Once the mechanism of instability is understood, corrective measures can be introduced allowing the increase of the beam current.

The study and understanding of the cause and nature of instability is important to allow the appropriate design of corrective measures which reduce the effect to the level of a mere perturbation or less. Accelerator design and development is done with the goal to eliminate or at least minimize collective effects through self imposed limitation on the performance, feedback systems, control mechanisms, etc. We will not be able to discuss space charge dependent instabilities here in detail postponing such studies to Vol. II while concentrating here only on some basic concepts of collective phenomena.

12.1 Linear Space-Charge Effects

A very basic collective effect in accelerator physics results from the repelling forces between equally charged particles within a beam. One of the goals of accelerator development is to reach a high particle beam brightness which implies that strong *self fields* within the beam itself due to space-charges must be expected. In Sect. 1.4.1 we found a balance of electric and magnetic fields giving a relativistic beam stability against being blown up by repelling electrostatic forces between particles. The compensation, however, is not perfect specifically for lower energy particles and a more detailed discussion of such space-charge effects is required. Specifically, a beam of uniform particle density, for example, generates Lorentz forces which increase linearly from the center of the beam causing a tune shift and may move the beam into a resonance as the beam intensity is increased.

12.1.1 Self Field for Particle Beams

The self fields of a beam depend on beam parameters like particle type, particle distribution, bunching, and energy of the particle. We will derive the nature and effect of these self fields for most common beam parameters.

To determine self fields we consider a continuous beam of particles with a line charge λ, or a volume charge $\rho(x, y)$. The electric fields within a beam are derived from a potential V defined by *Poisson's equation*

$$\Delta V = -4\pi \cdot \rho(x, y), \tag{12.1}$$

where ρ, being the electric charge density in the beam, is finite within and vanishes outside the beam. Similarly, the magnetic vector potential is defined by

$$\Delta \mathbf{A} = -4\pi \frac{\mathbf{v}}{c} \cdot \rho(x, y). \tag{12.2}$$

For a particle beam we may set $\mathbf{v} = (\approx 0, \approx 0, v)$ and the vector potential therefore contains only a longitudinal component $\mathbf{A} = (0, 0, A_z)$.

Generally, particle beams have an elliptical cross section and the solution to (12.1) for such a beam with constant charge density, $\rho = \text{const}$, has been derived by Teng [12.1, 2]. Within the elliptical beam cross section, where $x \leq a$ and $y \leq b$, the electric potential is

$$V(x, y) = -2\pi \rho \frac{ab}{a+b} \cdot \left[\frac{x^2}{a} + \frac{y^2}{b} \right] \tag{12.3}$$

and a, b are the horizontal and vertical half axis respectively. The vector potential for the magnetic field is from the discussions above

$$A_z(x, y) = -2\pi \rho \frac{v}{c} \frac{ab}{a+b} \cdot \left[\frac{x^2}{a} + \frac{y^2}{b} \right] \tag{12.4}$$

and both the electric and magnetic field can be derived by simple differentiations

$$\mathbf{E} = -\nabla V \quad \text{and} \quad \mathbf{B} = \nabla \times \mathbf{A} \quad \text{for} \tag{12.5}$$

$$E_x = \frac{4e\lambda}{a \cdot (a+b)} \cdot x, \qquad E_y = \frac{4e\lambda}{b \cdot (a+b)} \cdot y, \quad \text{and} \tag{12.6}$$

$$B_x = -\frac{4e\lambda\beta}{b \cdot (a+b)} \cdot y, \qquad B_y = \frac{4e\lambda\beta}{a \cdot (a+b)} \cdot x, \tag{12.7}$$

where $\beta = v/c$ and the linear charge density λ is defined by

$$\lambda = \pi ab \cdot \rho(x, y). \qquad (12.8)$$

Comparison of (12.7,8) reveals the relationship between electric and magnetic self fields of the beam to be

$$B_x = -\beta \cdot E_y, \qquad B_y = +\beta \cdot E_x. \qquad (12.9)$$

The electric as well as the magnetic field scales linearly with distance from the beam center and therefore both cause focusing and a tune shift in a circular accelerator.

In many applications it is not acceptable to assume a uniform transverse charge distribution. Most particle beams either have a bell shaped particle distribution or a gaussian distribution as is specially the case for electrons in circular accelerators. We therefore use in the transverse plane a gaussian charge distribution given by

$$\rho(x, y) = \frac{\lambda}{2\pi \, \sigma_x \sigma_y} \cdot \exp\left[-\frac{x^2}{2\sigma_x^2} - \frac{y^2}{2\sigma_y^2} \right]. \qquad (12.10)$$

Although many particle beams, but specifically electron beams, come in bunches with a gaussian distribution in all degrees of freedom, we will only introduce a bunching factor for the longitudinal variable and refer the interested reader for the study of a fully six dimensional gaussian charge distribution to ref. [12.3].

The potential for a transverse bigaussian charge distribution (12.10) can be expressed by [12.2]

$$V(x, y) = -e\lambda \int\limits_0^\infty \frac{1 - \exp\left[-\frac{x^2}{2(\sigma_x^2 + t)} - \frac{y^2}{2(\sigma_y^2 + t)} \right]}{\sqrt{(\sigma_x^2 + t) \cdot (\sigma_y^2 + t)}} \cdot dt \qquad (12.11)$$

which can be verified by back insertion into (12.1). From this potential we obtain for example the vertical electric field component by differentiation

$$E_y = -\frac{\partial V(x, y)}{\partial y} = e\lambda y \int\limits_0^\infty \frac{\exp\left[-\frac{x^2}{2(\sigma_x^2 + t)} - \frac{y^2}{2(\sigma_y^2 + t)} \right]}{(\sigma_y^2 + t) \cdot \sqrt{(\sigma_x^2 + t) \cdot (\sigma_y^2 + t)}} \cdot dt. \qquad (12.12)$$

No closed analytical expression exists for these integrals unless we restrict ourselves to a symmetry plane with $x = 0$ or $y = 0$ and small amplitudes $y \ll \sigma_y$ or $x \ll \sigma_x$ respectively. These assumptions are appropriate for most space-charge effects and the potential in the vertical midplane becomes

$$V(x = 0, y \ll \sigma_y) = -\frac{\lambda}{\sigma_y \cdot (\sigma_x + \sigma_y)} \cdot y^2. \qquad (12.13)$$

For reasons of symmetry a similar expression can be derived for the horizontal mid plane by merely interchanging x and y in (12.13). The associated electric fields are for $x = 0$ and $y \ll \sigma_y$

$$E_x = \frac{2\lambda}{\sigma_x \cdot (\sigma_x + \sigma_y)} \cdot x, \qquad E_y = \frac{2\lambda}{\sigma_y \cdot (\sigma_x + \sigma_y)} \cdot y, \qquad (12.14)$$

and the magnetic fields according to (12.9) are from (12.14)

$$B_x = -\frac{2\lambda\beta}{\sigma_y \cdot (\sigma_x + \sigma_y)} \cdot y, \qquad B_y = +\frac{2\lambda\beta}{\sigma_x \cdot (\sigma_x + \sigma_y)} \cdot x. \qquad (12.15)$$

All fields increase linearly with amplitude and we note that the field components in the horizontal midplane are generally much smaller compared to those in the vertical midplane because most particle beams in circular accelerators are flat and $\sigma_y \ll \sigma_x$.

12.1.2 Forces from Space-Charge Fields

The electromagnetic self fields generated by the collection of all particles within a beam exert forces on individual particles of the same beam or of another beam. The *Lorentz force* due to these fields can be expressed by

$$\mathbf{F} = e\,\mathbf{E} \cdot f_e + \frac{e}{c}[\mathbf{v} \times \mathbf{B}] \cdot f_e \cdot f_v, \qquad (12.16)$$

where we have added to the usual expression for the Lorentz force the factors f_e and f_v. Because the fields act differently depending on the relative directions and charge of beam and individual particle distinct combinations occur. We set $f_e = 1$ if both the beam particles and the individual particle have the same sign of their charge and $f_e = -1$ if their charges are of opposite sign. Similarly we set $f_v = 1$ or $f_v = -1$ depending on whether the beam and individual particle have the same or opposite direction of movement with respect to each other.

The vertical force from the self field, for example, of a proton beam on an individual proton within the same beam moving with the same velocity is from (12.16)

$$F_y(\uparrow\uparrow,++) = +e(1 - \beta^2) \cdot E_y. \qquad (12.17)$$

An antiproton moving in the opposite direction through a proton beam would feel the vertical force

$$F_y(\uparrow\downarrow,+-) = -e(1 + \beta^2) \cdot E_y. \qquad (12.18)$$

Expansion to other combinations of particles and directions of velocities are straightforward. For ions the charge multiplicity Z must be added to the fields or the individual particle or both depending on the case. The possible combinations of the force factors $\pm(1 \pm \beta^2)$ are summarized in Table 12.1.

Table 12.1. Self field force factors

++ ↑↑	+− ↑↑	++ ↑↓	+− ↑↓
−− ↑↑	−+ ↑↑	−− ↑↓	−+ ↑↓
$+(1-\beta^2)$	$-(1-\beta^2)$	$+(1+\beta^2)$	$-(1+\beta^2)$

The ±-signs in Table 12.1 indicate the charge polarity of beam and particle and the arrows the relative direction. We note a great difference between the case, where particles move in the same direction, and the case of beams colliding head on.

12.2 Beam – Beam Effect

In colliding beam facilities two counter rotating beams within one storage ring or counter rotating beams from two intersecting storage rings are brought into collision to create a high center of mass energy at the collision point which transforms into known or unknown particles to be studied by high energy experimentalists. The event rate is given by the product of the cross section for the particular event and the *luminosity* which is determined by storage ring operating conditions. By definition, the luminosity is the density of collision centers in the target multiplied by the number of particles colliding with this target per unit time. In the case of a colliding beam facility one beam is the target for the other beam. For simplicity we assume here that both beams have the same cross section and defer the general case to Vol. II. We also assume that each beam consists of B bunches. In this case the luminosity is

$$\mathcal{L} = \frac{N_1}{B A} N_2 \, \nu_{\text{rev}} \, , \tag{12.19}$$

where N_1 and N_2 are the total number of particles in each beam, A the cross section of the beams, and ν_{rev} the revolution frequency in the storage ring. In most storage rings the transverse particle distribution is gaussian or bell shaped and since only the core of the beam contributes significantly to the luminosity we may define standard beam sizes for all kinds of particles. For a gaussian particle distribution the effective beam cross section is

$$A_g = 4\pi \, \sigma_x \sigma_y \tag{12.20}$$

and the luminosity

$$\mathcal{L} = \frac{N_1}{4\pi \, \sigma_x \sigma_y \, B} N_2 \, \nu_{\text{rev}} \, . \tag{12.21}$$

The recipe for high luminosity is clearly to maximize the beam intensity and to minimize the beam cross section. This approach, however, fails because of the *beam – beam effect* which, due to electromagnetic fields created by the beams themselves, causes a tune shift and therefore limits the amount of beam that can be brought into collision in a storage ring. The beam – beam effect has first been recognized and analyzed by Amman and Ritson [12.4]

From Sect. (12.1) we find, for example, for two counter rotating beams of particle and antiparticle a vertical beam beam force of

$$F_y = -\frac{e \cdot (1 + \beta^2) 2 \lambda}{\sigma_y (\sigma_x + \sigma_y)} y. \tag{12.22}$$

This force is attractive and therefore focusing, equivalent to that of a quadrupole of strength

$$k = -\frac{F_y/y}{c^2 \beta^2 \gamma m} \tag{12.23}$$

causing a vertical tune shift of

$$\delta\nu_y = \frac{1}{4\pi} \int_{\text{coll}} \beta_y \, k \, ds. \tag{12.24}$$

Integrating over the collision length which is equal to half the bunch length ℓ because colliding beams move in opposite directions, we note that the linear charge density is $\lambda = eN/B/\ell$, where N is the total number of particles per beam and B the number of bunches per beam. With these replacements the beam beam tune shift becomes finally

$$\delta\nu_y = \frac{r_o N \beta_y}{2\pi B \sigma_y (\sigma_x + \sigma_y)}, \tag{12.25}$$

where r_o is the classical particle radius of that particle which is being disturbed. Obviously, the tune shift scales linear with particle intensity or particle beam current and inversely with the beam cross section. Upon discovery of this effect it was thought that the particle beam intensity is limited when the tune shift is of the order of $\approx 0.15 - 0.2$ which is the typical distance to the next resonance. Experimentally, however, it was found that the limit is much more restrictive with maximum tune shift values of $\approx 0.04 - 0.06$ for electrons [12.5– 8] and less for proton beams .

A definitive quantitative description of the actual beam – beam effect has not been possible yet due to its highly nonlinear nature. Only particles with very small betatron oscillation amplitudes will experience the linear tune shift derived above. For betatron oscillations larger than one σ, however, the field becomes very nonlinear turning over to the well known $1/r$-law at large distances from the beam center. In Vol. II we will discuss in more detail the phenomenology of the beam – beam effect, some empirical scaling as well as some theoretical models.

In spite of the inability to quantitatively describe the beam – beam effect by the linear tune shift it is generally accepted practice to quantify the beam – beam limit by the value of the linear tune shift. This is justified since the nonlinear fields of a particle beam are strictly proportional to the linear field and therefore the linear tune shift is a good measure for the amount of nonlinear fields involved.

12.3 Wake Fields

Electron storage rings are being planned, designed, constructed, and operated for a variety of applications. While in the past such storage rings were optimized mostly as colliding beam facilities for high energy physics, it becomes more and more apparent that in the future most applications for storage rings will be connected with the production of synchrotron radiation. Some of these radiation sources will be designed for high energy particle beams to produce hard x-rays while others have moderate to low beam energies to, for example, produce VUV and soft x-rays or to drive free electron lasers.

Since the radiation intensity produced is directly proportional to the stored electron beam current, it is obvious that the usefulness of such a radiation source depends among other parameters on the maximum electron beam current that can be stored in the storage ring. We will mainly point out limiting phenomena for the storable beam current specifically those that are determined by the particular design of the vacuum chamber and delay a more detailed discussion of instabilities to Vol. II.

The beam in an electron storage ring is composed of bunches which are typically a few centimeters long and are separated by a distance equal to one or more rf wavelengths. The total number of bunches in a storage ring can range from one bunch to a maximum of h bunches, where h is the *harmonic number* for the storage ring system.

Depending on the particular application and experiment conducted it may be desirable to store only a single bunch with the highest intensity possible. In other cases the maximum total achievable intensity is desired in as many bunches as possible and the particular bunch distribution around the ring does not matter. In either case the ultimate electron beam intensity will most probably be limited by instabilities caused by electromagnetic interaction of the beam current with the environment of the vacuum chamber. We ignore here technical limitations due to, for example, insufficient available rf power or inability to cool the radiation heating of the vacuum chamber.

12.3.1 Parasitic Mode Losses and Impedances

Due to tight particle bunching by the rf system to about (1/20)th of the rf wavelength, we have large instantaneous currents with significant amplitudes of Fourier components at harmonics of the revolution frequency up to about 20 times the rf frequency or down to wavelength of a few centimeters. Strong electromagnetic interaction between electron bunches and cavity like structures as part of the vacuum enclosure must therefore be expected. Any but the smallest steps in the cross section of the vacuum chamber constitute cavity like structures. A bunch passing by such a structure deposits electromagnetic energy which in turn causes heating of the structure and can act back on particles in a later segment of the same bunch or in a subsequent bunch. Schematically such fields, also called *wake fields*, are shown in Fig. 12.1 where the beam passes by a variation in the cross section of the vacuum chamber [12.10]. We will discuss the nature and the frequency spectrum of these wake fields to determine the effect on the stability of the beam and to develop counter measures to minimize the strength and occurrence of these wake fields.

We distinguish *broad band parasitic losses* where the quality factor Q is of the order of unity from narrow band losses with higher Q values. Fields

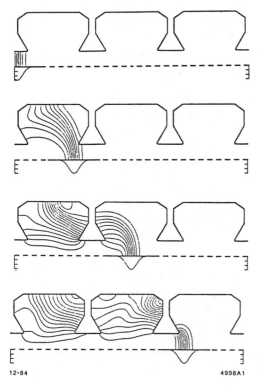

12-84   4998A1   **Fig. 12.1.** Parasitic mode fields [12.10]

from broad band losses last only a very short time of the order of one period and are mainly responsible for *single bunch instabilities*, where the fields generated by electrical charges in the head of the bunch act back on the particles in the tail of the same bunch. Due to the low value of the quality factor, $Q \approx 1$, these broad band wake fields decay before the next bunch arrives. These kind of losses are mainly due to sudden changes of the vacuum chamber cross section. Since these fields last only for a short time, they are responsible for *single bunch instabilities*.

Other structures, shaped like resonant cavities, can be excited by a passing particle bunch in *narrow band field modes* with relatively high values for the quality factor Q. Such fields persist a longer time and can act back on some or all subsequent bunches as well as on the same bunch after one or more revolutions. Due to their persistence they may cause *multibunch instabilities*.

Both types of fields can appear as longitudinal or transverse modes and cause correspondingly longitudinal or transverse instabilities. Obviously, a perfect vacuum chamber would have a superconducting surface and be completely uniform around the ring. This is not possible in reality because we need rf systems which by their nature are not smooth, injection/ejection components, synchrotron light ports, bellows, and beam position monitors. While we cannot avoid such lossy components we are able by proper design to minimize the detrimental effects of less than ideal components.

The loss characteristics of a particular piece or of the vacuum chamber for the whole ring is generally expressed in terms of an impedance Z or in terms of a loss factor k. To illustrate the different nature of wake fields we assume a cavity like change in the cross section of the vacuum chamber as shown in Fig. 12.2.

A bunch passing through such a structure on axis excites in lowest order a longitudinal electrical field and a transverse magnetic field as shown. Such

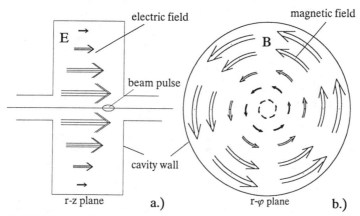

Fig. 12.2. Longitudinal parasitic mode

a field pattern will not cause a transverse deflection of the whole beam since the electrical field is strictly longitudinal and the transverse magnetic field is zero on axis and out of phase. For this situation we define a longitudinal impedance $Z_\parallel$ by

$$Z_\parallel(\omega) = -\frac{\int \mathbf{E}(\omega)\,\mathrm{d}s}{I(\omega)}, \tag{12.26}$$

where $\mathbf{E}(\omega)$ is the electric field at the frequency ω and $I(\omega)$ the Fourier transform of the bunched beam current. The r.h.s. of (12.26) is the energy gained per unit charge and is equivalent to an accelerating voltage divided by the current, where the actual frequency dependence depends on the specific physical shape of the "resonating" structure.

In a similar way we can define a transverse impedance. A beam passing off axis through a "cavity" excites asymmetric fields, as shown in Fig. 12.3, proportional to the moment of the beam current $I(\omega)\,\Delta x$, where Δx is the displacement of the beam from the axis.

Such an electrical field is connected through Maxwell's equation with a finite transverse magnetic field on axis, as shown in Fig. 12.3 which causes a transverse deflection of the beam. Consistent with the definition of the longitudinal impedance we define a transverse impedance by

$$Z_\perp(\omega) = \mathrm{i}\,\frac{\int \big(\mathbf{E}(\omega) + \frac{1}{c}\left[\mathbf{v}\times\mathbf{B}(\omega)\right]\big)_\perp\,\mathrm{d}s}{I(\omega)\,\Delta x}, \tag{12.27}$$

where $\mathbf{v}$ is the velocity of the particle and $\mathbf{B}(\omega)$ the magnetic field component of the electromagnetic field at frequency ω. In general the impedances are complex

$$Z(\omega) = Z_{\mathrm{Re}}(\omega) + \mathrm{i}\,Z_{\mathrm{Im}}(\omega). \tag{12.28}$$

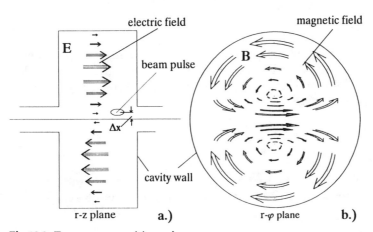

Fig. 12.3. Transverse parasitic mode

The resistive part of the impedance can lead to a shift in the betatron oscillation frequency of the particles while the reactive or imaginary part may cause damping or antidamping.

The impedance is a function of the frequency and its spectrum depends on the specific design of the vacuum chambers in a storage ring. The longitudinal impedance of vacuum chambers has been measured in SPEAR and in other existing storage rings and has been found to follow a general spectrum as a consequence of similar design concepts of storage ring components. SPEAR measurements, shown in Fig. 12.4, demonstrate the general form of the frequency spectrum of the vacuum chamber impedance [12.11].

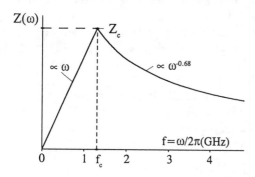

Fig. 12.4. SPEAR impedance spectrum [12.11]

Characteristic for the spectrum is the cutoff frequency f_c at which the linear impedance function reaches a maximum and above which the fields are able to propagate in the vacuum chamber. This cutoff frequency obviously is determined by the aperture of the vacuum chamber and therefore occurs at different frequencies for different rings with different vacuum chamber apertures. For the longitudinal broad band impedance at high frequencies above the cutoff frequency f_c we have the simple power law

$$Z_\parallel(\omega) = Z_c \, \omega^{-0.68}, \qquad (\omega > \omega_c). \tag{12.29}$$

To simplify comparisons between different storage rings we define a normalized impedance Z/n as the impedance at the cut off frequency divided by the mode number n which is the ratio of the cutoff frequency f_c to the revolution frequency f_{rev}

$$\left| \frac{Z}{n} \right|_c = \left| \frac{Z_c}{f_c/f_{rev}} \right|. \tag{12.30}$$

This definition of the normalized impedance can be generalized to all frequencies and together with (12.29) the impedance spectrum becomes

$$\left| \frac{Z_\parallel}{n} \right|_{eff} = \left| \frac{Z_\parallel}{n} \right|_c \left(\frac{\omega}{\omega_c} \right)^{-1.68}. \tag{12.31}$$

394

The interaction of the beam with the vacuum chamber impedance leads to an energy loss which has to be compensated by the rf system. We characterize this loss through the loss factor k defined by

$$k = \frac{\Delta U}{q^2},\tag{12.32}$$

where ΔU is the total energy deposited by the passing bunch and q is the total electrical charge in this bunch. This definition is a generalization of the energy loss of a single particle passing once through a resonator where $k = -(\omega/4)(R_s/Q)$ and R_s is the shunt impedance of this resonator. The loss factor is related to the real part of the impedance by

$$k = \frac{2}{q^2} \int_o^\infty \mathrm{Re}[Z(\omega)]\, I^2(\omega)\, \mathrm{d}\omega\tag{12.33}$$

and depends strongly on the bunch length as can be seen from measurements of the loss factor in SPEAR [12.12] shown in Fig. 12.5. Specifically, we find the loss factor to scale with the bunch length like

$$k(\sigma_\ell) \sim \sigma_\ell^{-1.21}.\tag{12.34}$$

Similar to the definitions of impedances we also distinguish a longitudinal and a transverse loss factor. Where only one is known, we can make an estimate of the other one through the approximate relation which is correct only for cylindrically symmetric structures [12.13, 14]

$$Z\perp = \frac{2R}{b^2} \frac{Z_\parallel}{n},\tag{12.35}$$

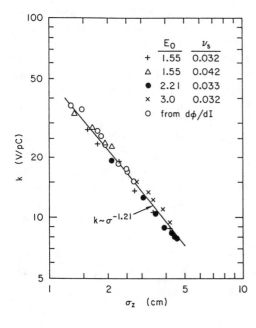

Fig. 12.5. Parasitic loss factor k in the storage ring SPEAR [12.12]

395

where $2\pi R$ is the ring circumference and b the typical vacuum chamber radius. The longitudinal impedance of the whole storage ring vacuum system including rf cavities can be determined by measuring the energy loss of particles in a high intensity bunch compared to the energy loss for particles in a low intensity bunch. Such loss measurements are performed by observing the shift in synchronous phase for the low and high intensity beam. The parasitic losses of rf cavities can be calculated very accurately with computer programs or are known from laboratory measurements. From the separate knowledge of cavity and total ring losses we derive the vacuum chamber losses by simple subtraction.

12.4 Beam Instabilities

Longitudinal or transverse impedances in a storage ring are responsible for a number of instabilities limiting the maximum attainable beam current. We distinguish two classes of instabilities, *single bunch instabilities* and *multibunch instabilities*. Single bunch current limitations or instabilities are mostly caused by the broad band, low Q impedances of the beam enclosure. These instabilities can occur in longitudinal as well as in transverse phase space.

The longitudinal broad band impedance can cause an instability which leads to internal bunch oscillations resulting in bunch lengthening and an increase in the beam energy spread. The onset of this turbulent bunch lengthening instability has been measured at SPEAR and is shown in Fig. 12.6 [12.12].

The increase in the bunch length and energy spread is obvious from these measurements for beam currents exceeding the threshold current at which point the first mode of internal turbulence, the quadrupole mode, becomes excited. This quadrupole oscillation manifests itself in the appearance of a strong frequency signal from the beam at twice the synchrotron oscillation frequency. Higher oscillation modes are excited at higher beam currents.

Internal bunch oscillations can be excited only by fields with wavelength of the order of the bunch length or shorter. For bunches shorter than the vacuum chamber dimensions only the part above the cutoff frequency of the impedance spectrum is relevant. If the impedance spectrum for a particular ring is known, like (12.31) for SPEAR, we can calculate the onset of the *bunch lengthening instability* or *micro wave instability* [12.15]. The threshold current per bunch for this instability to appear is given by

$$I_{\parallel \mathrm{th}} \approx \frac{\sqrt{2\pi}\, \alpha_{\mathrm{c}}\, (E/e)\, \sigma_\ell}{R\, (Z_\parallel/n)_{\mathrm{eff}}} \left(\frac{\sigma_E}{E}\right)^2 , \tag{12.36}$$

where α_{c} is the momentum compaction factor, σ_E/E the energy spread in the beam, σ_ℓ the bunch length, and R the average storage ring radius.

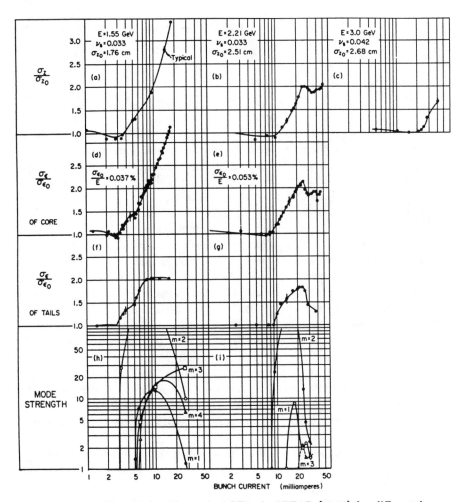

Fig. 12.6. Onset of longitudinal beam instability in SPEAR [12.12] for different beam energies. The graphs show the bunch length, the energy spread in the core and in the tails, and the mode signal as a function of bunch current for different energies

The microwave instability becomes particularly effective when very short bunches are desired. It does, however, not necessarily lead to beam loss because the instability is self stabilizing through Landau damping effected by the increased energy spread in the beam. It should be noted that the impedance in (12.34) must also include the broad band impedance of the rf cavity. We note that a large value for the broad band impedance should be avoided to reach high beam currents at short bunch lengths.

The transverse broad band impedance can lead to a *fast head tail instability* for positive chromaticities. This instability has been observed in many storage rings and manifests itself by a beam current dependent shift in the betatron tune. According to theory the current limit is reached when the betatron oscillation frequency has shifted by one half the synchrotron frequency. The functional dependencies of the threshold bunch current is given by [12.16]

$$I_{\perp th} \approx \frac{\nu_s (E/e) \sigma_\ell b^2}{(Z_\parallel/n)_{eff} \langle \beta_b \rangle}, \tag{12.37}$$

where ν_s is the synchrotron tune, $2b$ the smaller of the average horizontal or vertical vacuum chamber aperture, σ_ℓ the bunch length, and $\langle \beta_b \rangle$ the average value of the betatron function in the plane of the smaller aperture. This instability limits the single bunch current and is inversely proportional to the broad band impedance of the whole ring.

If the storage ring is filled with many bunches, there is a possibility that these bunches interact via the finite impedances with each other and instabilities may occur. Total energy considerations will not provide any saturation of such instabilities since the rf system provides a virtually un-limited source of energy for the stored beam. For such an interaction between bunches to happen we must require that the fields in resonating structures of the vacuum chamber must persist for many oscillations to still be present when subsequent bunches pass by. This excludes fields generated through broad band impedances since these occur at very high frequencies where the quality factor Q is low. For multibunch instabilities we are therefore look-ing for cavity like structures which have a larger value for the quality factor Q. The most obvious of such structures are of course the rf cavities in the storage ring. Indeed, in a well designed vacuum system for a storage ring the narrow band impedances of the vacuum system are negligible compared to those of the rf cavities.

A train of n bunches can oscillate in a number of modes with frequencies given by

$$\omega_p = \omega_{rev} \left[p \cdot n \pm (m + \nu) \right], \tag{12.38}$$

where $m = 1, 2, 3, \ldots (n-1)$, p is an integer, ω_{rev} the revolution frequency and ν the tune. The tune may be the synchrotron tune for longitudinal oscil-lations or the betatron tune for transverse betatron oscillations. Interaction of such a multibunch beam with an impedance leads to a complex shift of the oscillation frequency. The imaginary part of this frequency shift can cause damping or antidamping of specific mode frequencies. The complex shift of the frequency is proportional to [12.17]

$$\Delta\omega \sim i\, I_b \frac{\alpha_c\, Z(\omega_p)}{\nu_s\, (E/e)}, \tag{12.39}$$

were I_b is the bunch current and $Z(\omega_p)$ the impedance at any of the mode frequencies ω_p. The imaginary part of the frequency shift is the damping or antidamping rate of the instability, $1/\tau = \text{Im}(\Delta\omega)$.

Multibunch instabilities lead to increasing oscillations between bunches and eventually as the beam current is being further increased to loss of part of the beam, thus limiting the maximum attainable beam current. As in the cases for the single bunch instabilities, we find again that the complex frequency shift is proportional to the impedance. Here, however, we are looking for narrow band impedances which are large at frequencies equal to some multibunch oscillation mode frequency ω_p.

In summary we note a number of different instabilities which may limit the single bunch current as well as the multibunch current in a storage ring. All these instabilities result from the electromagnetic interaction of the particle beam with its environment. The strength of the interaction is governed by an impedance function which depends on the particular shape of the interacting environment. Theoretically, the onset of these instabilities can be derived and its parametrical dependence becomes obvious.

All limits on the beam current are proportional to the energy E of the beam, to the bunch length σ_ℓ and inversely proportional to the longitudinal or transverse impedance Z of the vacuum chamber. These dependencies are important guidelines for the design of new storage rings and specifically for the design of vacuum chambers and rf cavities. The importance of specific corrective measures depends much on the particular goals and applications of the storage ring.

The energy dependence is easily understandable since the reaction of particles to electromagnetic fields is inversely proportional to the energy of the particle. This aspect of the beam current limitation becomes important when one tries to accumulate beam into a storage ring at a low beam energy to save the cost of a full energy injector. In this case it is clear that the maximum achievable current in the storage ring is given by the lower current thresholds at the low injection energy.

It also has become apparent that instabilities are worse for short bunches than for long bunches. Nothing specific can be done to improve the instability situation where short bunches are desired. For many applications, however, the bunch length is irrelevant and it is clear that for such cases a longer bunch length would allow the storage of higher beam current. A longer bunch length can be obtained utilizing a second rf system operating at a higher rf frequency which can be tuned to lengthen the bunch through potential well distortion. Another solution is the use of a lower frequency main rf system which is the most economic solution for small, low energy storage rings.

Finally the maximum achievable beam current greatly depends on the impedance of the vacuum enclosure and special effort must be employed to minimize the impedance characteristics of both the vacuum chamber and the rf cavities. The optimum solution for the vacuum enclosure would have a perfectly uniform cross section all around the storage ring. It is therefore of great importance for the current capability of any storage ring that the vacuum chamber be designed such as to avoid as much as possible any change in the cross section. Obviously, we cannot do this perfectly since we have to provide for components like flanges, bellows, injection lines, position monitors to name just some of the components that introduce a finite impedance. The challenge for the design engineer will be to design the vacuum chamber components such that they perform their function while introducing the minimum possible impedance.

In addition to beam instabilities caused by vacuum chamber impedances we also observe significant heating losses due to the energy contained in the fields which become absorbed in the vicinity of cavity like structures. The higher order mode heating can be very severe and can lead to malfunctioning or catastrophic destruction of vacuum chamber elements. At the least the elevated temperatures of vacuum components due to higher order mode losses will cause higher thermal gas desorption than expected.

Measurements have been performed in many laboratories on basic vacuum chamber components (for example [12.18]). Some components show large unavoidable impedances, like electrostatic separating plates for high energy colliding beam facilities, and therefore appropriate cooling must be provided. In other cases the measured impedance of individual pieces are rather small and we need not be concerned about heating effects but the total impedance for many such pieces in the ring can become important with respect to instabilities. Open pump ports for example provide enough material to absorb the heating losses but constitute a significant impedance because there are many such components in a storage ring.

For high intensity storage rings we find therefore two reasons to minimize the impedance of components, beam instabilities and heating of the components. For this reason the gaps between flanges must be bridged by a metallic seal, the bellows must be shielded from the beam by an additional metallic enclosure, and the pump ports must be screened. Where a change in the cross section of the vacuum chamber cannot be avoided a gentle transition must be provided. With these precautions the loss factor or impedance can be reduced by as much as an order of magnitude to such low values that the vacuum chamber impedance becomes much smaller than that of the rf cavity.

A careful design of a storage ring will include the measurement, calculation or estimation of the impedance for each vacuum chamber component. Electrical models should be built and the impedance measured. The impedance for more regularly shaped components, particularly rotationally symmetric components can be calculated by computer programs. Estimates of component impedances determine the power losses and design engineers can then apply special cooling measures. For the whole storage ring an impedance budget including measured, calculated or estimated values for the impedance for every single component of the vacuum system should be established. Based on this impedance budget beam current limits and instability thresholds can be calculated.

We have summarily scanned over the problems of higher order mode losses and associated problems of instability and heating. A more detailed evaluation of wake fields and beam interaction with its environment will be pursued in Vol. II together with a discussion of related beam instabilities.

Problems

Problem 12.1. Verify that (12.3, 11) are indeed solutions of the respective Poisson equation. Prove that (12.13) is the potential for small amplitudes and $x = 0$.

Problem 12.2. Calculate the linear beam – beam tune shift for each beam under the following head on colliding beam conditions:
a) a 250 GeV proton beam colliding with a fully ionized 30 GeV/u Au ion beam. (proton emittance $\epsilon_{x,y} = 20\pi$ mm mrad, gold ion emittance $\epsilon_{x,y} = 33\pi$ mm mrad, $\beta^*_{x,y} = 2.0$ m, proton intensity 10^{11} p/bunch, a total of 60 bunches per beam, gold ion intensity 10^9 Au ions/bunch) [12.19].
b) a 250 GeV proton beam colliding with a fully ionized 100 GeV/u Au ion beam (parameters same as in a) but gold ion emittance $\epsilon_{x,y} = 10\pi$ mm mrad) [12.19].
c) a 30 GeV electron beam colliding with a 820 GeV proton beam [12.20–22]. The circumference of the rings is 6336 m, there are $2.1 \cdot 10^{13}$ protons and $0.8 \cdot 10^{13}$ electrons in 210 bunches and the beam sizes at the collision point are $\sigma_x/\sigma_y = 0.29/0.07$ mm for the proton beam and $0.26/0.02$ mm for the electron beam.
d) a 1.5 GeV electron beam colliding with a 1.5 GeV positron beam at a collision point with $\epsilon_x = 0.67 \, 10^{-6}$ rad-m, emittance coupling 27.7%, $\beta^*_x = 1.3$ m, $\beta^*_y = 0.1$ m and a beam current of 66 mA [12.23].

Problem 12.3. Estimate the strength of the octupole field component of the proton beam in RHIC at the collision point. Would an octupole be technically feasible to compensate for the beam – beam octupole term?

Problem 12.4. At the Stanford Linear Collider, SLC, an electron beam collides with a positron beam at up to 50 GeV per beam. Each bunch contains $5 \, 10^{11}$ particles and is focused to a beam diameter of 2.0 μm at the collision point where the betatron functions in both planes are $\beta^* = 0.005$ m. Calculate the beam – beam tune shift and the focal length of the beam lens for a bunch length of $\ell = 1$ mm. Compare with beam – beam limits in storage rings. Why can we tolerate a much greater beam – beam tune shift in a linear collider compared with a storage ring?

13 Beam Emittance and Lattice Design

The task of lattice design for proton and ion beams can be concentrated to a purely particle beam optics problem. Transverse as well as longitudinal emittances of such beams are constants of motion and therefore do not depend on the particular design of the beam transport or ring lattice. This situation is completely different for electron and positron beams in circular accelerators where the emission of synchrotron radiation determines completely the particle distribution in six-dimensional phase space. The magnitude and characteristics of synchrotron radiation effects can, however, be manipulated and influenced by an appropriate choice of lattice parameters and characteristics. We will discuss optimization and scaling laws for the transverse beam emittance of electron or positron beams in circular accelerators. A similar discussion on the manipulation of longitudinal phase space will be postponed to Vol. II due to the need to include nonlinear beam dynamics.

Originally electron storage rings have been designed, optimized, and constructed for the sole use as colliding beam facilities for high energy physics. The era of electron storage rings for experimentation at the very highest particle energies has, however, reached a serious limitation due to excessive energy losses into synchrotron radiation at very high particle energies. Of course, such a limitation does not exist for proton and ion beams with particle energies up to the order of some tens of TeV's and storage rings are therefore still the most powerful and productive research tool in high-energy physics. At lower and medium-energies electron storage rings with specially high luminosity still serve as an important research tool in high energy physics to study more subtle phenomena which could not be detected on earlier storage rings with lower luminosity.

To overcome the energy limitation in electron colliding beam facilities, the idea of *linear colliders* which avoids energy losses into synchrotron radiation [13.1, 2] becomes increasingly attractive to reach ever higher center of mass energies for high-energy physics. Even though electron storage rings are displaced by this development as the central part of a colliding beam facility they play an important role for linear colliders in the form of damping rings to prepare very small emittance particle beams.

The single purpose of electron storage rings for high-energy physics has been replaced by a multitude of applications of synchrotron radiation from such rings in a large variety of basic and applied research disciplines. It is therefore appropriate to discuss specific design and optimization criteria for electron storage rings.

Whatever the applications, in most cases it is the beam emittance which will ultimately determine the usefulness of the storage ring design for a particular application. We will derive and discuss physics and scaling laws for the equilibrium beam emittance in storage rings while using basic phenomena and processes of accelerator physics as derived in previous sections.

13.1 Equilibrium Beam Emittance in Storage Rings

The equilibrium beam emittance in electron storage rings is determined by the counteracting effects of quantum excitation and damping as has been discussed in Chap. 10. Significant synchrotron radiation occurs only in bending magnets and we find that the radiation from each bending magnet contributes independently to both quantum excitation and damping. The contribution of each bending magnet to the equilibrium beam emittance can be determined by calculating the average values for $\left\langle \frac{\mathcal{H}}{|\rho^3|} \right\rangle$ and $\left\langle \frac{1}{\rho^2} \right\rangle$ by

$$
\left\langle \frac{\mathcal{H}}{|\rho^3|} \right\rangle = \frac{1}{C} \int_0^C \frac{\mathcal{H}(s)}{|\rho^3(s)|} \, ds \,, \tag{13.1}
$$

where $\mathcal{H}$ is defined by (10.44) and C is the circumference of the storage ring. Obviously, this integral receives contributions only where there is a finite bending radius and therefore the total integral is just the sum of individual integrals over each bending magnet.

In evaluating the integral (13.1) we must be careful to include all contributions. The emission of photons depends only on the bending radius regardless of whether the bending occurs in the horizontal or vertical plane and since for the calculation of equilibrium beam emittances we are interested only in the energy loss due to the emission of photons; it is irrelevant in which direction the beam is bent. The effect of the emission of a photon on the particle trajectory, however, is different for both planes because in general the horizontal and vertical lattice and dispersion functions are different resulting in a different quantum excitation factor $\mathcal{H}$. For a correct evaluation of the equilibrium beam emittances in the horizontal and vertical plane (13.1) should be evaluated for both planes by determining $\mathcal{H}_x$ and $\mathcal{H}_y$ separately but including in both calculations all bending magnets in the storage ring.

The integral in (13.1) can be evaluated for each magnet if the values of the lattice functions at the beginning of the bending magnet are known. With these initial values the lattice functions at any point within the bending magnet can be calculated assuming a pure dipole magnet. With the definitions of parameters from Fig. 13.1 we find the following expressions

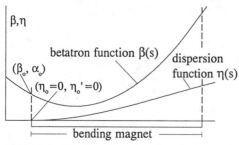

Fig. 13.1. Lattice functions in a bending magnet

for the lattice functions in a bending magnet where s is the distance from the entrance to the magnet

$$\beta(s) = \beta_{\mathrm{o}} - 2\alpha_{\mathrm{o}}s + \gamma_{\mathrm{o}}s^2\,,$$
$$\alpha(s) = \alpha_{\mathrm{o}} - \gamma_{\mathrm{o}}s\,,$$
$$\gamma(s) = \gamma_{\mathrm{o}}\,, \tag{13.2}$$
$$\eta(s) = \eta_{\mathrm{o}} + \eta'_{\mathrm{o}}s + \rho\,(1 - \cos\theta)\,,$$
$$\eta'(s) = \eta'_{\mathrm{o}} + \sin\theta\,.$$

Here the deflection angle is $\theta = s/\rho$ and $\beta_{\mathrm{o}}, \alpha_{\mathrm{o}}, \gamma_{\mathrm{o}}, \eta_{\mathrm{o}}, \eta'_{\mathrm{o}}$ are t
he values of the lattice functions at the beginning of the magnet. Before we use these equations we assume lattices where $\eta_{\mathrm{o}} = \eta'_{\mathrm{o}} = 0$. The consequences of this assumption will be discussed later. Inserting (13.2) into (13.1) we get after integration over one dipole magnet an expression of the form

$$\langle\mathcal{H}\rangle_{\mathrm{b}} = \beta_{\mathrm{o}}\,B + \alpha_{\mathrm{o}}\rho\,A + \gamma_{\mathrm{o}}\,\rho^2\,C\,, \tag{13.3}$$

where we have assumed the bending radius to be constant within the length ℓ_{b} of the magnet and where A, B, and C are functions of the full dipole deflection angle $\Theta = \ell_{\mathrm{b}}/\rho$ defined by

$$B = \tfrac{1}{2}\left(1 - \frac{\sin 2\Theta}{2\Theta}\right),$$
$$A = 2\,\frac{1 - \cos\Theta}{\Theta} - \tfrac{3}{2}\,\frac{\sin^2\Theta}{\Theta} - \tfrac{1}{2}\Theta + \tfrac{1}{2}\sin 2\Theta\,, \tag{13.4}$$
$$C = \tfrac{3}{4} + 2\cos\Theta + \tfrac{5}{4}\,\frac{\sin 2\Theta}{2\Theta} - 4\,\frac{\sin\Theta}{\Theta} + \tfrac{1}{6}\Theta^2 - \tfrac{1}{4}\Theta\sin 2\Theta$$
$$\quad + \tfrac{3}{2}\sin^2\Theta\,.$$

In a storage ring with dipole magnets of different strength all contributions must be added to give the average quantum excitation term for the whole ring

$$\left\langle \frac{\mathcal{H}}{|\rho^3|} \right\rangle = \frac{1}{C} \sum_i \left\langle \frac{\mathcal{H}_i}{|\rho_i^3|} \right\rangle \ell_{\mathrm{b},i} . \tag{13.5}$$

Here we sum over all magnets i with length $\ell_{\mathrm{b},i}$.

For small bending angles and an *isomagnetic ring* where all bending magnets have the same strength and length the expressions (13.4) can be approximated by

$$\begin{aligned}
B &\approx \tfrac{1}{3} \Theta^2 \left(1 - \tfrac{1}{5} \Theta^2 \right), \\
A &\approx -\tfrac{1}{4} \Theta^3 \left(1 - \tfrac{5}{18} \Theta^2 \right), \\
C &\approx \tfrac{1}{20} \Theta^4 \left(1 - \tfrac{5}{14} \Theta^2 \right).
\end{aligned} \tag{13.6}$$

The factor $\langle \mathcal{H}/|\rho^3| \rangle / \langle 1/\rho^2 \rangle$ becomes for an isomagnetic ring simply $\langle \mathcal{H} \rangle / |\rho|$ and the equilibrium beam emittance is with $J_x = 1$

$$\epsilon_{\mathrm{iso}} = C_q \gamma^2 \Theta^3 \frac{\langle \mathcal{H} \rangle}{|\rho|} . \tag{13.7}$$

Inserting (13.3) with (13.6) into (13.7) we get for the beam emittance in the lowest order of approximation

$$\epsilon_{\mathrm{iso}} = C_q \gamma^2 \Theta^3 \left[\frac{1}{3} \frac{\beta_{\mathrm{o}}}{\ell_{\mathrm{b}}} - \frac{1}{4} \alpha_{\mathrm{o}} + \frac{1}{20} \gamma_{\mathrm{o}} \ell_{\mathrm{b}} \right] + \mathcal{O}(\Theta^4) , \tag{13.8}$$

where $\gamma_{\mathrm{o}} = \gamma(s_{\mathrm{o}})$ is one of the lattice functions not to be confused with the particle energy γ.

Here we have assumed a completely separate function lattice where the damping partition number $J_x = 1$. For strong bending magnets this assumption is not quite justified due to the edge focusing of the bending magnets and the damping partition number should be corrected accordingly.

The result (13.8) shows clearly a cubic dependence of the beam emittance on the deflection angle Θ of the bending magnets which is a general lattice property since we have not yet made any assumption on the lattice type yet. Eqs. (13.3) and (13.8) have minima with respect to both α_{o} and β_{o}. By solving the derivation $\partial \langle \mathcal{H} \rangle / \partial \alpha_{\mathrm{o}} = 0$ for α_{o} we get for the optimum value

$$\alpha_{\mathrm{o\,opt}} = -\frac{1}{2} \frac{A}{C} \frac{\beta_{\mathrm{o}}}{\rho} . \tag{13.9}$$

After inserting (13.9) into (13.3) the optimum value for β_{o} can be derived from the derivative $\partial \langle \mathcal{H} \rangle / \partial \beta_{\mathrm{o}} = 0$ and is given by

$$\beta_{\mathrm{o,opt}} = \frac{2 C \rho}{\sqrt{4BC - A^2}} . \tag{13.10}$$

We insert (13.10) into (13.9) for

$$\alpha_{\mathrm{o,opt}} = \frac{-A}{\sqrt{4BC - A^2}} \tag{13.11}$$

and get for the optimum quantum excitation term (13.3) the simple expression

$$\langle \mathcal{H} \rangle_{\min} = \sqrt{4BC - A^2}\, \rho. \tag{13.12}$$

For small deflection angles $\Theta \ll 1$ the expansions (13.6) can be used to give the optimum lattice parameters at the entrance to the bending magnets where $\eta_o = \eta_o' = 0$

$$\alpha_{o,\mathrm{opt}} \approx \frac{(1 - \frac{5}{18}\,\Theta^2)\,\sqrt{15}}{\sqrt{1 - \frac{61}{105}\,\Theta^2}} \approx \sqrt{15}\,(1 + \tfrac{4}{315}\,\Theta^2),$$

$$\beta_{o,\mathrm{opt}} \approx \frac{\sqrt{12}\,(1 - \frac{5}{14}\,\Theta^2)\,L}{\sqrt{5}\,\sqrt{1 - \frac{61}{105}\,\Theta^2}} \approx \sqrt{\tfrac{12}{5}}\,L\,(1 - \tfrac{7}{105}\,\Theta^2), \tag{13.13}$$

$$\langle \mathcal{H} \rangle_{\min} \approx \frac{\Theta^3 \rho}{4\sqrt{15}}\,\sqrt{1 - \tfrac{61}{105}\,\Theta^2} \approx \frac{\Theta^3 \rho}{4\sqrt{15}}\,(1 - \tfrac{61}{210}\,\Theta^2).$$

The results are very simple for very small deflection angles where we may neglect the quadratic terms in the brackets of (13.13) and get

$$\alpha_{o,\mathrm{opt}} \approx \sqrt{15} \qquad \beta_{o,\mathrm{opt}} \approx \sqrt{\tfrac{12}{5}}\,\ell_{\mathrm{b}} \qquad \langle \mathcal{H} \rangle_{\min} \approx \frac{\Theta^3 \rho}{4\sqrt{15}}. \tag{13.14}$$

The minimum achievable equilibrium beam emittance in an isomagnetic ring is finally

$$\epsilon_{\mathrm{iso}} = C_q\,\gamma^2\,\frac{\Theta^3}{4\sqrt{15}}. \tag{13.15}$$

This approximation to small deflection angles is sufficiently accurate for most storage ring design applications. The error in the beam emittance calculation reaches only 10% for deflection angles of about 33° and therefore only in smaller storage rings with few bending magnets must the quadratic correction terms be taken into account. It is interesting to note, however, from the last equation in (13.13) that these quadratic correction terms lead to a smaller beam emittance compared to the lowest order of approximations although still higher order terms will quickly increase the beam emittance again as the bending angle is increased.

From (13.13) we note that the optimum betatron function must enter into the bending magnet with a strong convergence reaching a minimum at a distance $\ell_{\min} = \alpha_o/\gamma_o = {}^3\!/_8\ell_{\mathrm{b}}$ or about one third into the bending magnet as shown in Fig. 13.1. The dispersion functions η_o and η_o' were set to zero in (13.3). Numerical methods must be used to find the optimum solutions for finite values of the dispersion functions. In the following only very small values $\eta_o \ll 1$ and $\eta_o' \ll 1$ are assumed to determine the correction for a finite dispersion on the beam emittance. Retaining only linear terms in η_o, η_o', Θ and using (13.13) we get

$$\langle \mathcal{H} \rangle_{\eta\,\mathrm{min}} = \langle \mathcal{H} \rangle_{\mathrm{min}} + \frac{1}{\sqrt{5}} \left(\tfrac{5}{3}\eta_{\mathrm{o}} + 6\,\eta_{\mathrm{o}}'\,\ell_{\mathrm{b}} \right) \Theta + \mathcal{O}(\Theta^2). \qquad (13.16)$$

Obviously a further reduction in the beam emittance can, indeed, be obtained for negative values of η_{o} and η_{o}'. Nonlinear terms, however, quickly cause an increase again in the beam emittance. In summary it has been demonstrated that there are certain optimum lattice functions in the bending magnets to minimize the equilibrium beam emittance. No assumption about a particular lattice has been made yet. Another observation is that the beam emittance is proportional to the third power of the magnet deflection angle ($\epsilon \sim \Theta^3$) and proportional to the square of the beam energy ($\epsilon \sim E^2$). This point leads immediately to small deflection angles in order to achieve a small beam emittance. Low emittance storage rings, therefore, are characterized by many short magnet lattice periods.

13.2 Beam Emittance in Periodic Lattices

To achieve a small particle beam emittance a number of different basic magnet storage ring lattice units are available and in principle most any periodic lattice unit can be used to achieve as small a beam emittance as desired. More practical considerations, however, will limit the minimum beam emittance achievable in a lattice. While all lattice types to be discussed have been used in existing storage rings and work well at medium to large beam emittances, differences in the characteristics of particular lattice types become more apparent as the desired equilibrium beam emittance is pushed to very small values.

Of the large variety of magnet lattices that have been used in existing storage we will select three basic types and variations thereof to derive and discuss basic scaling laws for the beam emittance. The chosen lattices are among those most frequently used in storage ring design and therefore are a good representation of existing accelerator design capabilities. The three lattice types are

> the double bend achromat lattice (DBA) ,
> the triple bend achromat lattice (TBA) ,
> the triplet achromat lattice (TAL) ,
> the FODO lattice .

In the remainder of this section we discuss briefly emittance determining characteristics of these lattice types.

13.2.1 The Double Bend Achromat Lattice (DBA)

The *double bend achromat* or *DBA lattice* is designed to make full use of the minimization of beam emittance by the proper choice of lattice functions as discussed earlier. In Fig. 13.2 the basic layout of this lattice is shown.

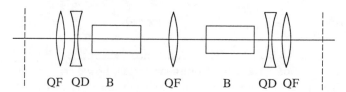

QF QD B   QF   B   QD QF

Fig. 13.2. Double bend achromat (DBA) lattice (schematic) [13.3]

A set of two or three quadrupoles provides the matching of the lattice functions into the bending magnet to achieve the minimum beam emittance. The central part of the lattice between the bending magnets may consist of one or more quadrupoles and its only function is to focus the dispersion function such that it is matched again to zero at the end of the next bending magnet resulting necessarily in a phase advance from bending magnet to bending magnet of close to 180°. This lattice type has been proposed first by Panofsky [13.3] and later by Chasman and Green [13.4] as an optimized lattice for a synchrotron radiation source and is therefore also called the *Chasman Green lattice*.

In Fig. 13.3 an example of a synchrotron light source based on this type of lattice is shown representing the solution of the design study for the European Synchrotron Radiation Facility ESRF [13.5].

The ideal minimum beam emittance in this lattice type has been derived in (13.15) for small bending angles and an isomagnetic ring with $J_x = 1$ as

$$\epsilon_{DBA} = \frac{C_q}{4\sqrt{15}} \gamma^2 \Theta^3 \tag{13.17}$$

or in more practical units

$$\epsilon_{DBA}(\text{rad m}) = 5.036 \times 10^{-13} E^2(\text{GeV}^2) \Theta^3(\text{deg}^3). \tag{13.18}$$

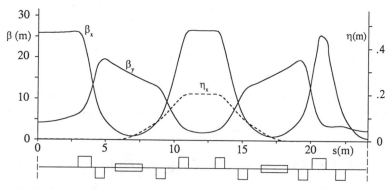

Fig. 13.3. Double bend achromat lattice of the European Synchrotron Radiation Facility, ESRF [13.5] (one half of 16 superperiods). The lattice is asymmetric to provide a mostly parallel beam in one insertion and a small beam cross section in the other.

In the actual ESRF lattice the parameters $(\alpha, \beta, \epsilon)$ achieved while preserving a reasonable range of beam stability, fall short of the optimum parameters $(\alpha_o^*, \beta_o^*, \epsilon_{min}^*)$ from (13.14, 18) as shown in the following table

Design Values:

$\alpha = 1.27$

$\beta = 2.49$ m

$\epsilon = 28.8 \times 10^{-11} \dfrac{\text{rad m}}{\text{GeV}^2}$

Ideal Values:

$\alpha_o^* = 3.873$

$\beta_o^* = 3.04$ m

$\epsilon_{min}^* = 8.96 \times 10^{-11} \dfrac{\text{rad m}}{\text{GeV}^2}$

Obviously, the actual lattice functions and in particular the slope of the betatron function differ from the optimum values as defined by (13.14) and consequently the beam emittance is larger by a factor of three than it could be theoretically. The reason for this are higher order chromatic and geometric aberrations which become large for the ideal solution and would severely limit the transverse stable area available for the beam.

The optimum choice for $\alpha_o^* = \sqrt{15}$ causes the betatron function to reach a sharp minimum at about one third into the bending magnet and then to increase from there on to large values causing nonlinear aberration problems in subsequent quadrupoles. In general it is therefore not possible to reach the optimum conditions of (13.17). For arbitrary values of α_o, however, there is still an optimum value for the initial betatron function β_o which can be derived from (13.8) by differentiation with respect to β_o and solving for

$$\frac{\beta_o}{\ell_b} = \sqrt{\frac{3}{20}} \sqrt{1 + \alpha_o^2}. \tag{13.19}$$

Inserting into (13.8) the beam emittance in units of the minimum beam emittance ϵ_{DBA}^* from (13.17) we obtain

$$\frac{\epsilon_{DBA}}{\epsilon_{DBA}^*} = \tfrac{1}{2} \gamma_o \beta^* + 8 \frac{\beta_o}{\beta^*} - \sqrt{15}\, \alpha_o. \tag{13.20}$$

This minimum beam emittance as defined by (13.20) is plotted in Fig. 13.4 as a function of the betatron function β_o for various parameters α_o.

It is apparent from Fig. 13.4 that more moderate values for α_o can be used without much loss in beam emittance. This weak dependence can be used to lessen the problem caused by nonlinear aberrations. Actually the maximum value of the betatron function $\beta(\ell_b)$ at the end of the bending magnet can be minimized for $\alpha_o = 4\sqrt{15}/17$ at the expense of a loss in beam emittance of about a factor of two. For this condition we have the lattice parameters

$$\frac{\alpha_o}{\alpha_o^*} = \frac{4}{17}, \qquad \frac{\beta(\ell_b)}{\beta_o^*(\ell_b)} = \frac{17}{32},$$

$$\frac{\beta_o}{\beta_o^*} = \frac{23}{68}, \qquad \frac{\epsilon_{DFA}}{\epsilon_{DFA}^*} = \frac{32}{17}. \tag{13.21}$$

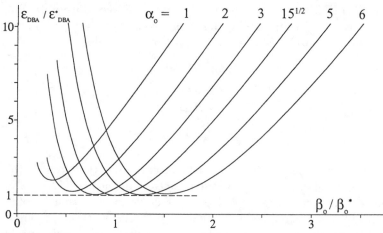

Fig. 13.4. Scaling of beam emittance and lattice functions in a double bend achromat (DBA) lattice

This lattice type can be very useful for synchrotron light sources where many component and dispersion free straight sections are required for the installation of insertion devices. For damping rings this lattice type is not quite optimum since it is a rather "open" lattice with a low bending magnet *fill factor* and consequently a long damping time. Other more compact lattice types must be pursued to achieve in addition to a small beam emittance also a short damping time.

13.2.2 The Triple Bend Achromat Lattice (TBA)

The *triple bend achromat* lattice shown in Fig. 13.5 is a variation of the double bend achromat lattice, where the dispersion function is rather small in the case of a single central quadrupole.

Since the central part of the double bend achromat is the only place where the chromaticity can be corrected we find that the small values of the dispersion function require rather strong sextupole fields.

To avoid the need for longer bending magnet free space and let the dispersion function reach larger values it has been proposed to use some of this extra space for an additional bending magnet although the larger dispersion function in this dipole contributes more to the quantum excitation than the other magnets [13.6, 7].

13.2.3 The Triplet Achromat Lattice (TAL)

The *triplet achromat* lattice provides a very compact lattice resulting in a small ring circumference. The basic structure of this lattice type is shown in Fig. 13.6 in its practical realization at the storage ring ACO [13.8].

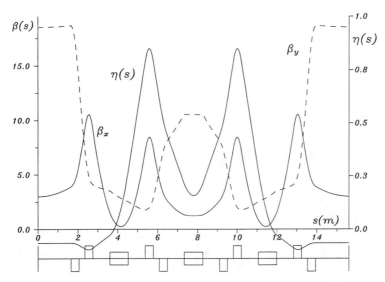

Fig. 13.5. Triple bend achromat lattice (TBA) of the storage ring BESSY [13.6, 7]

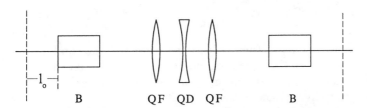

Fig. 13.6. Triplet achromat lattice (TAL) of the storage ring ACO [13.8]

The lattice functions in the bending magnets are determined by their values in the middle of the insertion $\beta = \beta^*$ and $\alpha = \alpha^* = 0$, since there are no quadrupoles in the insertion straight sections. At the entrance to the bending magnet we have $\beta_o = \beta^* + (\ell_o^2/\beta^*)$, $\alpha_o = -\ell_o/\beta^*$ and $\gamma_o = 1/\beta^*$ where ℓ_o is the distance of the magnet entrance to the middle of the insertion. With this (13.3) becomes

$$\langle \mathcal{H} \rangle_{TAL} = \left(\beta^* + \frac{\ell_o^2}{\beta^*} \right) B - \frac{\ell_o \rho}{\beta^*} A + \frac{\rho^2}{\beta^*} C . \tag{13.22}$$

Again there is an optimum value for β^*. Using the expressions (13.6) one obtains for small bending angles

$$\langle \mathcal{H} \rangle_{TAL} = \rho \Theta^3 \left[\frac{1}{20} \frac{\ell_b}{\beta^*} + \frac{1}{3} \frac{\beta^*}{\ell_b} + \frac{1}{3} \frac{\ell_o^2}{\ell_b^2} \frac{\ell_b}{\beta^*} + \frac{1}{4} \frac{\ell_o}{\ell_b} \frac{\ell_b}{\beta^*} \right] \tag{13.23}$$

411

and the optimum value of the betatron function β^* derived from the derivative $\mathrm{d}\langle\mathcal{H}\rangle_{TAL}/\mathrm{d}\beta^* = 0$ is given by

$$\left(\frac{\beta^*}{\ell b}\right)^2_{\mathrm{opt}} = \frac{3}{4}\left(\frac{1}{5} + \frac{\ell_o}{\ell_b} + \frac{4}{3}\frac{\ell_o^2}{\ell_b^2}\right). \tag{13.24}$$

The minimum value for the quantum excitation factor becomes with (13.24) from (13.23)

$$\langle\mathcal{H}\rangle_{\mathrm{min}} = \tfrac{2}{3}\rho\,\Theta^3\left.\frac{\beta^*}{\ell_b}\right|_{\mathrm{opt}} \tag{13.25}$$

and the minimum beam emittance in an isomagnetic triplet achromat lattice with $J_x = 1$ is finally

$$\epsilon_{TAL,\mathrm{min}} = \frac{1}{\sqrt{3}}C_q\gamma^2\,\Theta^3\left(\frac{1}{5} + \frac{\ell_o}{\ell_b} + \frac{4}{3}\frac{\ell_o^2}{\ell_b^2}\right)^{1/2} \tag{13.26}$$

or in practical units

$$\epsilon_{TAL,\mathrm{min}}\,(\mathrm{rad\ m}) = 52.01\times10^{-13}\left.\frac{\beta^*}{\ell_b}\right|_{\mathrm{opt}}E^2(\mathrm{GeV}^2)\,\Theta^3(\mathrm{deg}^3). \tag{13.27}$$

Similar to the result in case of the double bend achromat the equilibrium beam emittance scales like the square of the beam energy and on the cube of the deflection angle in the bending magnets, while the difference in the lattice type shows up in a geometric factor.

The beam emittance varies only little in the vicinity of the optimum betatron function β^*. With $b = \beta^*/\beta^*_{\mathrm{opt}}$ and (13.24) this variation can be expressed by

$$\frac{\epsilon_{TAL}}{\epsilon_{TAL,\mathrm{min}}} = \tfrac{1}{2}\frac{1+b^2}{b}. \tag{13.28}$$

A further reduction in the beam emittance can be obtained in this lattice if the dispersion function in the magnet free insertion is not matched to zero, $\eta^* \neq 0$. It can be shown that in this case

$$\langle\mathcal{H}\rangle = \rho\,\Theta^3\left\{\left[\frac{1}{20}\frac{\ell_b}{\beta^*} + \frac{1}{3}\frac{\beta^*}{\ell_b} + \frac{1}{3}\frac{\ell_o^2}{\ell_b^2}\frac{\ell_b}{\beta^*} + \frac{1}{4}\frac{\ell_o}{\ell_b}\frac{\ell_b}{\beta_o}\right]\right.$$
$$\left. - \eta^*\frac{\rho}{\beta_o\ell_b}\left(\frac{1}{7} + \frac{\ell}{\ell_b} - \frac{\rho}{\ell_b^2}\eta^*\right)\right\} \tag{13.29}$$

and looking for the optimum value of the dispersion function in the insertion we get from $\partial\langle\mathcal{H}\rangle/\partial\eta^* = 0$

$$\left.\frac{\eta^*}{\ell_b}\right|_{\mathrm{opt}} = \frac{\ell_o}{2\rho}\left(\frac{1}{3} + \frac{\ell_o}{\ell_b}\right). \tag{13.30}$$

Using $\eta^* = \eta^*_{\text{opt}}$ in (13.29) we may derive also an optimum value for the betatron function

$$\left(\frac{\beta^*}{\ell_b}\right)^2_{\text{opt}} = \frac{1}{4}\left(\frac{4}{15} + \frac{\ell_o}{\ell_b} + \frac{\ell_o^2}{\ell_b^2}\right), \tag{13.31}$$

and the minimum beam emittance in an isomagnetic triplet achromat lattice becomes finally

$$\epsilon_{TAL} = \tfrac{1}{3}C_q\gamma^2\Theta^3\left(\frac{4}{15} + \frac{\ell_o}{\ell_b} + \frac{\ell_o^2}{\ell_b^2}\right)^{1/2}. \tag{13.32}$$

With the optimization of the dispersion function the beam emittance can be reduced by another factor 1.5 – 2 depending on the ratio ℓ_o/ℓ_b being zero or very large respectively. It is interesting to note that the optimum dispersion η^* must be positive for minimum beam emittance.

13.2.4 The FODO Lattice

The FODO Lattice, shown schematically in Fig. 13.7 is the most commonly used and best understood lattice in storage rings optimized for high-energy physics colliding beam facilities where large beam emittances are desired. This choice is obvious considering that the highest beam energies can be achieved while maximizing the fill factor of the ring with bending magnets.

The FODO lattice provides the most space for bending magnets compared to other lattices. The usefulness of the FODO lattice, however, is not only limited to high-energy large emittance storage rings. By using very short cells very low beam emittances can be achieved as has been demonstrated in the first low emittance storage ring designed [13.9] and constructed [13.10] as a damping ring for the linear collider SLC to reach an emittance of 11×10^{-9} m at 1 GeV.

The lattice functions in a FODO structure have been derived and discussed in detail in Chap. 6 and are generally determined by the focusing parameters of the quadrupoles. Since FODO cells are not achromatic the dispersion function is in general not zero at either end of the bending magnets. An example of a FODO cell with lattice functions is shown in Fig.13.7.

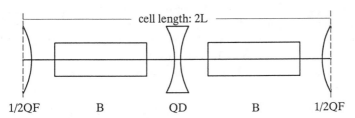

Fig. 13.7. FODO lattice (schematic)

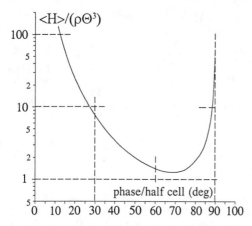

Fig. 13.8. Beam emittance of a FODO lattice as a function of the betatron phase advance per half cell in the deflecting plane

The beam emittance can be derived analytically for thin lens approximation by integrating the quantum excitation factor along the bending magnets. The result is shown in Fig. 13.8 where the function $[\langle\mathcal{H}\rangle/(\rho\Theta^3)]$ is plotted as a function of the betatron phase advance per FODO half cell ϕ which is determined by the focal length of the quadrupoles. The beam emittance for an isomagnetic FODO lattice is given by [13.11]

$$\epsilon_{FODO} = C_q\,\gamma^2\,\Theta^3\,\frac{\ell_b}{\ell_{b,o}}\,\frac{\langle\mathcal{H}\rangle}{\rho\,\Theta^3}\,, \tag{13.33}$$

where $\ell_{b,o}$ is the actual effective length of one bending magnet and $2\ell_b$ the length of a FODO cell. From Fig. 13.8 it becomes apparent that the minimum beam emittance is reached for a betatron phase of just below $140°$ per FODO cell. In this case $\langle\mathcal{H}\rangle/(\rho\,\Theta^3) \approx 1.25$ and the minimum beam emittance in such a FODO lattice in practical units is

$$\epsilon_{FODO}(\mathrm{m\,rad}) = 97.53 \cdot 10^{-13}\,\frac{\ell_b}{\ell_{b,o}}\,E^2(\mathrm{GeV}^2)\,\Theta^3(\mathrm{deg}^3)\,. \tag{13.34}$$

Comparing the minimum beam emittance achievable in various lattice types as determined by (13.18, 26, 33) the FODO lattice seems to be the least appropriate lattice to achieve small beam emittances. This, however, is only an analytical distinction. FODO cells can be made much shorter than the lattice units of other structures and for a given circumference many more FODO cells can be incorporated than for any other lattice. As a consequence, the deflection angles per FODO cell can be much smaller. For very low emittance storage ring, therefore, it is not a priori obvious that one lattice is better than another. Additional requirements for a particular application must be included in the determination of the optimum storage ring lattice.

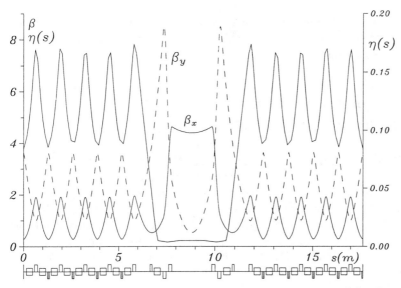

Fig. 13.9. Lattice structure for one of two superperiods of the SLC damping rings[13.9] with two insertions for injection and ejection as well as rf cavities

An appropriate example is that of the damping ring for the *Stanford Linear Collider*, SLC, see Fig. 13.9, where not only a small beam emittance is desired but also a short damping time. The short damping time requirement demands a ring circumference as small as possible to maximize the average synchrotron radiation power. This is most effectively done in a FODO lattice where a high percentage of the circumference is covered with dipole magnets. A double focusing achromat lattice could give an equally small emittance but because of the much more "open" magnet structure the damping time would be much longer, thus missing design requirements.

On the other hand, small emittance storage rings to serve as high brilliance photon sources require a significant number of magnet free insertions for the installation of insertion devices like wiggler or undulator magnets while the damping time is not a critical parameter. In this case a DFA, a TBA, or a TAL lattice is more appropriate since they provide magnet free insertion straight sections in a natural way.

A still smaller beam emittance than in a regular FODO lattice can be achieved if the defocusing quadrupole is eliminated by using a combined function bending magnet that is focusing in the vertical plane [13.12]. In this case the horizontal betatron function has a minimum in the bending magnet which is close to the optimum condition discussed earlier in this section. Since only the vertical focusing is combined with the bending field, $\vartheta < 0$, and the variation of the damping partition number leads to an additional damping in the horizontal plane, although, consistent with Robinson's damping criterion, at the expense of the longitudinal damping.

Lattices based on this combination of vertical focusing and horizontal bending have been worked out for synchrotron radiation sources proposed and under construction.

13.3 Optimum Emittance
for Colliding Beam Storage Rings

The single most important parameter of colliding beam storage rings is the *luminosity* and most of the design effort is aimed at maximizing the collision rate. Particularly, the beam emittance must be chosen properly for maximum luminosity. From the discussions in Sect. 12.2 we know that the luminosity must be optimized within the boundary condition of the beam – beam effect. Based on experimental observations we assume an empirical limit for the maximum allowable *beam – beam tune shift* $\delta\nu_{max}$, and select appropriate storage ring parameters such that the luminosity is maximized within this limit,

$$\delta\nu_y \leq \delta\nu_{max}. \tag{13.35}$$

Before we specify optimum lattice and beam size parameters we solve (12.25) for the particle number N and insert into (12.21) for

$$\mathcal{L} = \frac{\pi}{r_o^2} \left(\frac{\delta\nu_{max}}{\beta_y^*}\right)^2 B \frac{\sigma_y}{\sigma_x} (\sigma_x + \sigma_y)^2 \nu_{rev}. \tag{13.36}$$

In electron positron colliders the beams are very flat, $\sigma_y \ll \sigma_x$, and the luminosity is

$$\mathcal{L}_e = \frac{\pi}{r_o^2} \left(\frac{\delta\nu_{max}}{\beta_y^*}\right)^2 \sigma_x \sigma_y B \nu_{rev}. \tag{13.37}$$

The lattice for a *colliding beam storage ring* is selected such that at the collision points the vertical betatron function converges to a sharp minimum. The value for maximum luminosity of such a *low beta insertion* was first recognized and analyzed by Robinson and Voss [13.10].

To further maximize the luminosity large beam cross sections are desired, however, in such a way that the betatron function β_y^* at the collision point can be kept small at the same time. This is possible by choosing the equilibrium beam emittance as large as possible. We choose therefore a lattice, in most cases a FODO lattice, with weak focusing as discussed in the previous section. The limit is either given by single bunch instability or by economic factors since the apertures of the magnets must be made large as well.

Last but not least one should collide many bunches. This requires two intersecting rings or sophisticated beam separation controls. In a single ring two multibunch beams would cause collisions at many places where the betatron function is not made small and therefore the tune shift parameter would exceed the limit. Therefore, only half as many bunches as there are collision points can be stored in each beam and brought into safe collision only at regular collision points. More bunches can be used if both beams are kept separate through the use of electrostatic fields at the beginning of the arcs between collision points [13.14]. These electrostatic fields cause a kick in different directions for electrons and positrons and making use of the betatron oscillation both beams oscillate with respect of each other. With proper placement of the separating plates the beam separation can be made such that both beams are separated where they would otherwise collide outside the regular collision points. Such a scheme has been implemented at the Cornell storage ring CESR [13.15] where eight bunches per beam are stored while there are only two low beta collision points. Eight bunches would collide at 16 points, and all but two collisions are avoided by beam separation with electrostatic fields.

So far we have always assumed that the vertical tune shift is larger than the horizontal tune shift. This is true in most cases of electron – positron colliding beam storage rings. Where this is not true parameters must be chosen such that both tune shifts are below or at the limit.

Problems

Problem 13.1. For a circumference of your choice determine the lowest beam emittance achievable with a DBA, a TBA or FODO lattice. Assume for the DBA lattice the unit cell of Fig. 13.3, for the TBA lattice Fig. 13.5 and for the FODO lattice a bending magnet fill factor of 50%.

Problem 13.2. Verify Eq. (13.16)

Problem 13.3. Determine basic FODO lattice parameters for a 2 GeV e^+/e^- colliding beam storage ring with two collision points to reach a design luminosity of $\mathcal{L}_e = 10^{33}$ cm^{-2} sec^{-1}. The betatron functions at the collision point be $\beta_y^* = 5$ cm and $\beta_x^* = 1.3$ m and the emittance coupling 10%. Calculate beam sizes in the arc, aperture requirements, circumference and beam current. What is the total synchrotron radiation power? Adjust, if necessary, your design to keep the maximum synchrotron radiation power at the vacuum chamber wall below a practical limit of 5 kW/m.

Problem 13.4. The linear focusing of the beam – beam effect changes also the betatron function. Derive an expression that relates the change in the value of the betatron function β_y^* at the collision point to the beam – beam tune shift $\delta\nu$.

Problem 13.5. Compare both, the the PEP storage ring (Table 6.2, Fig. 6.19) at 6 GeV and the ESRF ring (Fig. 13.3, $\rho = 20\,\mathrm{m}$, $\epsilon_x = 9.0\,10^{-9}$ m-rad at 6 GeV, $C = 800$ m) as synchrotron light sources. Consider adding damping wigglers to reduce the beam emittances. Where are the optimum locations in the lattices for such damping wigglers? Calculate the reduction of the beam emittance in both rings per unit length of damping wiggler and determine the minimum emittance achievable in each ring. In doing so ignore the space availability for the length of wiggler magnets needed in each ring. Why are damping wigglers more effective in a large ring compared to a smaller ring?

Bibliography

The most recent developments of particle accelerators being pursued primarily in national laboratories are documented and published through an informal exchange of internal notes. This makes it extremely difficult for interested newcomers to absorb the current status of the field. Significant efforts are being made both in Europe and the United States to conduct yearly Accelerator Physics Schools and publish proceedings of the lectures and seminars. Even those proceedings are often available only from the organizing high energy physics laboratories. In very recent years, however, this situation has received great attention in the community and it is to be expected that the field of accelerator physics is now gaining the maturity of other fields by establishing regular publication channels.

Biannual accelerator physics conferences are held both in the United States and more recently since 1988 also in Europe. The US conference contributions are regularly published in the IEEE Nuclear Sciences Series and are therefore widely distributed and accessible throughout the technical library system.

In the following bibliography most commonly accessible literature is compiled for guidance.

General Books on Accelerator Physics

1   *Handbuch der Physik* edited by S. Flügge , Vol 44, Springer Verlag, Heidelberg, 1950

2   Livingood J., *Principles of Cyclic Accelerators*, Van Norstrand, 1961.

3   Blewett J., Livingston M.S., *Particle Accelerator*, McGraw-Hill, 1962.

4   Kolomensky A.A., Lebedev A.N., *Theory of Cyclic Accelerators*, North Holland, 1966.

5   Livingston M.S., *The Development of High Energy Accelerators*, Dover Publications, Inc., New York, 1966

6   Kollath R., *Teilchenbeschleuniger*, Dümmler, Bonn , 1967, translated as *Particle Accelerators*, Pittman, London, 1967.

7   Rosenblatt J., *Particle Acceleration*, Methuen, London, 1968.

8   Persico E., Ferrari E., Segre S.E., *Principles of Particle Accelerators*, W.A.Benjamin, 1968.

9   *The Stanford two Mile Accelerator* edited by R.B. Neal, W.A.Benjamin, 1968.

10   *Linear Accelerators* edited by P. Lapostolle and A. Septier, North Holland, 1970.

11   Scharf W., *Particle Accelerators and Their Uses*, Harwood Academic, New York, 1985

12   Humphries Jr S., *Principles of Charged Particle Accelerators*, Wiley and Sons, New York, 1986.

13   *High-Brightness Accelerators* edited by A. K. Hyder, M. F. Rose and A.H. Guenther, Plenum Press, New York and London, 1986.

14   *New Techniques for Future Accelerators* edited by M. Puglisi, S. Stipcich and G. Torelli, Plenum Press, New York and London, 1987

15   K. Wille: *Physik der Teilchenbeschleuniger und Synchrotron-strahlungsquelle*, Teubner, Stuttgart, 1992

16   A. Hofmann, *Physics of Synchrotron Radiation* Cambridge University Press, Cambridge, to be published 1993

17   D. Edwards, M. Syphers, *An Introduction to the Physics of High Energy Accelerators* Wiley and Sons, New York, 1993

18   A. Chao, *Theory of Collective Beam Instabilities in Accelerators* Wiley and Sons, New York, to be published 1993

Literature on Beam Dynamics

19   Steffen K.G., *High Energy Beam Optics*, Wiley and Sons, New York, 1965.

20   Banford A.P., *The Transport of Charged Particle Beams*, Spon, London, 1966.

21   *Focusing of Charged Particles* edited by A. Septier, Vols. I and II, Academic Press, 1967.

22   Bruck H., *Accelerateurs circulaire de particules* , Presses Universitaires de France, Paris, 1968.

23   Lichtenberg A.J., *Phase Space Dynamics of Particles*, Wiley and Sons, New York, 1969.

24   *Electron Storage Rings*, Intl. School of Physics, Varenna 1969, edited by B.Touschek, Academic Press, 1971.

25   Lawson J.D., *The Physics of Charged Particle Beams* ,Clarendon Press, Oxford, 1977.

26   Carey D.C., *The Optics of Charged Particle Beams*, Harwood Academic, New York, 1987

27   Hagedorn P., *Non-Linear Oscillations*, Clarendon Press, Oxford, 1988.

Journals and Conference Proceedings

28   "Particle Accelerators", Gordon Breach Pub. Company

29   "Physics of Particle Accelerators", US Summer School Series by the American Institute of Physics, AIP Conf.Proc. #87, #105, #127, #153, #184

30   Proceedings of CERN Accelerator Schools, CAS Reports since 1984, CERN, Geneva

31 "Frontiers of Particle Beams", Lecture Notes from CERN/US Accelerator Schools published in "Lecture Notes in Physics" by Springer Verlag

32 Proceedings of the biannual National US Accelerator Conferences, IEEE Trans. on Nucl. Sci.

33 Proceedings of the biannual European Accelerator Conference

34 Proceedings of the triannual International Conference on High Energy Accelerators

35 Proceedings of the biannual International Conference on the Applications of Accelerators in Research and Industry held in Denton, Texas and published in Nuclear Instruments and Methods

References

Chapter 1

1.1   M. Faraday: Poggendorf Annalen **48**, 430 (1839)
1.2   J. Plücker: Annalen der Physik und Chemie **13**, 88 (1858)
1.3   J.W. Hittorf, Annalen der Physik und Chemie, 5. Reihe, 16. Band, 1 (1869)
1.4   E. Goldstein: Monatsberichte der Königl. Akad. der Wiss. zu Berlin, 284 (1876)
1.5   E. Wiedemann: Zeitschrift für Elektrochemie **8**,155 (1895)
1.6   J.J. Thomson: *Conduction of Electricity Through Gases*, (Cambridge University Press, Cambridge 1906), ed. 2, p.602
1.7   J. Larmor: Philos. Mag. **44**, 503 (1897)
1.8   A. Liènard: L'Eclaire Electrique **16**, 5 (1898)
1.9   E. Wiechert: Archieves Neerlandaises 549 (1900)
1.10 G.A. Schott: Philos. Mag. [6] **13**, 194 (1907)
1.11 G.A. Schott: Annalen der Physik und Chemie **24**, 635 (1907)
1.12 H. Greinacher: Z.Physik **4**, 195 (1921)
1.13 G. Ising: Arkiv för Matematik, Astronomi och Fysik **18**, 1 (1924)
1.14 R. Wideroe: Archiv für Elektrotechnik **21**, 387 (1928)
1.15 R.J. Van de Graaff: Phys.Rev. **38**, 1919 (1931)
1.16 E.O. Lawrence, M.S. Livingston: Phys. Rev. **40**, 19 (1932)
1.17 J.D. Cockcroft, E.T.S. Walton: Proc.of the Royal Soc.of London, **A136**, 619 (1932)
1.18 D. W. Kerst, R. Serber: Phys.Rev. **60**, 53 (1941)
1.19 D. Ivanenko, I. Ya. Pomeranchouk: DAN (U.S.S.R.) **44**, 343 (1944)
1.20 J. Schwinger: Phys.Rev., **75**, 1912 (1949)
1.21 V. I. Veksler: DAN (U.S.S.R.) **44**, 393 (1944)
1.22 E.M. McMillan: Phys.Rev., **68**, 143 (1945)
1.23 J.P. Blewett: Phys. Rev. **69**, 87 (1946)
1.24 L.W. Alvarez: Phys.Rev. **70**, 799 (1946)
1.25 E.L. Ginzton, W.W. Hansen, W.R. Kennedy: Rev.Sci.Inst. **19**, 89 (1948)
1.26 N. Christofilos: U.S. Patent No. 2,736,766 (1950)
1.27 H. Motz: J. Appl. Phys. **22**, 527 (1951)
1.28 M.S. Livingston, J.P. Blewett, G.K. Green, L.J. Haworth: Rev. Sci.Instrum. **21**, 7 (1950)
1.29 E. Courant, M.S. Livingston, H. Snyder: Phys.Rev. **88**, 1190 (1952)
1.30 M. Chodorow, E.L. Ginzton, W.W. Hansen, R.L. Kyhl, R. Neal, W.H.K. Panofsky: Rev.Sci.Instrum. **26**, 134 (1955)
1.31 M. Sands: Phys.Rev. **97**, 470 (1955)
1.32 E. Courant, H. Snyder: Ann. Phys. **3**, 1 (1959)
1.33 W. Scharf: *Particle Accelerators and their Uses* (Harwood Academic, New York 1985)
1.34 E.P. Lee and R.K. Cooper: Part.Accel. **7**, 83 (1976)
1.35 I. Borchardt, E. Karantzoulis, H. Mais, G. Ripken: Z.Phys. C - Particles and Fields **39**, 339 (1988)

1.36   *Conceptual Design of the Relativistic Heavy Ion Collider*, RHIC, BNL-Report 52195, (Brookhaven National Laboratory, Upton NY 1989)

1.37   A.D. Hansen, L.H. Lanzl, E.M. Lyman, and M.B. Scott: Phys.Rev. **84**, 634 (1951)

Chapter 2

2.1   M.S. Livingston: *The Development of High-Energy Accelerators* (Dover, New York 1966)

2.2   H. Greinacher: Z.Physik **4**, 195 (1921)

2.3   M. Schenkel: Elektrotechnische Zeitschrift **40**, 333 (1919)

2.4   J.D. Cockcroft, E.T.S. Walton: Proc. of the Royal Soc. of London **A136**, 619 (1932)

2.5   R.J. Van de Graaff: Phys.Rev., **38**, 1919 (1931)

2.6   R.J. Van de Graaff, J.G. Trump, W.W. Buechner: Rep. Prog.in Phys., **11**, 1 (1948)

2.7   A. Bouwers: *Elektrische Höchstspannungen* (J.W. Edwards, Ann Arbor, Michigan, 1944)

2.8   N.C. Christofilos, R.E. Hester, W.A.S. Lamb, D.D. Reagan, W.A. Sherwood and R.E. Wright: Rev.Sci.Instum. **35**, 866 (1964)

2.9   C.A. Kapetanakos and P. Sprangle: Phys.Today, **38**, No.2 (1985) 58

2.10   D. Keefe: *Induction Linacs*, in *High Brightness Accelerators* ed. by Hyder et al. (Plenum, New York, 1988)

2.11   G. Ising: Arkiv för Matematik, Astronomi och Fysik **18**, 1 (1924)

2.12   R. Wideroe: Archiv für Electrotechnik **21**, 387 (1928)

2.13   D.H. Sloan, E.O. Lawrence: Phys.Rev. **38**, 2021 (1931)

2.14   M. Chodorow, E.L. Ginzton, I.R. Neilsen, S. Sonkin: Proc.Inst. Radio Engrs. **41**, 1584 (1953)

2.15   L.W. Alvarez: Phys.Rev. **70**, 799 (1946)

2.16   E.L. Ginzton, W.W. Hansen, W.R. Kennedy: Rev.Sci.Instum. **19**, 89 (1948)

2.17   R. Neal (ed.) *The 2 Mile Linear Accelerator* (Benjamin, New York 1968)

2.18   P. Lapostolle, A. Septier (eds.) *Linear Accelerators* (North-Holland, Amsterdam 1970)

2.19   J.C. Slater: Rev. Mod. Phys. **20**, 473 (1948)

2.20   R.B.Neal: HEPL ML Report No. 513, (Stanford University, Stanford, CA 1958

2.21   M. Borland: *A High-Brightness Thermionic Electron Gun*, Ph.D. Thesis, (Stanford University, Stanford, CA 1991)

Chapter 3

3.1   *Physics of Particle Accelerators*, M. Month, M. Dienes (eds.), AIP Conf. Proc. #**184**, (Am. Inst. Phys. New York 1989) p.1829

3.2   J.M Peterson: in *Physics of Particle Accelerators*, M. Month, M. Dienes (eds.), AIP Conf.Proc. #**184**, (Am. Inst. Phys. New York 1989) p.2240

3.3   *Design Study of a 15 to 100 GeV e+ - e- Colliding Beam Machine (LEP)*, CERN/ISR-LEP 78-17 (CERN, Geneva 1978)

3.4   J. Schwinger: Phys.Rev. **75**, 1912 (1949)

3.5   M. Tigner: Nuovo Cimento, **37**, 1228 (1965)

3.6   *The SLAC Linear Collider, Conceptual Design Report*, SLAC-229, (SLAC, Stanford University, Stanford CA 1980)

3.7   J. Slepian: U.S. Patent 1,645,304 (1922)

3.8   G. Breit, M.A. Tuve: Carnegie Institution Year Book **27**, 209 (1927)

3.9   R. Wideroe: Archiv für Electrotechnik **21**, 387 (1928)

3.10   E.T.S. Walton: Proc.of the Cambridge Phil.Soc. **25**, 569 (1929)

3.11   M. Steenbeck: U.S. Patent 2,103,303 (1935)

3.12   D. W. Kerst, R. Serber: Phys.Rev. **60**, 53 (1941)

3.13   S.P. Kapitza and V.N. Melekhin: *The Microtron*, (Harwood Academic, London, 1987)

3.14   B.H. Wiik, P.B. Wilson: NIM **56**, 197 (1967)

3.15   M. Eriksson: NIM **203**, 1 (1982)

3.16   E.O. Lawrence, N.E. Edlefsen: Science **72**, 378 (1930)

3.17   E.O. Lawrence, M.S. Livingston: Phys. Rev., **40**, 19 (1932)

3.18   V. I. Veksler: DAN (U.S.S.R.) **44**, 393 (1944)

3.19   E.M. McMillan: Phys.Rev. **68**, 143 (1945)

3.20   J.R. Richardson, K.R. MacKenzie, E.J. Lofgren, B.T. Wright: Phys.Rev. **69**, 669 (1946)

3.21   W.M. Brobeck, E.O. Lawrence, K.R. MacKenzie, E.M. McMillan, R. Serber, D.C. Sewell, K.M. Simpson, R.L. Thornton: Phys.Rev. **71**, 449 (1947)

3.22   D. Bohm, L. Foldy: Phys.Rev., **72**, 649 (1947)

3.23   L.H. Thomas: Phys.Rev., **54**, 580 (1938)

3.24   1-2 GeV Synchrotron Radiation Source, Conceptual Design Report, Lawrence Berkeley Laboratory, University of California, Berkeley, California, PUB-5172 Rev. (1986)

3.25   D. Bohm, L. Foldy: Phys.Rev., **70**, 249 (1946)

3.26   M.S. Livingston, J.P. Blewett, G.K. Green, L.J. Haworth: Rev.Sci. Instrum. **21**, 7 (1950)

3.27   N. Christofilos: U.S. Patent No. 2,736,766, (1950)

3.28   E. Courant, M.S. Livingston, H. Snyder: Phys.Rev. **88**, 1190 (1952)

Chapter 4

4.1   E.R. Cohen, B.N. Taylor: Rev. of Mod. Phys. **59**, 1121 (1987)

4.2   J. Larmor: Philos. Mag. **44**, 503 (1897)

4.3   W.K.H. Panofsky, W.R. Baker: Rev. Sci.. Instrum. **21**, 445 (1950)

4.4   E.G. Forsyth, L.M. Lederman, J. Sunderland: IEEE Trans., NS-**12**, 872 (1965)

4.5   D. Luckey, Rev. Sci. Instrum. **31**, 202 (1960)

4.6   B.F. Bayanov, G.I. Silvestrov: Zh. Tekh. Fiz., **49**, 160 (1978)

4.7   B.F. Bayanov, J.N. Petrov, G.I. Silvestrov, J.A. Maclachlan, G.L. Nicholls: Nucl. Instrum. and Methods **190**, 9 (1981)

4.8   S. van der Meer: CERN 61-7 (1961); ibid. 62-16 (1962) CERN, Geneva

4.9   E. Regenstreif: CERN 64-41 (1964), CERN, Geneva

4.10   G.I. Budker, Proc. of the Int. Conf. on High Energy Accelerators, Dubna(USSR) (1963)

4.11   H. Wiedemann: Int. Rep. H-14 (1966) DESY, Hamburg, Germany

4.12   G.E. Fischer: "Iron Dominated Magnets" in AIP Conf. Proc. **153**, Vol 2, 1047 (1987), R. G. Lerner (ed.) (Am. Inst. Phys., New York 1987)

4.13   N. Christofilos: U.S. Patent No. 2,736,766, (1950)

4.14   E. Courant, M.S. Livingston, H. Snyder: Phys.Rev. **88**, 1190 (1952)

4.15   E. Courant, H. Snyder, Ann. Phys., **3**, 1 (1959)

Chapter 5

5.1   N. Christofilos: U.S. Patent No. 2,736,766, (1950)

5.2   E. Courant, M.S. Livingston, H. Snyder: Phys.Rev. **88**, 1190 (1952)

5.3   J. Moser: "Stabilitätsverhalten kanonischer Differentialgleichungssysteme", Nachr. der Akad. der Wiss. in Göttingen, **IIa**, No. 6, 87 (1955)

5.4   E. Courant, H. Snyder: Ann. Phys. **3**, 1 (1959)

5.5   K.G. Steffen: *High Energy Beam Optics*, (Wiley, New York, 1965)

5.6   G.E. Fischer: "Iron Dominated Magnets" in AIP Conf. Proc. **153**, Vol 2, 1047 (1987), R. G. Lerner (ed.) (Am. Inst. Phys., New York 1987)

5.7   R. Perin, S. van der Meer: "The program **MARE** for the computation of two-dimensional static magnetic fields", CERN 67-7 (1967), later expanded by Ch. Iselin under the name **MAGNET**, CERN, Geneva

5.8   K. Halbach: Lawrence Livermore National Laboratory Report, UCRL-17436, (1967) Livermore, CA

5.9   R.F.K. Herzog: Acta Phys. Austriaca, **4**, 431 (1951)

5.10   R.Q. Twiss, N.H. Frank: Rev. Sci. Instrum., **20**, 1 (1949)

5.11   W.K. Panofsky: in *High Energy Beam Optics* by K.G. Steffen, (Wiley, New York 1965), p. 117

5.12   R. Chasman, K. Green, E. Rowe: IEEE Trans., NS-**22**, 1765 (1975)

5.13   D. Einfeld, G. Mülhaupt: Nucl. Instrum. and Methods **172**, 55 (1980)

5.14   E. Rowe: IEEE Trans. NS-**28**, 3145 (1981)

5.15   A. Jackson: 1987 IEEE Part. Accel. Conf. Washington DC, IEEE Cat. No. 87CH2387-1, p.476

5.16   K. Siegbahn, N. Svartholm: Nature **157**, 872 (1946)

5.17   D.L. Judd: Rev. Sci. Instr., **21**, 213 (1950)

5.18   E.E. Chambers, R. Hofstadter: CERN Symposion 1956, Vol. 2, p.295

5.19   A.P. Banford: *The Transport of Charged Particle Beams*, (Spon, London, 1966)

5.20   D. L. Judd, S.A. Bludman: Nucl. Instrum. and Methods **1** , 46 (1956)

Chapter 6

6.1   E. Courant, H. Snyder: Ann. Phys. **3**, 1 (1959)

6.2   G. Hemmie: "Die Zukunft des Synchrotrons, (DESY II)", Int.Note DESY M-82-18, DESY, Hamburg (1982)

6.3   J. Roßbach, F. Willeke: "DESY II Optical Design of a New 10 GeV Electron-Positron Synchrotron", Int.Note DESY M-83-03, DESY, Hamburg (1983)

6.4   S. Kheifets, T. Fieguth, K.L. Brown, A. Chao, J.J. Murray, R.V. Servranckx, H. Wiedemann: Proc. 13th Int. Conf. on High Energy Accelerators, Novosibirsk, USSR (1986)

6.5   G.E. Fischer, K.L. Brown, F. Bulos, T. Fieguth, A. Hutton, J.J. Murray, N. Toge, W.T. Weng, H. Wiedemann: 1987 IEEE Part. Accel. Conf. Washington, DC, IEEE Cat. No. 87CH2387-1, p. 139

6.6   L. Emery: *A wiggler-based ultra-low emittance damping ring lattice and its chromatic correction* Ph.D. Thesis, (Stanford University, Stanford, CA 1990)

6.7   A.A. Garren, D.E. Johnson: 1987 IEEE Part. Accel. Conf. Washington, DC, IEEE Cat. No. 87CH2387-1, p. 163

6.8   J.J. Stoker: *Non Linear Vibrations*, Intersc. New York (1950) Chapter IV

6.9   H. Wiedemann, M. Baltay, J. Voss, K. Zuo, C. Chavis, R. Hettel, J. Sebek, H.D. Nuhn, J. Safranek, L. Emery, M. Horton, J. Weaver, J. Haydon, T. Hostetler, R. Ortiz, M. Borland, S. Baird, W. Lavender, P. Kung, J. Mello, W. Li, H. Morales, L. Baritchi, P. Golceff, T. Sanchez, R. Boyce, J. Cerino, D. Mostowfi, D.F. Wang, D. Baritchi, G. Johnson, C. Wermelskirchen, B. Youngman, C. Jach, J. Yang, R. Yotam, Proc. 1991 IEEE Part. Accel. Conf. San Francisco, IEEE Conf. Rec. 91CH3038-7, p.2688

6.10   H.G. Hereward, K. Johnsen and P. Lapostolle: Proc. CERN Symposium on High Energy Physics, **1**, 179 (1956)

6.11   A.J. Lichtenberg, K.W. Robinson: Int. Note CEA 13, Harvard University (1956)

6.12   K.L. Brown, F. Rothacker, D.C. Carey, C. Iselin: Laboratory Reports, SLAC-91 (1977), CERN 80-04 (1980), NAL-91

6.13   A.A. Garren, J.W. Eusebio: LBL Report UCID-10153 (1975)

6.14   M.D. Woodley, M.J. Lee, J. Jäger, A.S. King: IEEE Trans. NS-30, 2367 (1983)

6.15   F.C. Iselin and J. Niederer: CERN/LEP-TH/88-38

6.16   G.E. Fischer, W. Davis-White, T. Fieguth, and H. Wiedemann: Proc. of the 12th Int. Conf. on High Ener. Accel., Fermilab, Chicago, (1983)

6.17   E. Keil: in "Theoretical Aspects of the Behaviour of Beams in Accelerators and Storage Rings", ed. by M.H. Blewett, CERN 77-13, 29 (1977)

6.19   M. Bassetti, M. Preger: Int. Note T-14 (1972), Laboratori Nazionali di Frascati, Italy

6.18   M. Lee, P. Morton, J. Rees, B. Richter, N. Spencer: SPEAR-Note 102, (1971), SLAC, Stanford, CA

6.20   T.L. Collins: Int. Note Cambridge Electron Accelerator, CEA-86, (1961)

6.21   K. Robinson and G.A. Voss: Harvard University, CEAL-1029, (1966) Cambridge, USA

6.22   Summary of the Preliminary Design of Beijing 2.2/2.8 GeV Electron Positron Collider (BEPC), IHEP Academica Sinica, Dec 1982

Chapter 7

7.1   G. Lüders: Nuovo Cim.Suppl. 2, 1075 (1955)

7.2   J. Moser: Nachr. der Akad. der Wiss. in Göttingen, IIa, No. 6, 87 (1955)

7.3   R. Hagedorn, M.G.N. Hine, A. Schoch in Proc. of the CERN Symposium, (CERN, Geneva, 1956) p.237

7.4   E.D. Courant in Proc. of the CERN Symposium, (CERN, Geneva, 1956) p.254

7.5   M. Barbier in Proc. of the CERN Symposium, (CERN, Geneva, 1956) p.262

7.6   A.A. Kolomenski in Proc. of the CERN Symposium, (CERN, Geneva, 1956) p.265

7.7   L.J. Laslett, K.R. Symon in Proc. of the CERN Symposium, (CERN, Geneva, 1956) p.279

7.8   J. Moser in Proc. of the CERN Symposium, (CERN, Geneva, 1956) p.290

7.9   R. Hagedorn in Proc. of the CERN Symposium, (CERN, Geneva, 1956) p.293

7.10   E. Regenstreif in Proc. of the CERN Symposium, (CERN, Geneva, 1956) p.295

7.11   A. Schoch: CERN 57-23, (1958), CERN, Geneva

7.12   E. Courant, H. Snyder: Ann. Phys. 3, 1 (1959)

Chapter 8

8.1   G. Ising: Arkiv för Matematik, Astronomi och Fysik, 18, 1 (1924)

8.2   R. Wideroe: Archiv für Elektrotechnik, 21, 387 (1928)

8.3   M.S. Livingston: The Development of High-Energy Accelerators, (Dover, New York 1966)

8.4   V.I. Veksler: DAN (U.S.S.R.) 44, 393 (1944)

8.5   E.M. McMillan: Phys.Rev. 68, 143 (1945)

8.7   L.W. Alvarez: Phys.Rev. 70, 799 (1946)

8.8   K. Johnsen: CERN Symp. on High Energy Accel. CERN 56-25, 106 (1956), CERN, Geneva

8.9   G.K. Green: CERN 56-25, 103 (1956), CERN, Geneva

8.10   H. Goldstein: Classical Mechanics, (Addison-Wesley, Cambridge 1950)

8.11   D. Deacon: Theory of the Isochronous Storage Ring Laser, Ph.D. Thesis, (Stanford University, Stanford, CA 1979)

8.12   C. Pellegrini, D. Robin: Nucl. Instrum. and Methods A301, 27 (1991)

8.13   C.G. Lilliequist, K.R. Symon: MURA Internal Report, MURA-491, (1959)

Chapter 9

9.1   A. Liénard: L'Eclairage Electrique **16**, 5 (1898)

9.2   E. Wiechert: Archives Neerlandaises, 546 (1900)

9.3   G.A. Schott: Annalen der Physik **24**, 635 (1907)

9.4   G.A. Schott: Phil. Mag.[6] **13**, 194 (1907)

9.5   G.A. Schott: *Electromagnetic Radiation*, Cambridge University Press, Cambridge (1912)

9.6   D.W. Kerst, R. Serber: Phys. Rev. **60**, 53 (1941)

9.7   D. Ivanenko and I.Ya. Pomeranchouk: Phys. Rev. **65**, 343 (1944)

9.8   J.P. Blewett: Phys. Rev. **69**, 87 (1946)

9.9   describing work of C. Sutis: Sci. News Lett. **51**, 339 (1947)

9.10   describing work of F. Haber: Electronics **20**, 136 (1947)

9.11   F. Elder, A. Gurewitsch, R. Langmuir, H. Pollock: Phys. Rev. **71**, 829 (1947)

9.12   The author would like to thank Prof. M. Eriksson, Lund, Sweden for introducing him to this approach into the theory of synchrotron radiation

9.13   M. Sands: *Physics with Intersecting Storage Rings* B. Touschek (ed.), (Academic Press, New York 1971)

9.14   R. Coisson: Opt.Comm. **22**, 135 (1977)

9.15   R. Bossart, J. Bosser, L. Burnod, R. Coisson, E. D'Amico, A. Hofmann, J. Mann: Nucl. Instrum. and Methods, **164**, 275 (1979)

9.16   R. Bossart, J. Boser, L. Burnod, E. D'Amico, G. Ferioli, J. Mann and F. Meot: Nucl. Instrum. and Methods **184**, 349 (1981)

9.17   J.M Peterson: in AIP Conf. Proc. #184, M. Month, M. Dienes (eds.), Physics of Particle Accelerators, New York (1989) p.2240

9.18   J.D. Jackson: *Classical Electrodynamics* (Wiley, New York, 1975)

9.19   D. Ivanenko and A.A. Sokolov: DAN (USSR) **59**, 1551 (1972)

9.20   J. Schwinger: Phys.Rev. **75**, 1912 (1949)

9.21   D.H. Tomboulian and P.L. Hartman: Phys.Rev **102**, 102 (1956)

9.22   G. Bathow, E. Freytag, R. Haensel, J. Appl. Phys. **37**, 3449 (1966)

9.23   M. Abramowitz and I. Stegun: *Handbook of Mathematical Functions*, (Dover, New York, 1972)

9.24   L.I. Schiff: Rev. of Sci. Instrum. **17**, 6 (1946)

9.27   H. Wiedemann, P. Kung and H.C Lihn: Nucl. Instrum. and Methods **A319**, 1 (1992)

9.28   F.C. Michel: Phys. Rev. Let. **48**, 580 (1982)

9.25   T. Nakazato, M. Oyamada, N. Niimura, S. Urasawa, O. Konno, A. Kagaya, R. Kato, T. Kamiyama, Y. Torizuka, T. Nanba, Y. Kondo, Y. Shibata, K. Ishi, T. Oshaka, M. Ikezawa: Phys. Rev. Let. **63**, 1245 (1989)

9.26   E.B.Blum, U.Happek and A.J. Sievers: Nucl. Instrum. and Methods **A307**, 568 (1992)

9.29   I am thankful to Mrs. P. Kung and Mr. H.C. Lihn, who provided the data for this graph

9.30   M. Born, E. Wolf: *Principles of Optics* (Pergamon, Oxford, 1975)

9.31   G.B. Airy: Trans. Cambr. Phil. Soc. **5**, 283 (1835)

9.32   M. Berndt, W. Brunk, R. Cronin, D. Jensen, R. Johnson, A. King, J. Spencer, T. Taylor, H. Winick, IEEE Trans. NS-**26**, 3812 (1979)

9.33   E. Hoyer, T. Chan, J.W.G. Chin, K. Halbach, K.J. Kim, H. Winick, J. Yang, IEEE Trans. NS-30, 3118 (1983)

9.34   K. Halbach: J.de Physique, C1, suppl.no 2, Tome 44, February 1983

9.35   W. Heitler, *The Quantum Theory of Radiation*, Clarendon Press, Oxford (1954)

9.36   W.M. Lavender, *Observation and Analysis of X-Ray Undulator radiation from PEP*, PhD Thesis, Stanford University, 1988

9.37   R.H. Milburn: Phys.Rev.Let. **4**, 75 (1963)

9.38   F.A. Arutyunian, V.A. Tumanian: Phys. Rev. Let. **4**, 176 (1963)

9.39   F.A. Arutyunian, I.I. Goldman V.A Tumanian: ZHETF(USSR) eightbf 45, 312 (1963)

9.40   I.F. Ginzburg, G.L. Kotin, V.G. Serbo, V.I. Telnov: *Colliding γ e and γγ beams based on the single pass accelerators of VLEPP type* (Inst. of Nucl.Physics, Novosibirsk, USSR, Preprint 81-102, 1981)

Chapter 10

10.1   B. Touschek, Proc. 1963 Summer Study on Storage Rings, Accel. and Experim. at Super High Energies, Brookhaven, BNL-Report 7534, 171 (1963)

10.2   A. Piwinski: Proc. 9th Int. Conf. on High Energy Accelerators, Geneva (1974) 405

10.3   J.D. Bjorken, S.K. Mtingwa: Part. Accel. **13**, 115 (1983)

10.4   K.W. Robinson, Phys. Rev. **111**, 373 (1958)

10.5   M. Sands, Phys. Rev. **97**, 470 (1955)

10.6   D. Deacon: *Theory of the Isochronous Storage Ring Laser*, PhD Thesis, (Stanford University, Stanford, CA 1979

10.7   C. Pellegrini, D. Robin: Nucl. Instrum. and Methods, **A301**, 27 (1991)

10.8   J.M. Paterson, J.R. Rees, H. Wiedemann, Internal. PEP Note #125, SLAC (1975), Stanford, CA

10.9   B. Buras, S. Tazzari, *Report of the ESRP*, c/o CERN, Geneva (1984)

10.10 S. Tazzari: "Electron Storage Rings for the Production of Synchrotron Radiation" in *Frontiers of Particle Beams*, M. Month, S. Turner (eds.), Lect. Notes Phys. Vol. **296** (Springer Verlag Berlin, Heidelberg, 1986) p.140

10.11 H. Wiedemann: "Low Emittance Storage Ring Design" in *Frontiers of Particle Beams*, M. Month, S. Turner (eds.), Lect. Notes Phys. Vol. **296**, (Springer Verlag, Berlin, Heidelberg, 1986), p.390

10.12 H. Wiedemann: "Storage Ring Optimization" in *Handbook on Synchrotron Radiation*, Vol. 3, ed. by G. Brown and D.E. Moncton, (North-Holland, Amsterdam 1991) p.31

10.13 H. Wiedemann, "Scaling of FODO Cell Parameters", PEP Note #39, PEP Summer Study, Stanford/Berkeley, 1973

10.14 K. Robinson, G.A. Voss, Proc. Int. Symp. Electron and Positron Storage Rings, Saclay, III-4, (Presses Universitaire de France, Paris 1966

10.15 H. Wiedemann, "Enlargement of the Electron Beam Cross Section in a Storage Ring due to an Oscillating Synchrotron Radiation Damping Time Constant", PEP Note #48, SLAC, Stanford CA 1973

10.16 W.K.H. Panofsky, W.A. Wenzel, Rev. Sci. Instrum. **27**, 967 (1956)

Chapter 11

11.1   V.L. Highland: Nucl. Instrum. and Methods **129** (1975), 497 and ibid. **161** 171 (1979)

11.2   H. Bethe: Proc. Camb. Phil. Soc. **30**, 524 (1934)

11.3   H. Bethe, W. Heitler: Proc. Roy. Soc., A **146**, 83 (1934)

11.4   W. Heitler: *The Quantum Theory of Radiation*, (Clarendon Press, Oxford 1954) p.249

11.5   E. Garwin, Proc. of 4th Int. Vac. Congr., Manchester, England (1968), London Inst.of Phys. and Phys. Soc., 1968, Pt. 1, p.131, or SLAC Pub 0392 (1968)

11.6   J. Kouptsidis, A.G. Mathewson: DESY 76/49 (1976), DESY, Hamburg

11.7 O. Gröbner, A.G. Mathewson, P. Strubin : CERN-Report LEP-VA/AGM/sm, (1983), CERN, Geneva

11.9 D. Bostic, U. Cummings, N. Dean, B. Jeong, J. Jurow: IEEE Trans. NS-**22**, 1540 (1975)

Chapter 12

12.1 L. Teng: "Transverse Space Charge Effects", Argonne National Laboratory, Int. Report ANLAD-59, (1960), Chicago

12.2 L. Teng: " Primer on Beam Dynamics in Synchrotrons", in *Physics of Particle Accelerators*, AIP Conf.Proc. #**184**, M. Month (ed.), Am. Inst. Phys., New York, 1989)

12.3 J.E. Augustin: "Effect Faiceau-Faiceau", Laboratoire de L'Accelerateur Lineaire, Raport techn. 36-69 JEA-LN (1969) Orsay, France

12.4 F. Amman, D. Ritson: "Space Charge Effects in Electron-Electron and Positron-Electron Colliding Or Crossing Beam Rings", Int. Conf. High Ener. Accel., Brookhaven (1961) p.471

12.5 R. Belbeoch et al., Rapport Techn. 3-73 (1973), Orsay, France

12.6 F.Amman et al. Proc. VIII. Int. Conf. High Ener. Accel., Geneva, 132 (1971)

12.7 I.B.Vasserman et al., Proc.All-Union Conf.Char.Part.Accel., Dubna, USSR (1978)

12.8 H. Wiedemann, in "Nonlinear Dynamics and the Beam – Beam Interaction", M.Month, J.C. Herrera (eds.) AIP Conf.Proc. #**57**, p.84

12.9 G. Guignard, in "Nonlinear Dynamics and the Beam – Beam Interaction", M. Month, J.C. Herrera (eds.) AIP Conf.Proc. #**57**, p.69

12.10 T.Weiland: Part. Accel. Vol. **15**, 245 (1984)

12.11 A.W. Chao, J. Gareyte: SLAC Internal Note SPEAR-197 (1976)

12.12 P.B. Wilson, R. Servranckx, A.P. Sabersky, J. Gareyte, G.E. Fischer, A.W. Chao: IEEE Trans., NS-**24**, 1211 (1977)

12.13 W. Schnell, CERN Report CERN/ISR-RF/70-7, Geneva (1970)

12.14 F. Sacherer in Proc. 9th Int. Conf. High Ener. Accel., Stanford, USA, 347 (1974)

12.15 L.J.Laslett, K.V.Neil, A.M.Sessler: Rev. Sci. Instrum. **32** 279 (1961)

12.16 D.Kohaupt: DESY Report 80-22, (1980), DESY, Hamburg

12.17 F.Sacherer, IEEE Trans. NS-**20**, 825 (1973)

12.18 J.N. Weaver, P.B. Wilson, J.B. Styles, IEEE Trans. Nuc. Sci. Vol. NS-**26** 3971 (1979)

12.19 "Conceptual Design of the Relativistic Heavy Ion Collider, RHIC", Report BNL 52195 (1989), Brookhaven National Laboratory, Upton, USA

12.20 "HERA, A Proposal for a large Electron-Proton Colliding Beam Facility at DESY", DESY Report HERA 81-10, (1981)

12.21 G.A. Voss, Proc. of the 12th Int. Conf. High Ener. Accel., Fermilab, Batavia, USA, 1983

12.22 B.H. Wiik, IEEE Trans. NS-**32**, 1587 (1985)

12.23 "Summary of the Preliminary Design of Beijing 2.2/2.8 GeV Electron Positron Collider", IHEP, Academica Sinica, (1982)

Chapter 13

13.1 M. Tigner: Nuovo Cimento **37**, 1228 (1965)

13.2 J.E. Augustin, N. Dikanski, Y. Derbenev, J. Rees, B. Richter, A. Skrinski, M. Tigner, H. Wiedemann: Proc. of the Workshop on Possibilities and Limitations of Accelerators and Detectors, FNAL-Report (1979) p.87

13.3 W.H.K. Panofsky: in *High Energy Beam Optics* by K.G. Steffen, (Wiley, New York 1965), p.117

13.4 R. Chasman, K. Green, E. Rowe: IEEE Trans., NS-22, 1765 (1975)

13.5 "Design Report of the European Synchrotron Radiation Facility, ESRF", (1985) Grenoble, France

13.6 D. Einfeld and G. Mülhaupt: Nucl. Instrum. and Methods 172, 55 (1980)

13.7 E. Rowe: IEEE Trans. NS-28, 3145 (1981)

13.8 H. Bruck: "The Orsay Project of a Storage Ring for Electrons and Positrons of 450 MeV Maximum Energy", Proc. Int. Conf. on High Ener. Accel., Dubna (USSR), (1963) p.365

13.9 H. Wiedemann: "Scaling of Damping Rings for Colliding Linac Beam Systems", Proc. 11th Int. Conf. High Ener. Accel., (Birkhäuser Verlag, Basel 1980) p.693

13.10 G. Fischer, W. Davis-White, T. Fieguth, H. Wiedemann: Proc. 12th Int. Conf. High Ener. Accel., FNAL, Batavia, (1983)

13.11 R. Helm and H. Wiedemann: PEP-Note 303, (1979), SLAC, Stanford University

13.12 G. Vignola: Brookhaven Report BNL 36867, (1985)

13.10 K. Robinson and G.A. Voss: Harvard University, CEAL-1029, (1966) Cambridge, USA

13.14 H. Wiedemann, Proc. of the 9th Int. Conf. High Ener. Accel., Stanford (USA), (1974)

13.15 D. Rice, 1989 IEEE Particle Accelerator Conference, IEEE Cat. 89CH2669-0, p.444

Author Index

Subject Index

437

transit time 267
– factor 38, 40
transition energy 180, 278
– approximate 210
TRANSPORT 213
transport systems periodic, 81
transport systems symmetric, 81
transverse fields 82
transverse focusing 60
travel time 179
travelling wave 35
trim magnets 238
triple bend achromat 172, 410
triplet achromat 410
tune 200, 254
– measurement 201
– shift 242, 243
– amplitude dependence, 252
– field errors, 240
tune approximate, 201
tune synchrotron oscillation, 276
twin brother paradox 14
Twiss parameters 153

ultra high vacuum 380
undulator magnet 330
undulator 225
– helical 331
– wavelength 331
– flat 331
units cgs 10
units MKS 10

unity transformation matrix 199
unstable fixed point 286
upright magnets 92

vacuum chamber environment 273
Van de Graaff accelerator 3, 5, 29
vector relations 80
voltage breakdown 265
voltage multiplication 28
voltage multiplier circuit 29

wake fields 391
wall losses 39
wave guide 33, 35
wave length shifter 327
wave length undulator, 331
wave number 271
weak focusing 60, 68, 119
wedge magnet 114, 143
Wideroe linear accelerator structure 270
Wideroe principle 270
Wideroe structure 32
Wideroe $1/2$–condition 54, 56, 58
wiggler magnet 4, 173, 225, 327
– achromat 174
– dispersion function 174
wire lens 85
Wronskian 106

x–rays 3
– beams 57
– tube 28